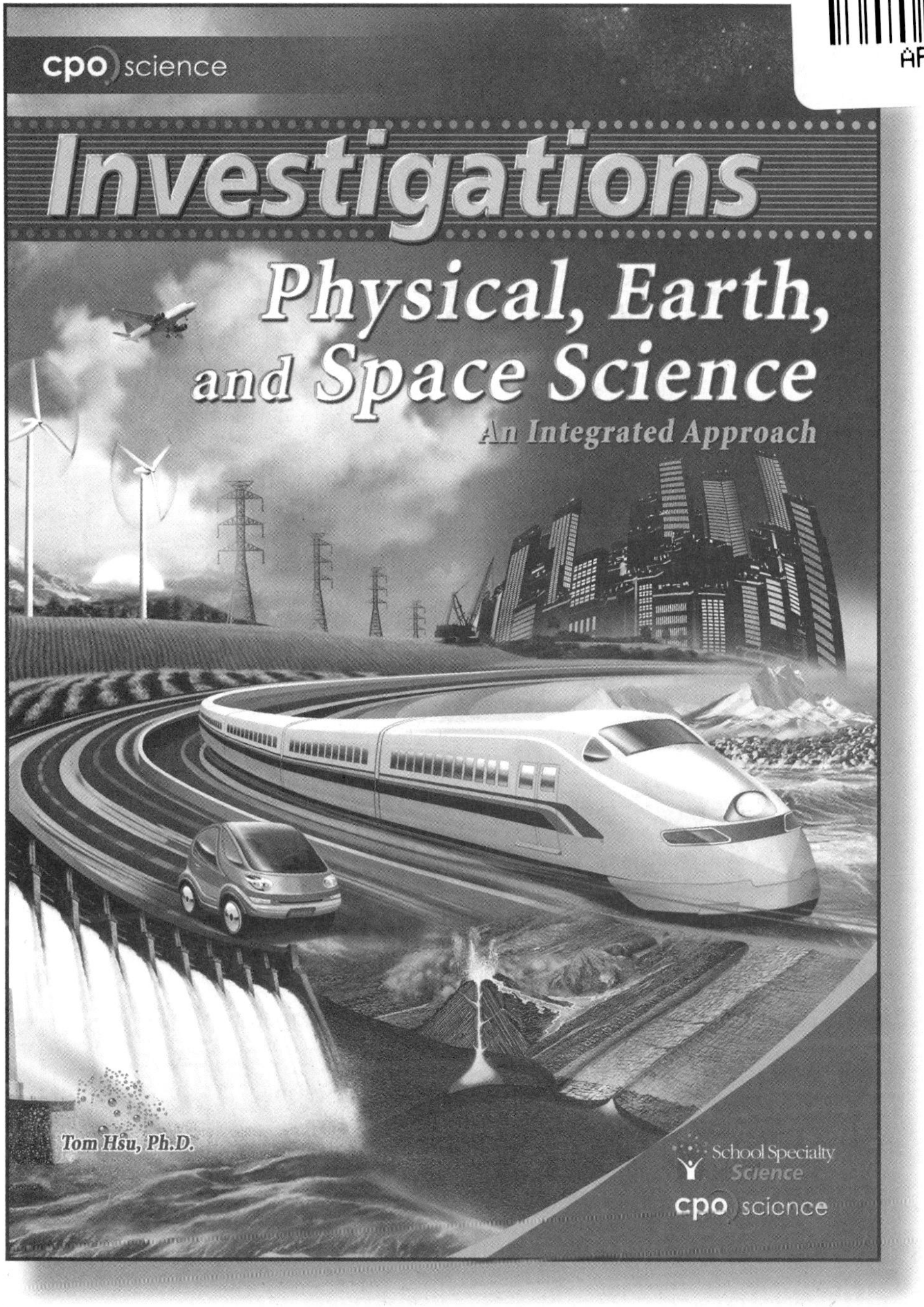

School Specialty Science

cpo science

Physical, Earth, and Space Science Investigations
First Edition

ISBN: 978-1-60431-098-6
Part Number: 492-3930
2 3 4 5 6 7 8 9 - WCE - 14 13 12 11 10

CPO Science
80 Northwest Boulevard
Nashua, NH 03063
(866)588-6951
http://www.cposcience.com

Printed and Bound in the United States of America

Credits

Principal Writers

Thomas C. Hsu, Ph.D. – Author

Nationally recognized innovator in science and math education and the founder of CPO Science. Holds a Ph.D. in Applied Plasma Physics from the Massachusetts Institute of Technology (MIT), and has taught students from elementary, secondary and college levels.

Erik Benton – Principal Investigation Writer

B.F.A., University of Massachusetts with minor in Physics

Taught for eight years in public and private schools, focusing on inquiry and experiential learning. Erik brings extensive teaching and technical expertise, ranging from elementary and adult education to wildlife research.

Scott Eddleman – Co-Author, Project Manager

B.S., Biology, Southern Illinois University; M.Ed., Harvard University

Taught for 13 years in urban and rural settings. Developed two successful science-based school-to-career programs. Worked on National Science Foundation-funded projects at TERC.

Mary Beth Abel – Co-Author

B.S., Marine Biology, College of Charleston; M.S., Biological Sciences, University of Rhode Island.

Taught science and math at an innovative high school and at the college level. Has expertise in scientific research, inquiry-based teaching methods, and science curriculum development.

Patsy Eldridge – Co-Author

B.S., Biology, Grove City College; M.Ed., Tufts University.

Experienced science teacher and national hands-on science trainer and presenter. As an adjunct professor for Endicott College in Beverly, MA, and the College of Charleston, developed content-intensive Physical Science courses for educators.

Sonja L. Taylor – Principal Teacher's Guide Writer

B.S., Chemistry, Stephen F. Austin State University; M.Ed., Divergent Learning, Columbia College

Worked as an analytical chemist and has taught high school chemistry, biology, and physical science.

Senior Editor

Lynda Pennell – Executive Vice President

B.A., English; M.Ed., Administration, Reading Disabilities, Northeastern University; CAGS Media, University of Massachusetts, Boston.

Nationally known in high school restructuring and for integrating academic and career education. Served as the director of an urban school for 5 years and has 17 years teaching/administrative experience in the Boston Public Schools.

Art and Illustration

Polly Crisman – Graphics Manager/Illustrator

B.F.A., University of New Hampshire

Worked as a designer and illustrator in marketing and advertising departments for a variety of industries. She created the CPO Science logo and supervises the graphic design image for CPO publications and media products.

Bruce Holloway – Senior Designer/Illustrator

Pratt Institute, N.Y., Boston Museum of Fine Arts

Expertise in illustration, advertising graphics, exhibits and product design. Commissioned throughout his career by The National Wildlife Federation's Conservation Stamp Campaign. Bruce has worked as senior designer with CPO Science since 2000 and collaborated with various teams to create all CPO book covers.

Jesse Van Valkenburgh – Designer/Illustrator

B.F.A. Illustration, Rochester Institute of Technology

Worked in prepress and design. Was responsible for creative design and prepress film production for computer catalogs, brochures and various marketing materials. Jesse completes photography and illustrations as a graphic designer for CPO book and media products.

Contributing Writers and Editors

Lainie Ives – Writer, Editor, Ancillary Materials, Connections

B.A., Gordon College and graduate coursework at Cornell University's Shoals Marine Laboratory and Wheelock College

Taught elementary and middle school science at an independent school and an environmental education center in New England. Taught middle and high school English overseas.

Stacy Kissel – Writer and Content Editor, Connections

B.S., Civil and Environmental Engineering, Carnegie Mellon University; M.Ed., Physics Education, Boston College

Teaches physics, math, and integrated science at Brookline High in Massachusetts.

James Sammons – Writer, Virtual Rocks CD and Rock Cards Developer, and Content Editor

B.S., Chemistry and Biology, University of Rhode Island.

Has taught middle school science for 30 years. Member of the Geological Society of America and the American Geophysical Union.

Mary Ann Erickson – Editor and Assessment Writer

B.S., Naval Architecture and Marine Engineering, Massachusetts Institute of Technology

Ran a technical writing consulting business, writing process control manuals for water treatment plants, software design and mining equipment.

David H. Bliss – Content Editor and Assessment Writer

B.S., Science, Cornell University; M.Ed., Zoology.

Has taught for 39 years in the science field: biology, chemistry, earth science, and physics.

Lawrence Sabbath – Content Editor

B.S. & M.A., Wayne State University (Detroit)

Science Consultant at the Oakland Schools Science, Math & Technology Center.

Pamela J. W. Gore, Ph.D.– Content Editor

B.S., University of Maryland, College Park; M.S., M.Phil., and Ph.D., The George Washington University.

Professor of Geology, Georgia Perimeter College.

Editorial Consultant

Christine Golden

B.A., Psychology, Gordon College: M.B.A., Rivier College

Project manager at *Imperial Communications* since 1999, with 22 years publishing experience. Owner and managing editor of *Big Dog Publishing Services*. Christine's work centers on editing of K-12 textbook material.

Equipment Design and Material Support

Thomas Narro – Senior Vice President

B.S., Mechanical Engineering, Rensselaer Polytechnic Institute

Accomplished design and manufacturing engineer; experienced consultant in corporate reengineering and industrial-environmental acoustics.

Danielle Dzurik – Mechanical Engineer

B.S., Industrial Design, Auburn University

Focuses her efforts in product development on creating new products and improving upon older designs.

Kathryn Gavin – Purchasing and Quality Control Manager

Responsible for all functions related to purchasing raw materials and quality control of finished goods. Works closely with product development and design.

Ancillary Materials

Laine Ives – Skill and Practice Worksheets, Projects, Connections

Mary Ann Erikson – ExamView® Assessment Suite

Laura Preston – Guided Readings, Connections

Leslie Sheen – Graphic Organizers, Connections

Cory and Laura Ort – Lesson Organizer

Project and Technical Support

Susan Gioia – Educational CPO Science Administrator

Expertise in office management. Oversees all functions necessary for the smooth product development of CPO products, including print and media.

Lynn L'Heureux – Technical Consultant

Owner of M&M Composition, LLC. Has worked in textbook composition for 10 years and specializes in math, computer, and science texts.

Reviewers

Mark Baker – Diamond, OH

Dr. Nicholas Benfaremo – South Portland, ME

Kip Bollinger – Carlisle, PA

Daryl Campbell – Needham, MA

Nancy Baker Cazan – Massillon, OH

Ann Cleary – Medina, OH

Jean A. Cyders – Canton, OH

Lou Fellner – Bayville, NJ

Dr. Gregorio Garcia – Brownsville, TX

Cort Gillen – Cypress, TX

Alan P. Gnospelius – San Antonio, TX

Lisa Q. Gothard – East Canton, OH

Liz Gregory – Austin, TX

Tracy Hollars – Cleveland, OH

William C. Huckeba – Irving, TX

Chrystal Brooke Johnson – Irving, TX

Marty Kilroy – White, GA

Daniel Klein – Navarre, OH

Kathleen Kuhn – Uniontown, OH

Jay Kurima – Fort Worth, TX

Ed Laubacher – Uniontown, OH

Matt Leatherberry – Minerva, OH

John C. Lineweaver – Abilene, TX

Anita K. Marshall – Cypress, TX

Jerry Martin – Florence, SC

Maria Estela Martinez – Irving, TX

Mark Mathews – McDonough, GA

Chandra R. Maxey – Irving, TX

Alyson Mazza – Salem, NH

Tom Melia – Prairie Village, KS

Thomas McArthur – North Canton, OH

Michael Mihalik – Uniontown, OH

William Miller – Fort Worth, TX

LaMoine Motz – Waterford, MI

Stacey L. Nunley – Detroit, MI

Joel C. Palmer – Mesquite, TX

Neil Parrot – Massillon, OH

Jim Perry – Mansfield, MA

Jeff Pickle – North Canton, OH

Lisa M. Polar – Corapolis, PA

Emily O. Price – Waxahachie, TX

Julie Randolph – Beaumont, TX

Steve Remenaric – Massillon, OH

Michael Vela – Concord, MA

Melissa Vela – Concord, MA

Scott Vetter – Shreveport, LA

Elizabeth Volt – Cleveland, OH

David Warner – Massillon, OH

Larry Weathers – Arlington, MA

Dan Williamson – Irving, TX

Kathleen Woodring – Edna, TX

Jaclyn L. Ziders – Uniontown, OH

Special Thanks To

Brookline High School Science Students

Brookline High School, MA

How to Read an Investigation

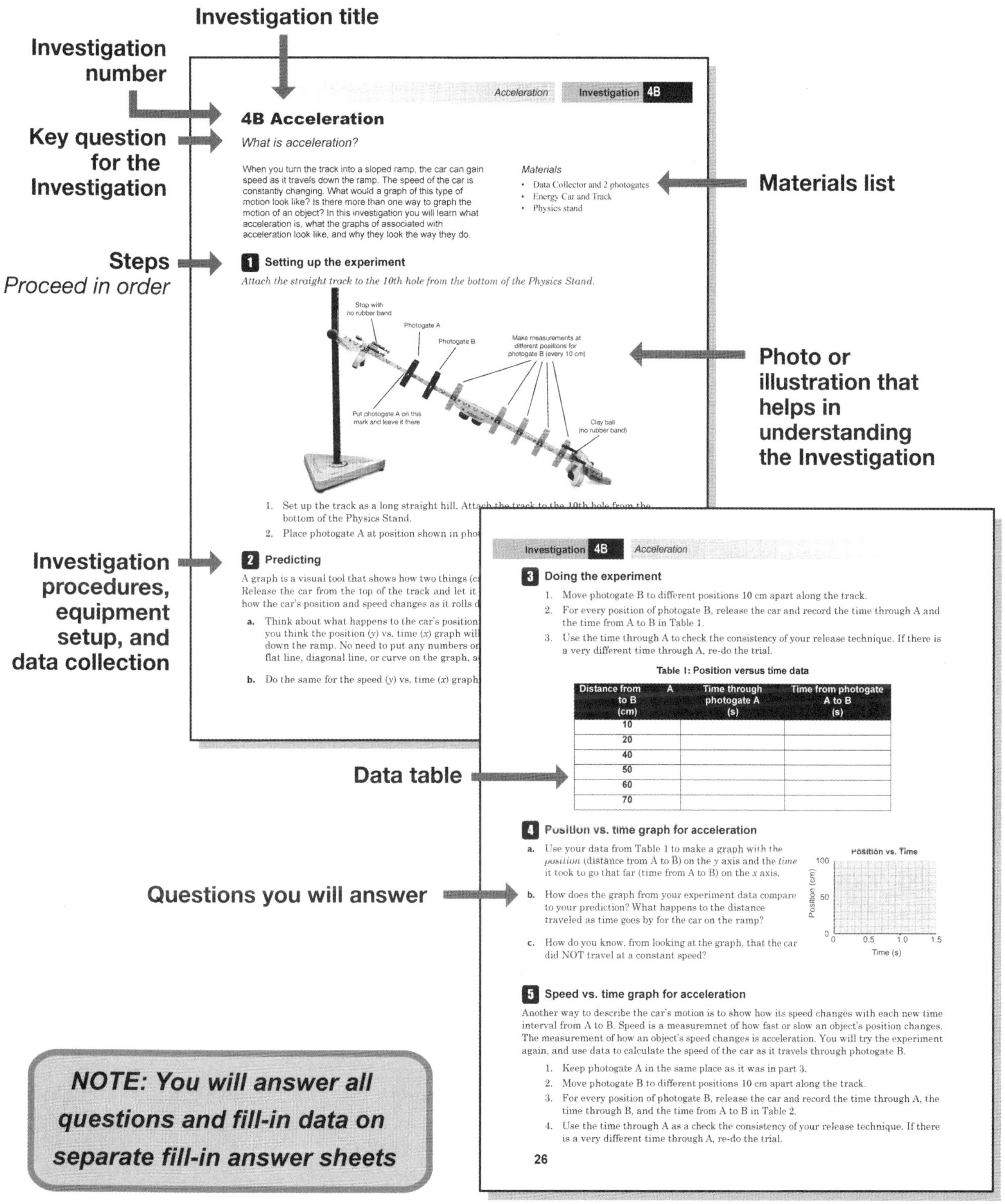

Acceleration | Investigation 4B

4B Acceleration

What is acceleration?

When you turn the track into a sloped ramp, the car can gain speed as it travels down the ramp. The speed of the car is constantly changing. What would a graph of this type of motion look like? Is there more than one way to graph the motion of an object? In this investigation you will learn what acceleration is, what the graphs of associated with acceleration look like, and why they look the way they do.

Materials
- Data Collector and 2 photogates
- Energy Car and Track
- Physics stand

1 Setting up the experiment

Attach the straight track to the 10th hole from the bottom of the Physics Stand.

1. Set up the track as a long straight hill. Attach the track to the 10th hole from the bottom of the Physics Stand.
2. Place photogate A at position shown in pho

2 Predicting

A graph is a visual tool that shows how two things (c
Release the car from the top of the track and let it
how the car's position and speed changes as it rolls d

a. Think about what happens to the car's position you think the position (y) vs. time (x) graph wil down the ramp. No need to put any numbers or flat line, diagonal line, or curve on the graph, a

b. Do the same for the speed (y) vs. time (x) graph

Investigation 4B | Acceleration

3 Doing the experiment

1. Move photogate B to different positions 10 cm apart along the track.
2. For every position of photogate B, release the car and record the time through A and the time from A to B in Table 1.
3. Use the time through A to check the consistency of your release technique. If there is a very different time through A, re-do the trial.

Table 1: Position versus time data

Distance from A to B (cm)	Time through photogate A (s)	Time from photogate A to B (s)
10		
20		
40		
50		
60		
70		

4 Position vs. time graph for acceleration

a. Use your data from Table 1 to make a graph with the *position* (distance from A to B) on the y axis and the *time* it took to go that far (time from A to B) on the x axis.

b. How does the graph from your experiment data compare to your prediction? What happens to the distance traveled as time goes by for the car on the ramp?

c. How do you know, from looking at the graph, that the car did NOT travel at a constant speed?

5 Speed vs. time graph for acceleration

Another way to describe the car's motion is to show how its speed changes with each new time interval from A to B. Speed is a measuremnet of how fast or slow an object's position changes. The measurement of how an object's speed changes is acceleration. You will try the experiment again, and use data to calculate the speed of the car as it travels through photogate B.

1. Keep photogate A in the same place as it was in part 3.
2. Move photogate B to different positions 10 cm apart along the track.
3. For every position of photogate B, release the car and record the time through A, the time through B, and the time from A to B in Table 2.
4. Use the time through A as a check the consistency of your release technique. If there is a very different time through A, re-do the trial.

26

NOTE: You will answer all questions and fill-in data on separate fill-in answer sheets

Lab Safety

Observing safety precautions is an extremely important practice while completing science investigations. Using science equipment and carrying out laboratory procedures always requires attention to safety. The purpose of learning and discussing safety in the lab is to help you learn how to protect yourself and others at all times.

The investigations in this book are designed to reduce safety concerns in the laboratory. The physics investigations use stable equipment that is easy to operate. The chemistry investigations use both household and laboratory chemicals. Although these chemicals might be familiar to you, they still must be used safely.

You will be introduced to safety by completing a skill sheet to help you observe the safety aids and important information in your science laboratory. In addition to this skill sheet, you may be asked to check your safety understanding and complete a safety contract. Your teacher will decide what is appropriate for your class.

Throughout this book, safety icons and words and phrases like "caution" and "safety tip" are used to highlight important safety information. Read the description for each icon carefully and look out for them when reading your book and doing investigations.

	**General safety:** Follow all instructions carefully to avoid injury to yourself or others.
	**Wear safety goggles:** Requires you to wear eye protection to prevent eye injuries.
	**Wear a lab apron or coat:** Requires you to wear a lab apron or coat to prevent damage to clothing and to protect from possible spills.
	**Wear gloves:** Requires you to protect your hands from injury due to heat or chemicals.
	**Poisonous chemicals:** Requires you to use extreme caution when working with chemicals in the laboratory and to follow all safety and disposal instructions from your teacher.
	**Skin irritant:** Requires you to use extreme caution when handling chemicals in the laboratory due to possible skin irritation and to follow all safety and disposal instructions from your teacher.
	**Respiratory irritant:** Requires you to perform the experiment under a laboratory hood and to avoid inhaling fumes while handling the chemicals.
	**Laser:** Requires you to use extreme caution while using a laser during investigations and to follow all safety instructions.

Lab safety is the responsibility of everyone! Help create a safe environment in your science lab by following the safety guidelines from your teacher as well as the guidelines discussed in this document.

Chapter Investigations

Optional Investigations

Equipment Setups

1A Measurement

How do you use scientific tools to make accurate measurements?

In everyday life we use tools to make our work easier. For example, it is difficult to pound a nail without a hammer, to wrap a gift without tape, or to open clear plastic packages without scissors. In science, it is important to be able to correctly choose and use laboratory equipment to make measurements. During the "Measurement Games" you and a partner will practice choosing the correct tool to make measurements accurately. The group with the best averages overall wins the "Measurement Games!"

Materials (per group)

- Small paper cup
- Pebbles
- Water
- Masking tape
- Graduated cylinder
- Electronic scale (or triple-beam balance)
- Meter stick
- Beaker
- Measuring tape
- Data Collector
- GeoBox

1 Stop and think

a. Are you familiar with the tools necessary to make measurements of length, mass, and volume? Look at the list of materials above and write down a tool used for measuring each property.

Property	Tool used
Length	
Mass	
Volume	

b. Do you understand what metric units coincide with length, mass, and volume? Discuss this with your partner and write an appropriate unit that corresponds to each property.

Property	Unit used
Length	
Mass	
Volume	

c. Read about the events listed below in Part 2 and predict the outcome of each event for yourself before you actually perform the task.

Event	Prediction
Straw Javelin	
Paper Cup	
Pebble grab	
Side Step	
Hoppity Hop	

2 Doing the activity

A. Perform the "Measurement Games" event, collect, and record data.

1. **Straw Javelin:** During this event, you will be throwing a straw as far as you can, like it is a javelin. Your front foot may not cross the start line, and you must throw the straw like a javelin with only one hand. Measure the distance of your throw in meters and centimeters.
2. **Paper Cup Challenge:** How much water can you move from a GeoBox to a beaker in 10 seconds using just one paper cup? Use a graduated cylinder to measure the volume of water you successfully transferred. Be careful so you don't spill any water!
3. **Pebble Grab:** Who can grab the greatest mass of pebbles? Use ONLY ONE HAND to grab as many pebbles as you can out of a container. Transfer them to an electronic balance to measure the mass. Be sure the balance is measuring in the correct units before you begin!
4. **Side Step:** How far is your leg span? From a starting point step as far as you can to the side. Your partner will measure the length of your step in meters and centimeters.
5. **Hoppity Hop:** Who can hop 10 meters the fastest on one foot? Measure and mark 10 meters on the floor with the masking tape. Use the Data Collector in stopwatch mode to time how long it takes your partner to hop 10 meters on one foot.

B. Results

1. Record your results below. Any result with missing or incorrect units will be automatically disqualified from the Measurement Games!
2. After you have recorded your results there will be a class discussion to see who the winners are in each event. Decide within your group who has the best score for each event. Use that score for the class data set. Record the data in the data table your teacher has drawn on the board. Determine the best overall score for the group winner. Record the individual winner's results for each event in the data table below.

Olympic event	My results	Winner's results
Straw Javelin		
Paper Cup Challenge		
Pebble Grab		
Side Step		
Hoppity Hop		

3 Thinking about what you observed

a. Calculate the difference between the winner's results and your results for each event. (Don't forget units!)

Event	Difference
Straw Javelin	
Paper Cup Challenge	
Pebble Grab	
Side Step	
Hoppity Hop	

b. In which event were you closest to the winner?

c. In which event were you the farthest away from the winner?

d. How close were you to your predictions?

e. Which measurement were you most familiar with before The Games? Why?

f. Which measurement did you find easiest to make during The Games? Why was it so easy for you?

g. Which measurement did you find to be the most difficult during the Games? Why?

4 Exploring on your own

a. Scientists work hard to be precise in their measurements when experimenting. In real life, most of us are accurate enough, but hardly precise, every time we perform an event that involves a measurement. Do you know the difference? Look it up and draw several examples that represent the difference between accuracy and precision.

b. Create a quiz that your classmates could take regarding the different tools used in a science lab and their appropriate units. Include an answer key.

c. Find a science article from a magazine, newspaper, or Internet website where scientists are measuring something. Explain in a one-paragraph summary what they are measuring, what tools they are using to make the measurements, and the units associated with the measurements.

1B Measuring Time

How is time measured accurately?

A measurement is a quantity with a unit that tells what the quantity means. For example, 3 seconds is a measurement of time that includes a quantity (3) and a unit (seconds). This investigation will explore time measurement.

Materials

- Data Collector and 2 photogates

1 Using the Data Collector as a stopwatch

A stopwatch measures a **time interval**. The Data Collector stopwatch shows time in seconds up to 60 seconds. The display shows **min:sec** for times longer than one minute.

1. Go to the Data Collector's timer function, and select stopwatch mode.
2. Practice starting and stopping the stopwatch.
3. Reset the stopwatch to zero.

2 Observing reaction time

The time it takes a signal from your brain to move a muscle is called **reaction time**.

1. This experiment takes two people. One person (the watcher) watches the stopwatch and the other person (operator) pushes the buttons without looking at the display. The watcher selects a time between 5 and 10 seconds and keeps the time secret.
2. The operator starts (and stops) the stopwatch *without looking at the display*. The watcher looks at the display and says STOP at the secret time. For example, if the secret time is 6 seconds, the watcher should say STOP when the display reaches 6.00 seconds.
3. Repeat the experiment several times, record your results in Table 1, and calculate reaction time by taking and average of the difference in times for all 5 trials.

Table 1: Reaction time

Trial	Secret time (s)	Measured time (s)	Difference (s)	Avg. difference reaction time (s)
1				
2				
3				
4				
5				

3 Mixed units for time

In physical science, you are usually going to measure time in seconds. However, time is often given in mixed units, which may include hours, minutes, and seconds. Consider the following three time intervals.

1. 16,000 seconds
2. 250 minutes
3. 4 hours, 23 minutes and 15 seconds (4:23:15)

a. Which one is in mixed units?

b. Can you tell which time is longest or shortest?

c. 1 minute = 60 seconds; how many seconds is 250 minutes?

d. 1 hour = 60 minutes; how many minutes are in 4 hours?

e. Use your answer from question d. to figure out how many seconds are in 4:23:15.

f. Arrange the three measurements from smallest to largest.

4 Using the photogates

A photogate allows us to use an infrared light beam to start and stop the Data Collector. When the timer function is in interval mode, it uses photogates to control the clock.

1. Connect a single photogate to the "**A**" input with a cord.
2. Select **interval mode** in the timer function of the Data Collector.
3. Try blocking the infrared beam with your finger and observe what happens on the timer display.

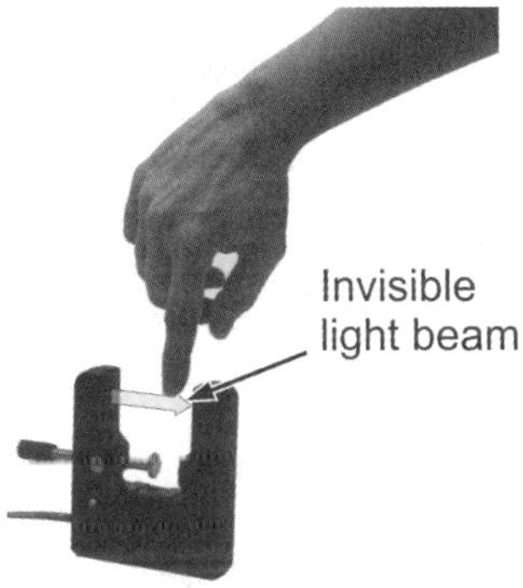

Try your own experiments until you can answer the following questions. Be very specific in your answer. Someone who has never used the Data Collector before should be able to read your answer and know what to do with the infrared beam to make the clock start and stop.

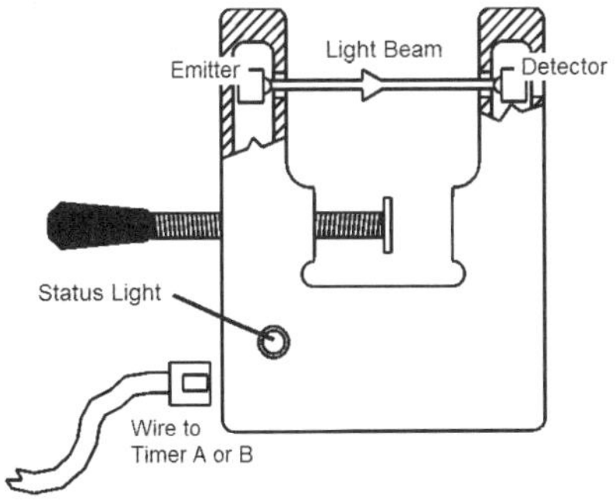

a. How do you start the clock?

b. How do you stop the clock?

c. What time interval has the clock measured?

5 Using two photogates

1. Connect a second photogate to the Data Collector.
2. Make sure the light on each photogate is green and press the reset button. Pressing reset clears the clocks and also tells the timer to look at its inputs to see which photogates are connected.
3. Do your own experiments and fill in the rest of Table 2.

Table 2: Timer and photogate rules

t_A	
How do you start the t_A clock?	
How do you stop the t_A clock?	

t_B	
How do you start the t_B clock?	
How do you stop the t_B clock?	

t_{AB}	
How do you start the t_{AB} clock?	
How do you stop the t_{AB} clock?	

2A Observation, Question, and Hypothesis

Is the flow rate constant no matter how much water is in the bucket?

When scientists study the world around them they make observations, think about what they observed, then pose a question based on what they observed and what they think will or will not happen. Once they come up with a question based on what they want to know about the world, the question is rephrased in the form of a statement called a hypothesis. Then, they perform an experiment to prove or disprove their *hypothesis* by examining and analyzing the data collected during the experiment.

In this investigation, you will answer the key question after you state a hypothesis. You will collect new data and learn how to organize this data in the form of a graph.

Materials

- Stream table stage
- Bucket with spigot
- Bucket without spigot
- Displacement tank
- Data Collector
- Water
- Graph paper

1 Getting started

In this investigation you will begin by making some observations about a system. The system you will use is a bucket filled with water, and another bucket used to catch the water as it is allowed to flow out of the first bucket through an open spigot. To get started, you will need to set up all the parts of your system.

2 Setting up

1. Turn the spigot on the bucket to the closed position. Then, fill the bucket halfway with water.
2. Put the bucket on the stream table stage. Place the second bucket under the spigot of the water-filled bucket on the stage.

Safety note: Do not stand on the stage! This piece of equipment can only support up to a mass of 10 kilograms.

3 Making observations

In this part of the investigation, you will observe the flow of the water as it empties out of the bucket. You won't measure the flow, but you and the other members of your team will watch the flow of the water. Try and be specific with your observations. Use these questions to guide your observations.

1. Is the flow rate getting faster as the bucket empties?
2. Is the flow rate getting slower as the bucket empties?
3. In this part of the investigation, you are collecting *qualitative* data. What would you need to do to collect *quantitative* data?

4 Thinking about what you observed

Now that you have observed the water emptying out of the bucket, it is time to form a hypothesis based on the key question.

a. Make a prediction based on the key question: ***Is the flow rate constant no matter how much water is in the bucket?***

b. How do you think the flow rate will change?

- Will the flow rate increase and let more water out for each 20-second interval as the bucket empties?
- Will the flow rate decrease and let less water out for each 20-second interval as the bucket empties?
- Will the flow rate remain constant for each 20-second interval as the bucket empties?
- Will the flow rate increase for some intervals and decrease with others as the bucket empties with no obvious pattern that we can see?

c. Combine your answers to questions a and b. This will be your hypothesis. State if you think the flow rate will change. If you think it will change, say how it will change.

5 Testing your hypothesis by doing an experiment

To test your hypothesis, you will measure the volume of water that flows out of the bucket for each 20-second interval until the bucket is empty. You will then calculate the flow rate during each of the 20-second intervals. From this data, you will decide if your hypothesis was correct or not.

1. Turn the spigot on the bucket to the closed position. Then, fill the bucket up to the fill line with water.
2. Put the bucket on the stream table stage. Place the displacement tank under the spigot of the bucket.
3. One partner will use the stopwatch mode of the Data Collector to time 20 seconds. Another partner will open, and then close the spigot when 20 seconds have elapsed. (Always open the spigot to the mark for each trial during this investigation).

4. Measure and record the amount of water (in mL) collected by the displacement tank in the second column of Table 1. Make sure the displacement tank is flat on the table when you read and record the volume in the data table.
5. Remember, do not pour the collected water back into the bucket on the stage. Pour the water into the collection bucket provided by your teacher.
6. Repeat the 20-second intervals until the bucket is empty, or no more water flows out of the spigot.

Note: Copy this data table into your notebook or on a separate piece of paper. You may not need all 12 interval rows on the data table.

Table 1: Volume when spigot is turned to the notch

Interval	Volume of water (mL)	Time (s)
1		
2		
3		
4		
5		
6		
7		
8		
9		
10		
11		
12		

6 Using your measurements

1. Once you have completed your measurements, use your data to calculate the flow rate for each interval.
2. Record the values in Table 2.

Table 2: Flow rate when spigot is turned to the notch

Interval	Flow rate (mL/s)
1	
2	
3	
4	
5	
6	
7	
8	
9	
10	
11	
12	

7 Making a graph

You have collected data. Now you need to interpret this data. One way to interpret data is to make a picture of it. A graph is a picture of data. It can help you identify patterns, trends, or other important information that may help to verify a hypothesis or make a prediction. A graph compares two types of information.

1. The two types of data in this investigation are flow rate and the interval number. You are going to make a graph of flow rate versus interval number.
2. There are many kinds of graphs. Each is useful in certain situations. You are going to use a bar graph for this investigation. A bar graph is useful to indicate specific amounts of things, like flow rate at different time intervals.
3. On the graph below, plot your data. You will put the flow rate on the *y*-axis, and the interval number on the *x*-axis. Draw each bar so that it is two squares wide. The height of each bar equals the flow rate for one interval.
4. Color in all the bars the same color with a pencil, pen, or colored pencil.

Title: ______________________________

y-axis: ______________________________

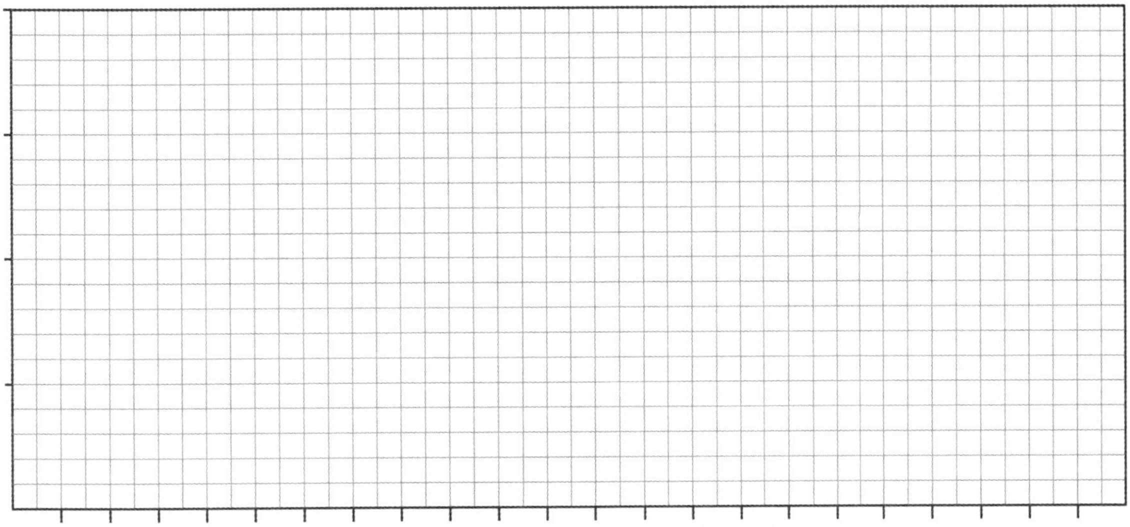

x-axis: ______________________________

8 Analyzing your data

a. Was your hypothesis correct? The answer to this question is known as your conclusion. Write a short paragraph to answer this question. Use your graph to help you answer this question.

b. Compare your data with the data collected by other lab groups. Did the data from the other lab groups support your hypothesis?

c. Did you find it easier to understand your data in number form on your data tables, or in picture form on your graph?

d. How can you make data presented in a table easier to understand? How can you make data presented in a graph easier to understand?

e. What condition was kept constant during each interval?

f. What condition changed and was not kept constant for each interval?

2B Experiments and Variables

How do you design a valid experiment?

Experiments help us collect evidence so we can unlock nature's puzzles. If an experiment is well planned, the results can provide an answer to a scientific question like "What would happen if I did this?" If the experiment is not well planned, you will still get results, but you may not know what they mean. In this investigation, you will experiment with a car on a ramp. Only by paying careful attention to the variables can you make sense of the results.

Materials

- Energy Car
- Data Collector and photogates
- Physics stand

1 Setting up the experiment

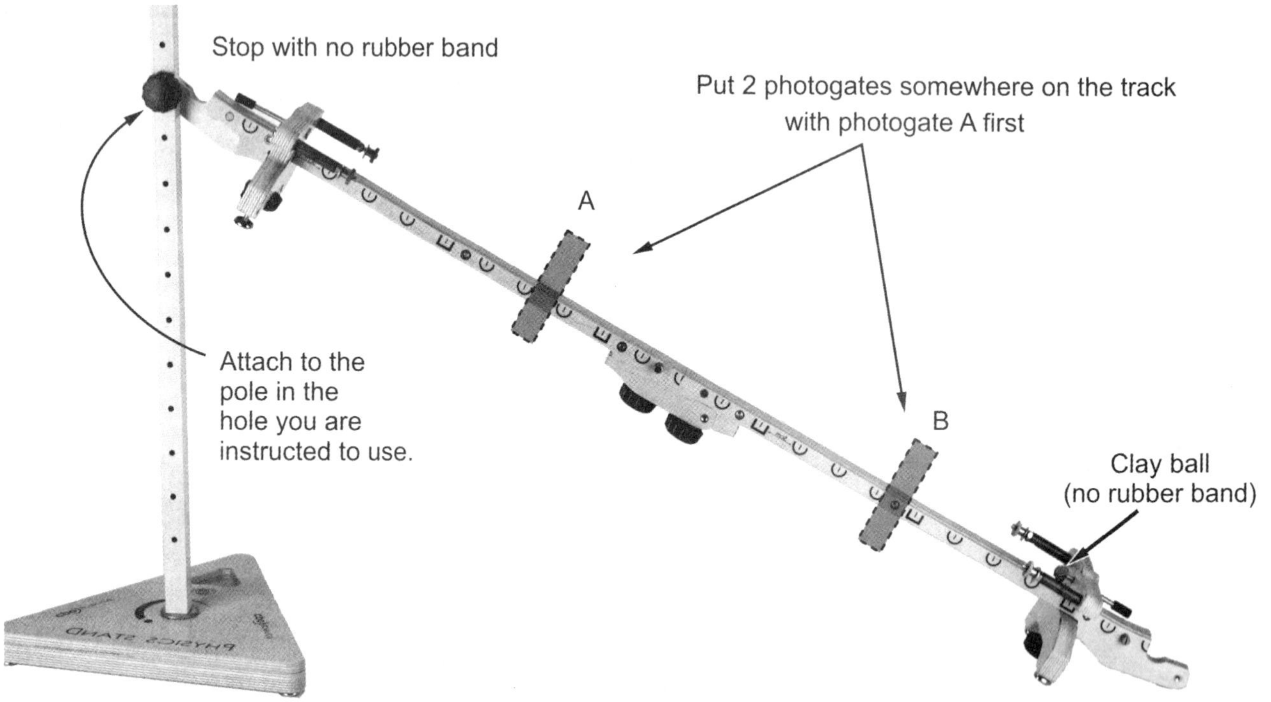

1. Set up the track with a long straight section. Your teacher will tell you which hole in the stand to attach the track. Each group will have a different angle.
2. Put a clay ball on the stop at the bottom.
3. Place two photogates on the track with photogate A higher than photogate B.
4. Roll the car down and record the time it takes the car to pass between the photogates (t_{AB}).

2 Stop and think

a. Which track should have the fastest car? Which track should have the shortest time between photogates?

b. Write a one sentence hypothesis that relates the time between photogates to the angle of the track.

c. Use Table 1 to record the results from each group in your class. Record the times in the column labeled "First Trial." Leave the column labeled "Second Trial" blank. How do the results compare with your hypothesis? Can you give a reason why they did or did not behave as you expected?

Table 1: Photogate times from A to B

	First trial	Second trial
Attachment hole (holes from bottom)	Time from A to B (s)	Time from A to B (s)

3 Variables

a. List at least six variables in your system which affect the time between photogates.

b. Which variable is the experimental variable in your class? How do you know?

c. What should be done with the remaining variables (other than the experimental variable)? Why should this be done?

d. Name two variables that should not be included in your system. These variables should not have much (or any) influence on the time from photogate A to B.

4 A controlled experiment

1. With your teacher and the rest of your class, decide on how to control the variables other than the experimental variable.
2. Practice rolling the car until you can get three consecutive times within 0.0010 second of each other.
3. Repeat the experiment using the experimental and controlled variables you discussed and decided upon. Record the new data in the column titled "Second Trial."

5 Applying what you have learned

a. Does the second trial of the experiment produce results that agree with your hypothesis?

b. Why does the second trial produce results that are more consistent with your hypothesis than the first trial did?

c. If something does not work, discuss what you should do to try and find the problem. List at least three steps that relate to variables, experiments, and controls.

3A Positive and Negative Position

How do we measure position in two dimensions?

We often use positive and negative numbers to tell right and left or forward and backward. In this investigation you will be measuring distances in two dimensions around your classroom.

Materials

- Meter stick
- Blank index cards
- Index cards with directions and questions prepared in advance

1 Describing direction in two dimensions

Your *position* is your location compared to a starting point or *origin*. The origin can be any point you choose. For example, when giving the position of your school, you may choose your house to be the origin. The position of your school might be 2 kilometers east of your house. When specifying a position, both the distance and the direction must be given.

The compass directions north, south, east and west can be used to tell direction. Positive and negative x- and y-coordinates can also be used to give directions. The x-coordinate tells the east–west position. East is in the positive direction, and west is in the negative direction. The y-coordinate tells the north–south position. North is in the positive direction, and south is in the negative direction. Coordinates can be written as an ordered pair (x, y).

For example, suppose the grocery store is 2 km east and 1 km south of your house. You would specify the position of the grocery store as (+2, –1) km.

2 Classroom scavenger hunt

1. Look around the floor of your classroom and locate the pieces of tape marked A, B, C, etc. Each of these letters will be used as an origin from which you will make measurements. The front of your classroom will correspond to north or $+y$ direction.
2. Your teacher has a pile of index cards that correspond to the letters. Each index card has a letter on one side and instructions on the other side. Choose one of the index cards. Find the card's letter on the floor of your classroom.
3. Read the instructions on the card. The instructions will tell you the distance and direction you must walk from the origin. Once you reach the final position, answer the question on the card. Write your answer in Table 1.

Origin point	Answer to question
A	
B	
C	
D	
E	
F	

4. Return the index card and choose another one. Complete cards A through F.

3 Thinking about position

a. Which compass direction corresponds with the positive x direction?

b. Which compass direction corresponds with the negative y direction?

c. Which compass direction corresponds with the negative x direction?

d. Which compass direction corresponds with the positive y direction?

e. You start at the origin and walk 3 meters north and 5 meters west. What are your coordinates in (x, y) form?

f. You start at the origin and walk 2 meters south and 1 meter east. What are your coordinates in (x, y) form?

g. How would you walk to get to the position (–1,–4)m?

h. How would you walk to get to the position (+2,–5)m?

4 Making your own scavenger hunt

1. Choose one of the lettered points in the classroom as your origin. Make a scavenger hunt card that gives directions from the origin to a destination in the classroom. Write a question to be answered at your destination.
2. Trade scavenger hunt cards with another group. Follow their directions and answer the question at the destination.

3B Topographic Mapping

How do you make a topographic map from a 3-dimensional surface?

Topography is the shape of the surface of an area and includes the elevations of land formations like mountains.The topography of a region is represented by a topographic map.

A topographic map is the two-dimensional representation of a three-dimensional land surface. Scientists use these types of maps to understand the effects of geologic processes on Earth's surface. Topographic maps show the difference in elevation through the use of contours. Contour lines connect points of equal elevation. Contour lines are drawn at specific intervals known as the contour interval. Once the contour lines are present, details are added to show land use using standard mapping symbols.

In this investigation, you will use a model land surface to make a contour map.

Materials

- GeoBox with the topo form (white land surface model) and topo lid (the flat lid)
- Container to hold water (or use a bucket from the stream table)
- Beaker to transfer water between containers
- Overhead projector markers (thin)
- Pencil
- Colored pencils
- Tracing paper
- Water colored with food coloring
- Metric ruler
- Topographic maps of local area

1 Making a topographic map

1. Note that the GeoBox has a sticker on the side. Each mark on this sticker represents one centimeter. Pour water into the GeoBox up to the first centimeter line.
2. Place the topo lid on the GeoBox. Stand over the GeoBox so that you are looking down on the topo form. With the overhead projector marker, first make a boarder for your map by outlining the lid. Then, outline the perimeter of the land surface onto the lid. This will be considered "sea level," or the 0 meter contour line.
3. Now, using your marker, number each centimeter above sea level. Use the sketch below to help you with this step.
4. Remove the topo lid and add water until the water level reaches the 1-centimeter mark. Replace the lid. Trace the "coastline," the line along which the water and land meet, onto the lid. All points on this line are 1 cm above sea level. They form a contour line, a line of equal elevation.

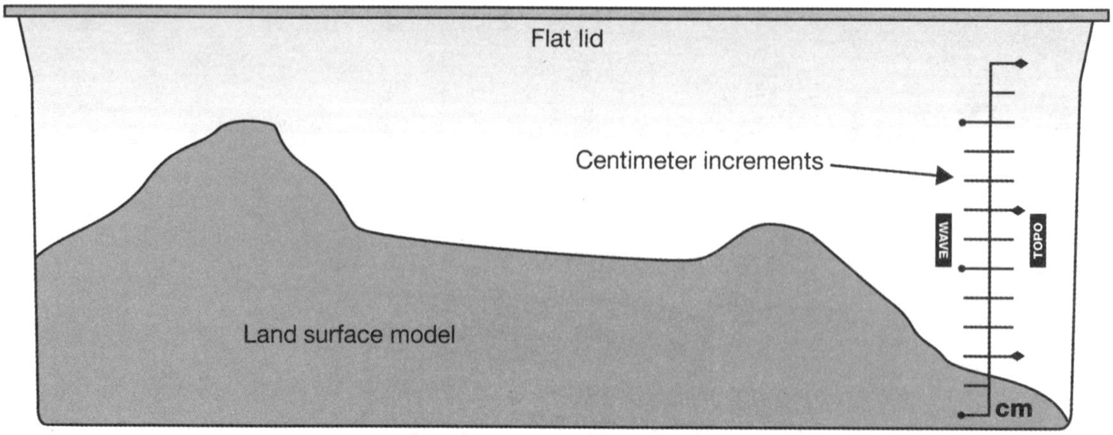

5. Add water to the level of the 2-centimeter mark. Replace the lid and again, trace the "coastline." All points on this line are 2 centimeters above sea level.
6. Continue this procedure until the topo form is covered with water.
7. Now you have a contour map of your land surface. Use the tracing paper to trace what is on the lid. Each partner makes his/her own contour map. The elevation for each contour represents 100 meters. Draw the contours as solid lines. Indicate the elevation as shown in the graphic at right.

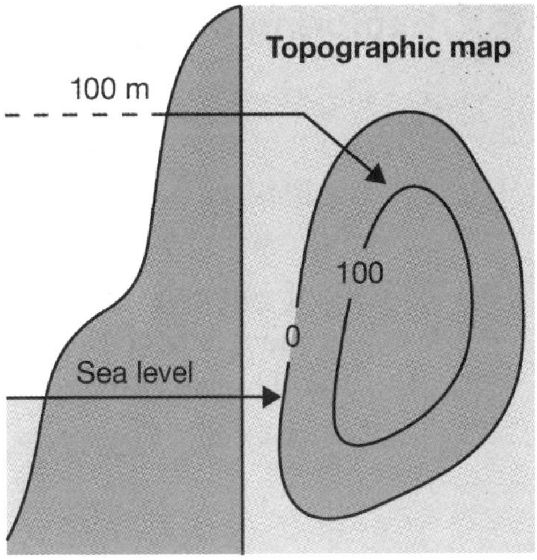

2 Labeling your map

1. Each partner will now finish his/her own map, which is now on tracing paper. Add details showing land-use using the USGS topographic map symbols (see next page). Make sure the following features are on your map. With your group or with the whole class, define any terms you do not understand.
   - Geographic north
   - A contour interval
   - A fractional scale
   - A verbal scale
   - A bar scale
   - A river
   - A depression
   - An airport
2. Color in areas on your map where appropriate. Water is shown in blue. Densely populated areas are shown in gray or pink. Wooded areas are in green and open areas in white. Individual buildings are solid black shapes.
3. When your group is finished, write a title on your map. Also, write your name or your group's name on the map.

3 Thinking about what you observed

a. What is the direction that your river is flowing?

b. What is the difference in elevation between the start and end of the river?

c. What is the overall total change in elevation in your map?

d. What is the highest elevation in your map?

e. By looking at your map, what area is the steepest? *Remember: Look at the contours to see how close or far away they are from each other.*

4 Exploring on your own

1. Obtain a topographic map of your local area. See if you can find where you live and some of the landmarks around you.
2. Besides scientists, who might be interested in a contour map? Why would it be useful to those people?
3. See if you can find out where the closest benchmark is in your local area. A benchmark is a place where the exact elevation is known. A marker is placed in the ground at that spot. There is an **X** on the marker with the exact elevation inscribed in it. (Hint: Benchmarks are noted on topographic maps.)

Topographic Map Symbols

Boundaries

- City
- Small Park

Buildings and related features

- Building
- School
- Athletic field
- Forest headquarters
- Airport
- Well (other than water), windmill or wind generator
- Picnic area
- Campground
- Winter recreation area

Contours

- Contour line (6000)
- Depression
- River mileage marker (Mile 69)

Railroads

- Railroad, single track
- Railroad, multiple track

Rivers, lakes and canals

- Perennial river
- Perennial lake/pond

Roads and related features

- Primary highway
- Secondary highway
- Trail
- Highway or road with median strip

Vegetation

- Woodland
- Shrubland
- Orchard
- Vineyard

U.S. Geological Survey used in compiling this information.

4A Speed

Can you predict the speed of the car as it moves down the track?

What happens to the speed of a car as it rolls down a ramp? Does the speed stay constant or does it change? In this investigation, you will measure the speed of a car at different points as it rolls down a ramp. Then you will make a graph that describes the motion, and predict the speed of the car somewhere on the ramp.

Materials

- Data Collector and 1 photogate
- Energy Car
- Physics stand

1 Describing speed

Suppose you ran in a race. What information do you need to describe your speed? Saying that you ran for 20 minutes would not be enough information. To describe your speed, you need two things:

1. The **distance** you traveled, and
2. The **time** it took you to travel that distance.

Example	Distance	Time	Speed
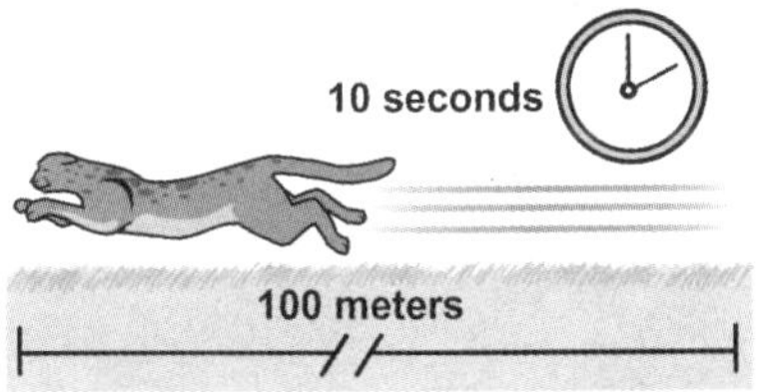	100 meters	10 seconds	10 m/s
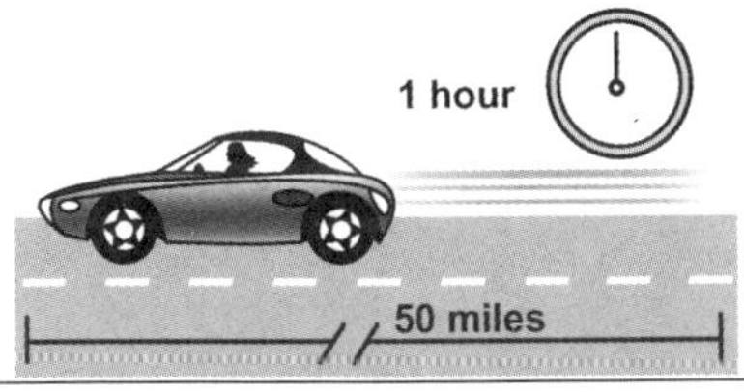	50 miles	1 hour	50 mi/hr
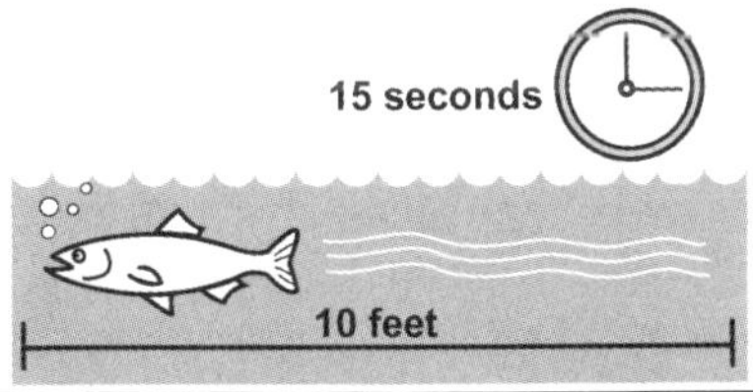	10 feet	15 seconds	0.67 ft/s

Based on the examples above, fill in the boxes to complete the *equation* for calculating speed.

$$\textbf{speed} = \frac{\square}{\square}$$

2 Making a hypothesis

Where is the car going the fastest on the ramp? Is it going fastest at the top, middle, or bottom?

a. Create a hypothesis about the car's speed on the ramp–where is it the fastest, and why do you think so?

3 Setting up the experiment

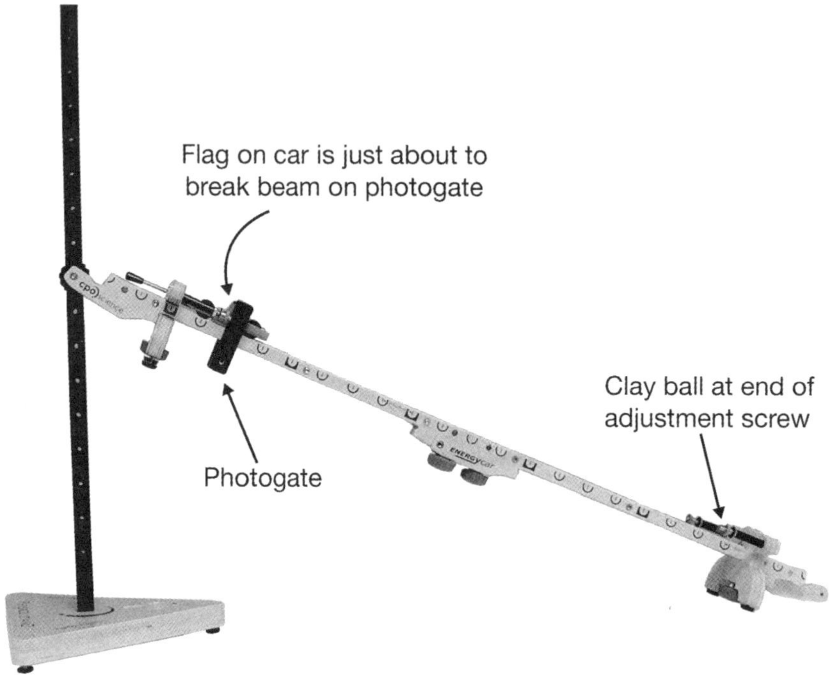

1. Join the two ends of the ramp that say "Energy Car" to each other. The ramp should be straight, without a bend in the middle. Use two threaded knobs to connect the two parts.
2. Attach the ramp to the physics stand using one threaded knob. Your teacher will tell your group which hole on the stand to use. Hole where ramp is attached: _____
3. Place a bumper at each end of the ramp. Place a clay ball on the end of the thumb screw at the bottom of the ramp to stop the car. The thumb screw at the top of the ramp will be the place where you start the car.
4. Attach a photogate to input A of the Data Collector.
5. Place the photogate so its screw is on the round mark that is closest to the front of the car.

4 Using the photogate to measure speed

As the car passes through the photogate, the Data Collector clock starts and stops. The Data Collector measures the length of time that the light beam is broken. Speed is equal to the distance traveled divided by the time taken. What distance do you use?

If you look at the car you will see a small “flag” on one side. This is the part of the car that blocks the photogate’s light beam. The distance the car moves while the light is blocked is the width of the flag, 1 cm.

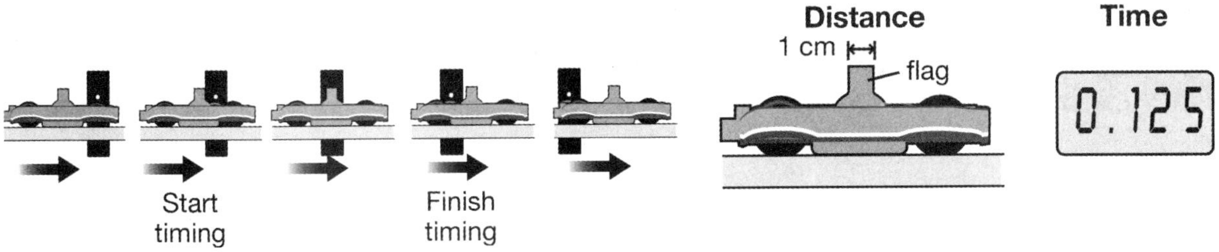

a. What is the *distance* traveled by the car in the example?

b. What is the *time* taken by the car in the example?

c. What is the *speed* of the car in the example?

5 Doing the experiment

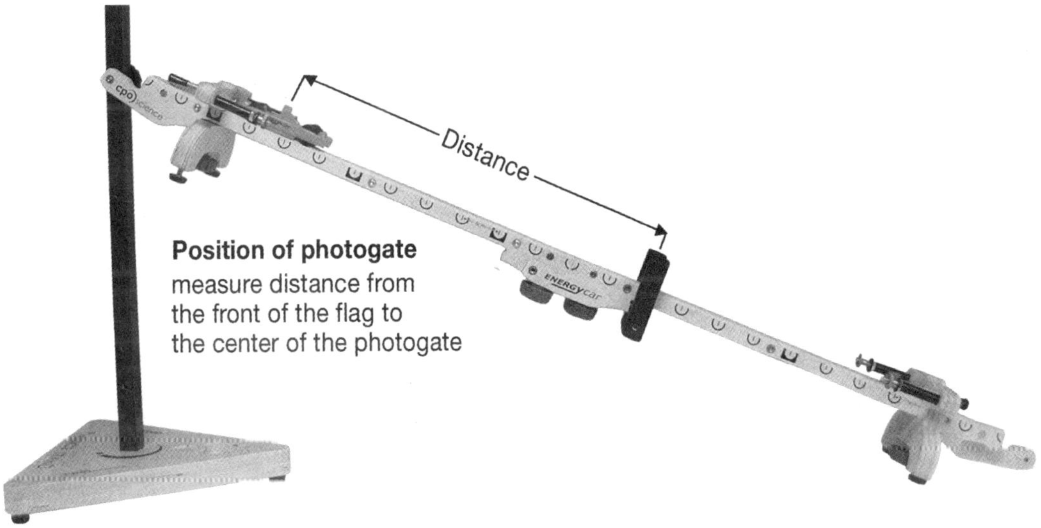

1. Place the car at the top of the ramp. Place the photogate 10 cm down the hill from the car. Measure the distance from the front edge of the car’s flag to the middle of the screw that holds the photogate to the track. If you line up the car’s flag over one of the circular markings, measuring is made easier, since each mark is 5 cm away from each other. Record the photogate’s position in Table 1.
2. Release the car without pushing it. Record the time through photogate A in the table.
3. Calculate the speed of the car using the distance traveled (1 cm) and the time at photogate A.
4. Move the photogate 10 cm down the track. Record the position, time, and speed.
5. Repeat the measurements of position, time, and speed for six different places spaced along the ramp. You will have to skip some places in the middle of the track where the photogate won’t attach.

Table 1: Position, time, and speed data

Position of photogate A (cm)	Time through photogate A (s)	Distance traveled by the car (cm)	Speed of the car (cm/s)
		1.00	
		1.00	
		1.00	
		1.00	
		1.00	
		1.00	

6 Analyzing the data

a. From your measurements, what can you say about the car's speed as it moves down the ramp?

b. Use your data to make a graph which shows how the car's speed changes as it rolls down the ramp. Put speed on the *y-axis* and position of the photogate on the *x-axis*. Be certain to label the axes with the correct variable and the proper unit of measurement. Give the graph a descriptive title. Include the number of the hole you used to connect the ramp to the stand in your title.

c. Describe what the graph shows about how the speed of the car is changing as it moves down the ramp.

d. Compare your graph with that of students who connected their ramps at different heights on the stand. Explain any differences you see.

7 Using your graph

Now that you have gathered, organized, and analyzed your data, it is time to use it to make a prediction. You measured the speed of the car at several places on the ramp as it rolled to the bottom. Now, you will predict what the speed of the car will be at a place you did not measure. There is a way to do this with the information represented by your graph.

1. In Table 2 record a position on the ramp where you did not measure the speed of the car. The position should be between two places where you did measure the speed.

Table 2: Predicted speed data

Selected position (cm)	Predicted speed at selected position (cm/s)	Actual speed at selected position (cm/s)	Percent correct of prediction

2. Use your graph to find the predicted speed of the car at the selected position. To do this, start on the x-axis at the position you have selected. Draw a line straight up until it intersects with the speed vs. position line on your graph. At the intersection point, draw a line horizontally over to the y-axis where the speed is recorded. This is the speed that corresponds to your predicted location. The graph to the right uses a position of 55 cm as an example. Use a different position. Record your predicted speed in Table 2.

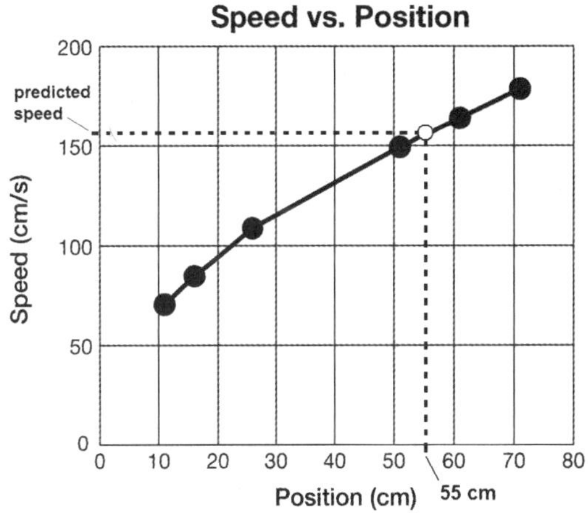

3. Place the photogate at the position you selected in step 1 and record the time it takes for the car to pass through the photogate.
4. Use the wing length (1.00 cm) and the time to calculate the speed. Record the actual speed in Table 2.
5. How does the predicted speed compare with the actual measured speed? What does this tell you about your experiment and measurements?

8 Calculating percent error

a. Find the difference between the predicted speed and the actual speed you calculated.

Predicted speed – Actual speed = Difference

b. Take this difference and divide it by the predicted speed, then multiply by 100.

(Difference ÷ Predicted speed) × 100 = Percent error

c. Use the percent error to calculate percent correct. Record percent correct in Table 2.

100 – Percent error = Percent correct

d. What do you think can account for any error you may have had?

4B Acceleration

What is acceleration?

When you turn the track into a sloped ramp, the car can gain speed as it travels down the ramp. The speed of the car is constantly changing. What would a graph of this type of motion look like? Is there more than one way to graph the motion of an object? In this investigation you will learn what acceleration is, what the graphs of associated with acceleration look like, and why they look the way they do.

Materials

- Data Collector and 2 photogates
- Energy Car
- Physics stand

1 Setting up the experiment

Attach the straight track to the 10th hole from the bottom of the Physics stand.

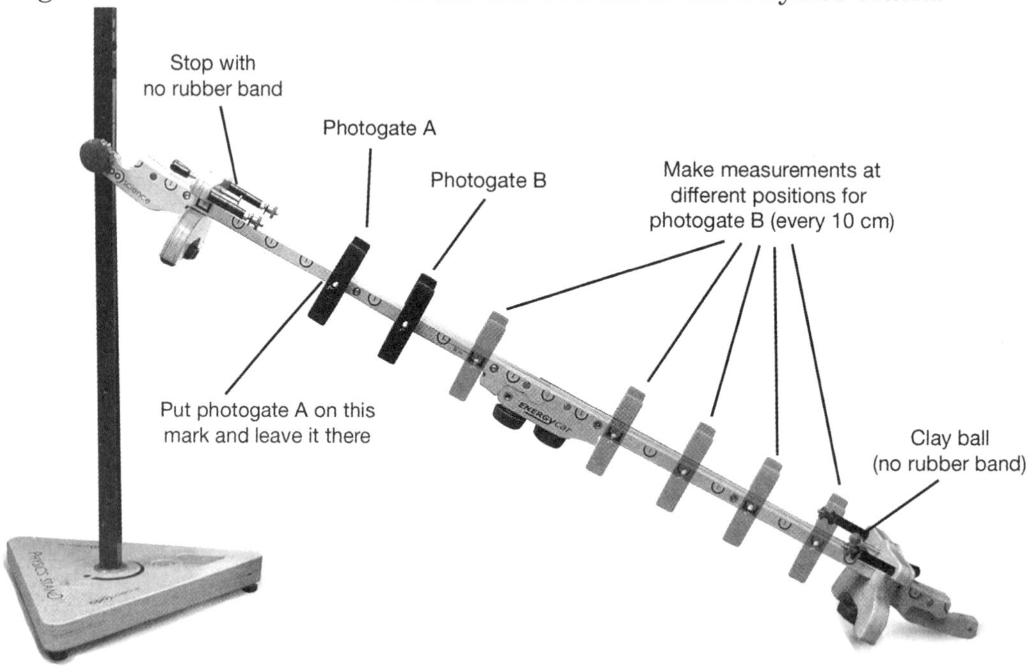

1. Set up the track as a long straight hill. Attach the track to the 10th hole from the bottom of the Physics Stand.
2. Place photogate A at position shown in photo. Keep photogate A in this position.

2 Predicting

A graph is a visual tool that shows how two things (called variables) are related to one another. Release the car from the top of the track and let it roll down the track a few times. Observe how the car's position and speed changes as it rolls down the track.

a. Think about what happens to the car's position as it rolls down the track. Sketch what you think the position (y) vs. time (x) graph will look like for the Energy Car as it moves down the ramp. No need to put any numbers on the axes; just label the axes and place a flat line, diagonal line, or curve on the graph, according to your prediction.

b. Do the same for the speed (y) vs. time (x) graph prediction.

3 Doing the experiment

1. Move photogate B to different positions 10 cm apart along the track.
2. For every position of photogate B, release the car and record the time through A and the time from A to B in Table 1.
3. Use the time through A to check the consistency of your release technique. If there is a very different time through A, re-do the trial.

Table 1: Position versus time data

Distance from A to B (cm)	Time through photogate A (s)	Time from photogate A to B (s)
10		
20		
40		
50		
60		
70		

4 Position vs. time graph for acceleration

a. Use your data from Table 1 to make a graph with the *position* (distance from A to B) on the *y* axis and the *time* it took to go that far (time from A to B) on the *x*-axis.

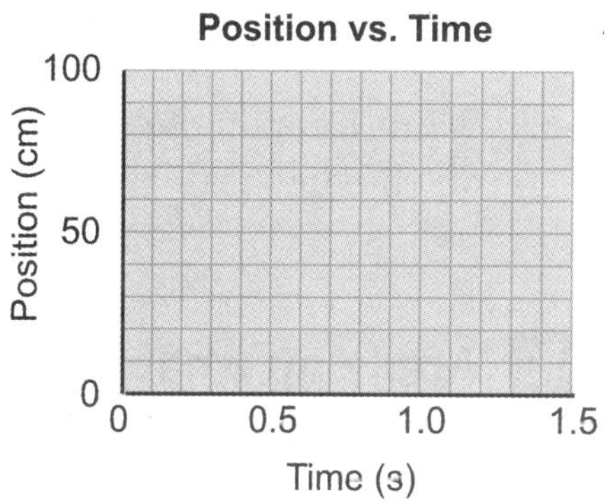

b. How does the graph from your experiment data compare to your prediction? What happens to the distance traveled as time goes by for the car on the ramp?

c. How do you know, from looking at the graph, that the car did NOT travel at a constant speed?

5 Speed vs. time graph for acceleration

Another way to describe the car's motion is to show how its speed changes with each new time interval from A to B. Speed is a measurement of how fast or slow an object's position changes. The measurement of how an object's speed changes is acceleration. You will try the experiment again, and use data to calculate the speed of the car as it travels through photogate B.

1. Keep photogate A in the same place as it was in part 3.
2. Move photogate B to different positions 10 cm apart along the track.
3. For every position of photogate B, release the car and record the time through A, the time through B, and the time from A to B in Table 2.
4. Use the time through A as a check for the consistency of your release technique. If there is a different time through A, do the trial again.

Table 2: Speed vs. Time Data

Distance from A to B (cm)	Time through photogate A (s)	Time through photogate B (s)	Time from photogate A to B (s)	Speed at photogate B (cm/s)
10				
20				
40				
50				
60				
70				

5. How fast is the car going as it travels through B each time? Calculate the speed of the car at each position of photogate B. Record your speeds in Table 2.

$$v = \frac{d}{t} \qquad v = \frac{1.00\ \mathbf{cm}}{\mathbf{time\ B}}$$

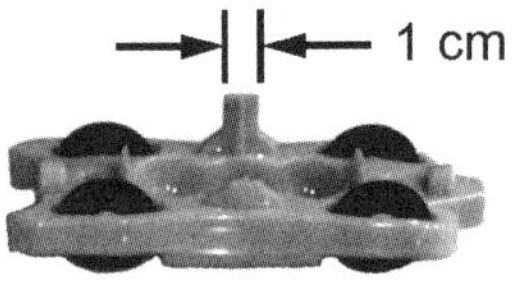

a. Draw a graph with the *speed* (speed through B) on the *y* axis and the *time* it took to go that far (time from A to B) on the *x*-axis.

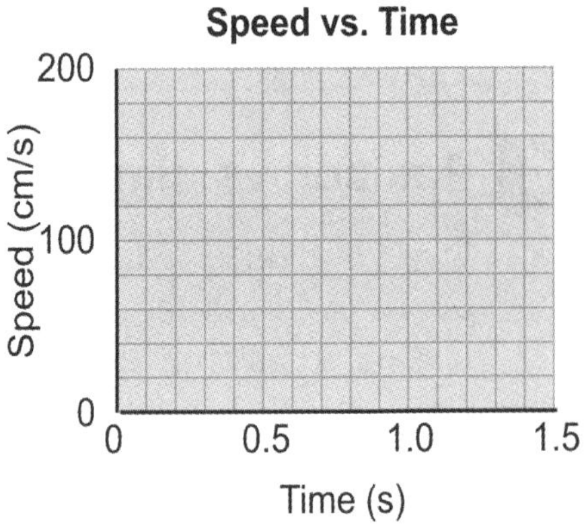

b. How does the graph from your experiment data compare to your prediction?

c. What happens to the speed as time goes by for the car on the downhill ramp?

6 Applying what you've learned

a. Did the car accelerate as it traveled down the hill? Justify your answer.

b. Use the formula for acceleration to find the average acceleration of your car. Use data from the trial in Table 2 where photogate B was 70 cm from A. The unit for acceleration will be cm/s/s.

$$acceleration = \frac{final\ velocity - initial\ velocity}{time\ elapsed} \qquad acceleration\ of\ car = \frac{speed\ \mathbf{B} - speed\ \mathbf{A}}{time\ \mathbf{A\ to\ B}}$$

c. Use your speed vs. time graph to predict how much time would elapse for the car to reach a speed of 200 cm/s.

d. Use your position vs. time graph to predict where on the track the car would reach a speed of 200 cm/s.

e. Test your predictions with a photogate at your predicted location on the track. Were your predictions accurate? Record your observations.

5A What is a Newton?

What is force and how is it measured?

You can think of force as a push or pull. Objects interact with each other (and you) through forces. It takes force to start an object's motion, and force to stop an object in motion. This investigation will explore the precise definition of force and measure the strength of forces.

Materials

- Spring scale (0 – 5 N)
- 15 round metal washers (1/2 inch inner diameter)
- Loop of string
- Electronic scale (or triple-beam balance)

1 Measuring forces

Forces have two important properties: strength and direction. In the English system of units, the strength of a force is measured in pounds. When you measure your own weight in pounds, you are measuring the force of gravity acting on your body. In the SI system, the strength of a force is measured in newtons (N). A quarter-pound hamburger has a weight of about 1 newton (1 lb = 4.448 N).

1. You can measure force with a spring scale. Before using the spring scale however, you must be sure it starts at zero. Calibrate the spring scale by turning the nut on the top until the plunger lines up with the zero mark.
2. Pull on the hook so the spring extends. When you pull, you are applying a force. Can you make a force of two newtons (2 N)?

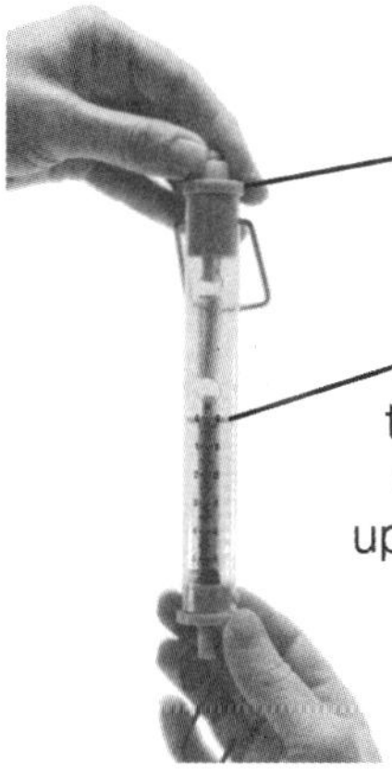

2 Weight: The force of gravity

Weight is a common force that you may be familiar with. On Earth, objects that have mass also have weight. Weight comes from the action of gravity on an object's mass.

1. Attach 3 steel washers to a loop of string.
2. Use a calibrated spring scale to measure the weight of the washers in newtons (N).
3. Use an electronic scale or triple beam balance to measure the mass in grams (g). Convert each mass in grams to kilograms (divide by 1000 or move decimal point three places to the left).
4. Repeat the measurements for 6, 9, 12, and 15 washers.

Table 1: Weight and Mass Data

Number of washers	Weight (N)	Mass (g)	Mass (kg)
3			
6			
9			
12			
15			

3 Stop and think

What do the results of your experiment tell you about the relationship between weight in newtons and mass in kilograms? Create a graph as described below to answer this question.

a. Make a graph of your data from Table 1. Place weight in newtons on the vertical (y) axis and mass in kilograms on the horizontal (x) axis.

b. Describe the graph. What does it tell you about the relationship between mass and weight?

c. Calculate the slope of your graph. The slope is equal to the strength of gravity (g) on Earth, measured in newtons per kilogram (N/kg).

d. Write an equation that relates weight in newtons, mass in kilograms, and the strength of gravity (g). It should be in the form: weight = ____________.

e. If an object has a mass of 10 kilograms, how much does it weigh in newtons?

4 Applying what you have learned

a. Explain how you could estimate the weight and mass of seven of your steel washers.

b. Find the weight and mass for seven of your steel washers. How close is the actual value to your estimated value? Explain some reasons why your value may not be perfectly accurate.

5B Friction

How does friction affect motion?

Friction is always present. Sometimes we want friction. For example, friction between tires and the road allows a car to be steered safely and to maintain its direction when moving. Other times we want to reduce friction. Putting oil on a bicycle chain allows it to work more efficiently with the gears. This investigation explores different effects of friction.

Materials

- Energy Car kit (including the sled)
- Physics stand
- Data Collector and photogates
- Tongue depressor
- Large paper plate
- Tape

1 Control setup

The first kind of friction you will be investigating is air friction. You will begin by finding out how the car moves before you add extra air friction.

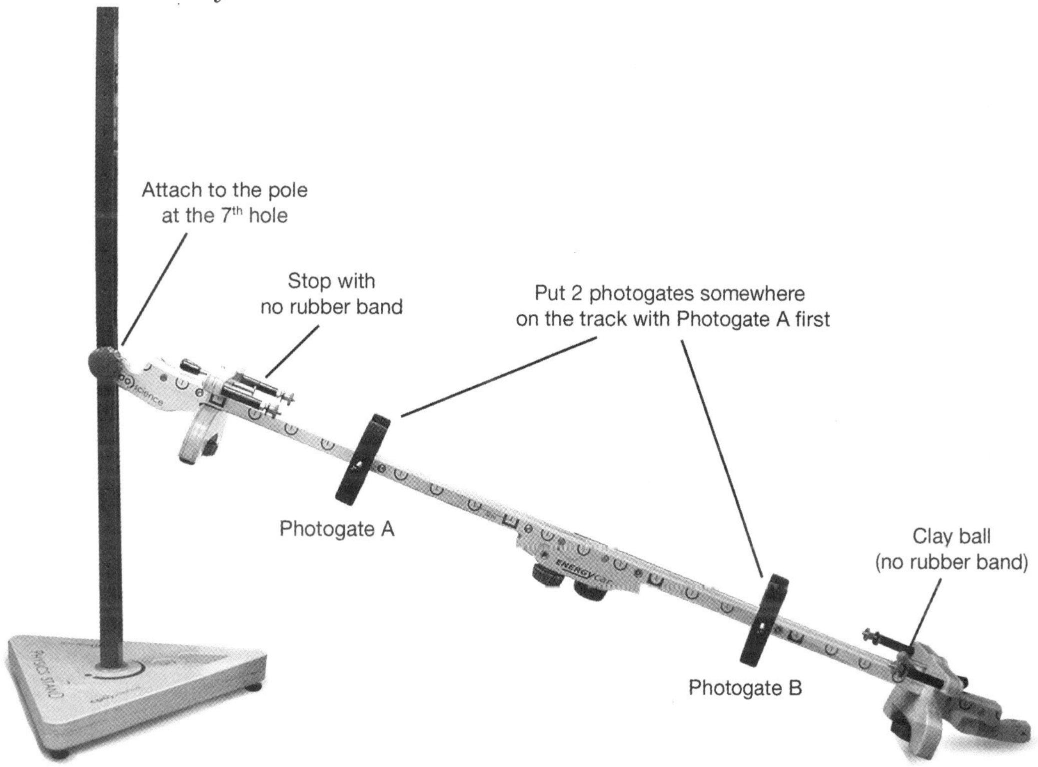

1. Set up the track as a long straight hill as shown above.
2. Attach the track to the stand at the seventh hole from the bottom.
3. Place one photogate near the top of the track and one near the bottom of the track.
4. Put a steel ball in the middle pocket of the car.
5. Let the car roll down the ramp, and record the time from A to B.
6. Measure the distance between the photogates.
7. Use the distance to calculate the average speed of the car.
8. Repeat two more times, for a total of three trials.
9. Calculate the average speed from your three trials.

Table 1: Control speeds

Trial	Time A to B (s)	Distance between A and B (cm)	Speed (cm/s)
1			
2			
3			
		Average Speed	

2 Create the "sail" car

A paper plate "sail" adds air friction (drag) to the car.

1. Tape a tongue depressor to the flag on the side of the car.
2. Tape a paper plate to the tongue depressor. Use enough tape to make sure it is securely attached.

3 Your hypothesis

a. Write a hypothesis that compares the speeds of the "sail car" and the normal car.

b. Explain the reasoning behind your hypothesis.

4 Do the experiment

1. The track and photogates should be set up as in part 1.
2. Put a steel ball in the middle pocket of the car.
3. Let the sail car roll down the ramp, and record the time from A to B.
4. Calculate the average speed of the car.
5. Repeat two more times, for a total of three trials.
6. Calculate the average speed from your three trials.

Table 2: Experimental speeds; sail car

Trial	Time A to B (s)	Distance between A and B (cm)	Speed (cm/s)
1			
2			
3			
		Average Speed	

5 Stop and think

a. Did your results confirm your hypothesis? Explain.

b. How did air friction affect the car's motion?

6 Applying what you have learned

a. Friction is a force that opposes motion. Explain where the friction force on the sail comes from.

b. How could you increase the air friction on the car? How could you decrease it?

c. Is the sail the only source of friction? Does the car have any friction forces acting on it other than air friction? Explain.

7 Setting up to measure rolling and sliding friction

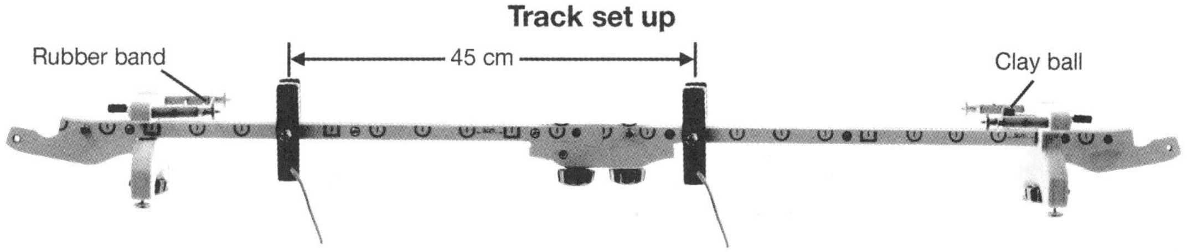

1. Set up the track so is exactly level. Put a clay ball on the stopper at one end of the track. Put a rubber band on the other end of the track. As you attach it, twist it once so it makes an X. You will be using the rubber band to launch the car and sled.
2. Adjust the stopper near at the launching end of the track so it is approximately 4 cm from the rubber band.
3. Place the sled on the track so the nose of the sled is touching the rubber band.
4. Place photogate A on the mark just ahead of the flag on the sled. The flag should not be blocking the photogate beam. Place photogate B 45 cm from photogate A.

8 Do the experiment

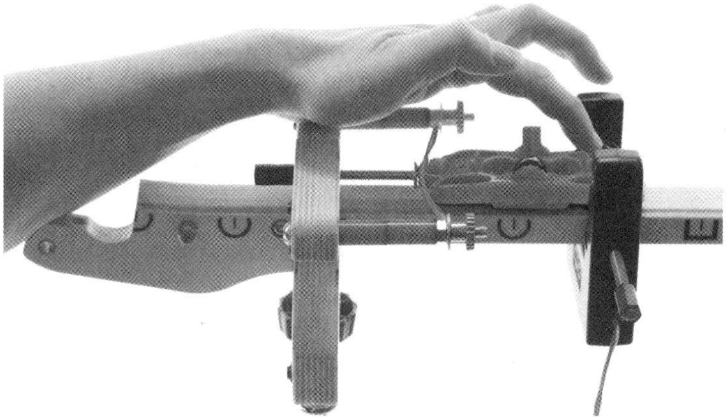

1. Launch the sled by resting your hand on the wooden track support and placing your index finger on the finger grip near the front of the sled. Practice a few times.
2. The sled should make it through both photogates. If it stops too soon, adjust the stopper so you can pull the rubber band back farther.
3. Record the time through each photogate and the time from photogate A to B.
4. Repeat for a total of three trials.

Table 3: Sliding friction data

Trial	Time through A (s)	Time through B (s)	Time from A to B (s)	Speed through A (cm/s)	Speed through B (cm/s)	Acceleration (cm/s^2)
1						
2						
3						
					average	

5. The speed of the sled through photogate A is the width of the flag (1 cm) divided by the time through A. Calculate the speed through A. Repeat for photogate B.
6. Use the two speeds and the average time from A to B to calculate the acceleration. Then find the average acceleration.
7. Repeat steps 3 through 6 using the car instead of the sled.

Table 4: Rolling friction data

Trial	Time through A (s)	Time through B (s)	Time from A to B (s)	Speed through A (cm/s)	Speed through B (cm/s)	Acceleration (cm/s^2)
1						
2						
3						
					average	

9 Thinking about your data

a. Were your accelerations positive or negative? Why?

b. Which decelerated more, the sled or the car?

c. What does your answer to the previous question tell you about the rolling friction of the car compared to the sliding friction of the sled?

d. How could you increase the sliding friction between the sled or track? How could you decrease it?

e. How could you increase the rolling friction between the car and track? How could you decrease it?

f. Compare rolling friction, air friction, and sliding friction. Which do you think has the greatest effect on the car's motion? Which has the least effect?

6A Newton's First and Second Laws

What is the relationship between force and motion?

The relationships between force and motion are known as Newton's laws. These are among the most widely used relationships in all of physics. The first law explains what happens when there is NO net force on an object, and the second law explains what happens when there IS a net force. The two laws are closely related. We will focus in this investigation on what happens to the Energy Car's motion when you change the force and the mass separately. Both laws apply!

Materials

- Energy Car
- Data Collector and photogate
- 3 same-size rubber bands
- 3 steel marbles
- Digital balance (or triple-beam balance)

1 Making predictions

Newton's second law explains what happens to the acceleration of an object when you change the force applied or change the mass of the object. For this first part of the investigation you will use different amounts of force to launch the car. You will not measure the resulting acceleration, because you would have to measure it while the car is still being acted upon by the rubber band, and this is impossible. Instead, you will measure the RESULT of the acceleration caused by the force of the rubber band. The RESULT will be measured by looking at changes in the car's speed as it rolls along the track after the launch.

a. Make a prediction: What will happen to the speed of the car as the force gets larger?

2 Changing force with constant mass

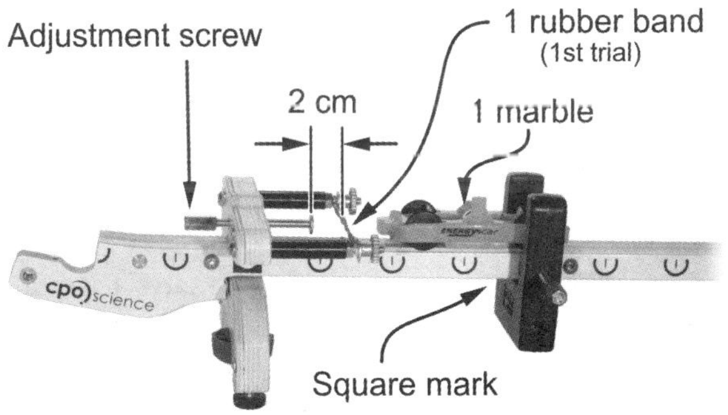

Rest your palm on the wood and pull the car against the screw with your finger on the tab nearest the far end of the car.

1. Set up the long straight track with a rubber band on one end and a clay ball on the other end. Adjust the screw so the rubber band deflects about two centimeters. Be sure to stretch the rubber band once or twice before using.
2. Put one photogate on the first square mark after the rubber band.
3. Put one marble in the center of the car. Use the screw to launch the car using the same deflection of the rubber band each time. This means the same force is applied to each launch. Try to get three launches with times that are within 0.0015 seconds. Average these times and record the result in Table 1.
4. Repeat the experiment with two and three rubber bands. Stretch before using!

5. Calculate the car's speed. Remember to use the tab width, 1.00 cm, for the distance.

Table 1: Changing force data

Number of rubber bands	Time through photogate (s)	Speed (cm/s)

3 Stop and think

a. During which portion of the car's motion is the rubber band affecting its speed?

b. Make a graph showing the speed of the car on the *y*-axis and the number of rubber bands on the *x*-axis. As the force was increased, what happened to the speed of the car?

c. Why was the same mass used for all trials (with different force)?

4 Changing mass with constant force

Use 1 rubber band to launch cars with 0, 1, 2, and 3 marbles.

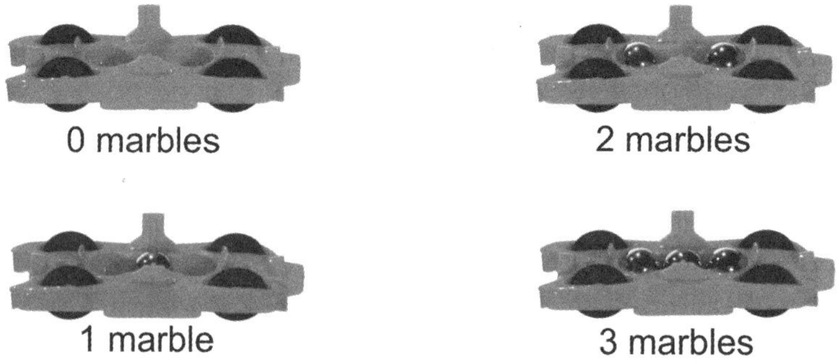

1. Put a single rubber band on the launching end of the track. Stretch it before using. Leave the photogate in the same place as it was (on the first square mark after the rubber band).
2. With the screw in the same place, launch cars of four different masses. Record the times in Table 2.
3. Measure the mass of the car with 0, 1, 2, and 3 steel marbles.
4. Calculate the car's speed. Remember to use the tab width, 1.00 cm, for the distance.

Table 2: Changing mass data

Mass of car (g)	Time through photogate (s)	Speed (cm/s)

5 Applying what you have learned

a. Use Table 2 to graph the speed of the car (y) against the mass (x). Does your graph show a direct relationship or an inverse relationship?

b. Why did the speed change when the same launching force from the rubber band was applied to cars of different mass? How do your observations support your answer?

c. Newton's First Law is often called "The Law of Inertia". How does this law apply to the car's motion when you changed the mass but kept the force constant?

d. Do you think the force applied to an object causes speed itself or causes changes in speed? Support your answer with at least one sentence of explanation for why you believe your answer is correct.

When an object's speed changes we say the object *accelerates*. Acceleration occurs whenever speed changes. To be precise, acceleration means the "change in speed divided by the change in time".

$$\text{Acceleration} = \frac{\text{change in speed}}{\text{change in time}}$$

e. Based on your experimental results, propose a mathematical relationship between the variables F (force), a (acceleration), and m (mass).

6 Fun challenge

Try this additional mini-experiment to see something interesting. You will launch the car with one rubber band and no marbles. Then you will launch the car with TWO rubber bands and TWO marbles in the car. This means you will double both the force and the mass at the same time. How will the times compare? Make a prediction.

1. Launch the car with one rubber band. Stretch the rubber band before using. There should be no marbles in the car. Get three times that are within 0.0015 s of each other and average. Record the time average in Table 2.
2. Launch the car with two rubber bands (be sure to stretch first!) and 2 marbles (placed in front and back of car). Get three times that are within 0.0015 s of each other and average. Record the time average in Table 2.

Table 3: Doubling Force and Mass

	Time through photogate (s)
1 rubber band, no marbles	
2 rubber bands, 2 marbles	

a. What do you notice about the times?

b. Explain your result. How does it compare to your prediction?

6B Newton's Third Law

What happens when equal and opposite forces are exerted on a pair of Energy Cars?

When you apply a force to throw a ball you also feel the force of the ball against your hand. That is because all forces come in pairs called action and reaction. This is Newton's Third Law of motion. There can never be a single force (action) without its opposite (reaction) partner. Action and reaction forces always act in opposite directions on two different objects. You can set up two Energy Cars to study Newton's Third Law.

Materials

- Energy Car
- Data Collector and photogates
- 1 Rubber band
- 2 Steel marbles
- Energy Car Link

1 Setting up and starting the experiment

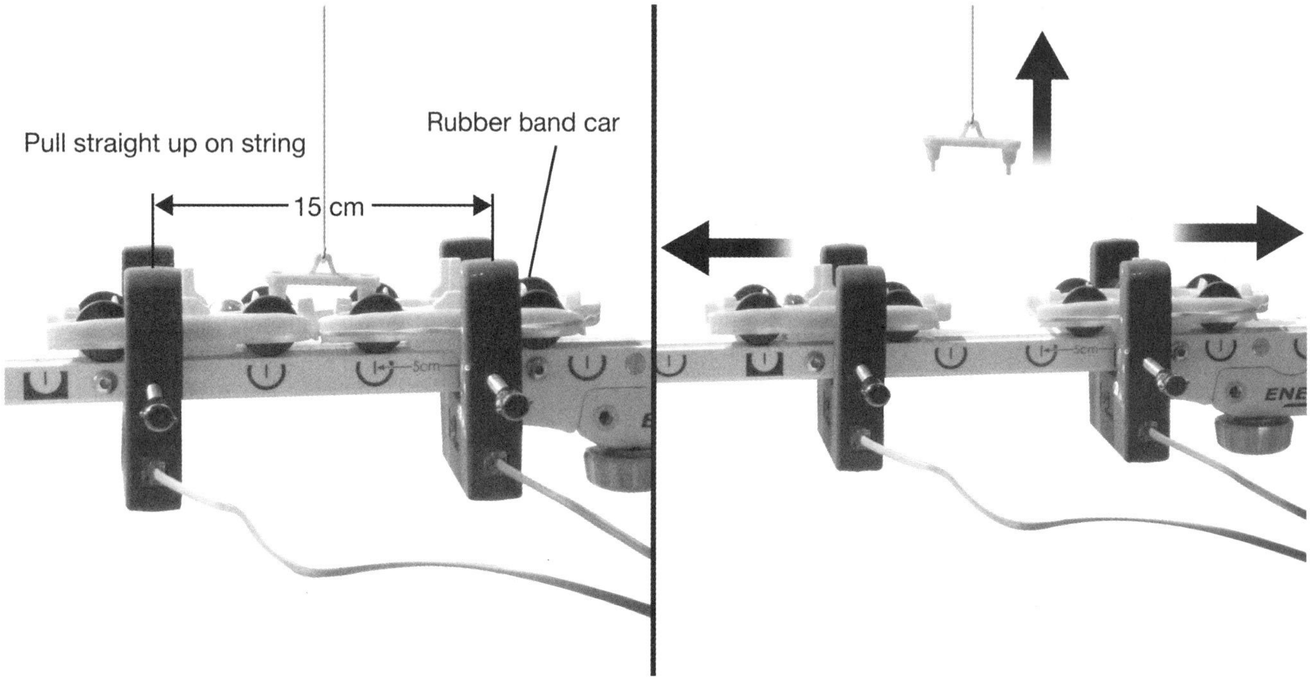

1. Set up the long straight track with a ball of clay on each stop. Use the bubble level to set the track level.
2. Place one steel marble in each car, and wrap one car with a rubber band.
3. Place two photogates 15-cm apart as shown in the photo.
4. Place the 2 cars, "nose to notch" between the photogates.
5. Squeeze the cars together and attach them with the Energy Car link.
6. Center the attached car pair between the photogates so each is about to break the photogate's beam, but do not actually break the beam. Check that both photogate indicator lights are still green. Make sure all 4 wheels of both cars are on the track.
7. With a *very quick* upward motion, pull the link *straight up* and out from the cars. ***CAUTION: Wear eyeglasses or safety glasses to avoid injury.***
8. Observe the time through each photogate. Repeat several times.

2 Thinking about what you observed

a. What caused the two cars to move when you took out the link?

b. According to Newton's third law, the cars experienced equal and opposite forces. How can you tell this is true by looking at the time through each photogate?

c. If one car was twice the mass of the other, would the cars still experience equal and opposite forces? Why or why not?

3 Changing the masses

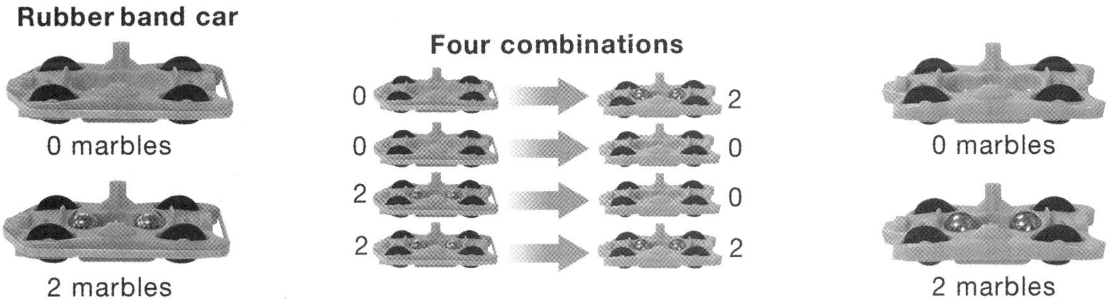

NOTE: Adding two steel marbles to the Energy Car doubles its mass.

1. Try the experiment with the four combinations of mass shown above. Take the average of three trials for each and record your data in Table 1. ***CAUTION: Wear eyeglasses or safety glasses to avoid injury.***
2. Calculate the average speed for each trial and record in Table 1.

Table 1: Energy Car Action–Reaction Data

Marble pairings for connected cars		Time through photogate (s)		Speed (cm/s)	
A	B	A	B	A	B
0 marbles	2 marbles				
0 marbles	0 marbles				
2 marbles	0 marbles				
2 marbles	2 marbles				

4 Applying what you have learned

a. How does the speed of each car pair compare when masses are equal?

b. How does the speed of each car compare when one of the pair has twice the mass?

c. Explain how your speed data supports the idea that there are equal and opposite action and reaction forces acting on the cars.

d. If the action and reaction forces are equal in strength, when the cars separate, why does one car move at a different speed than the other car when the masses are unequal? Hint: the answer involves the second law of motion.

7A Energy in a System

How is energy related to motion?

A system is a group of objects that interact with each other. Energy measures the ability of a system to change itself or other systems. This investigation is about systems and energy.

Materials

- Energy Car
- Data Collector and one photogate
- Rubber band
- Physics stand

1 Making a system

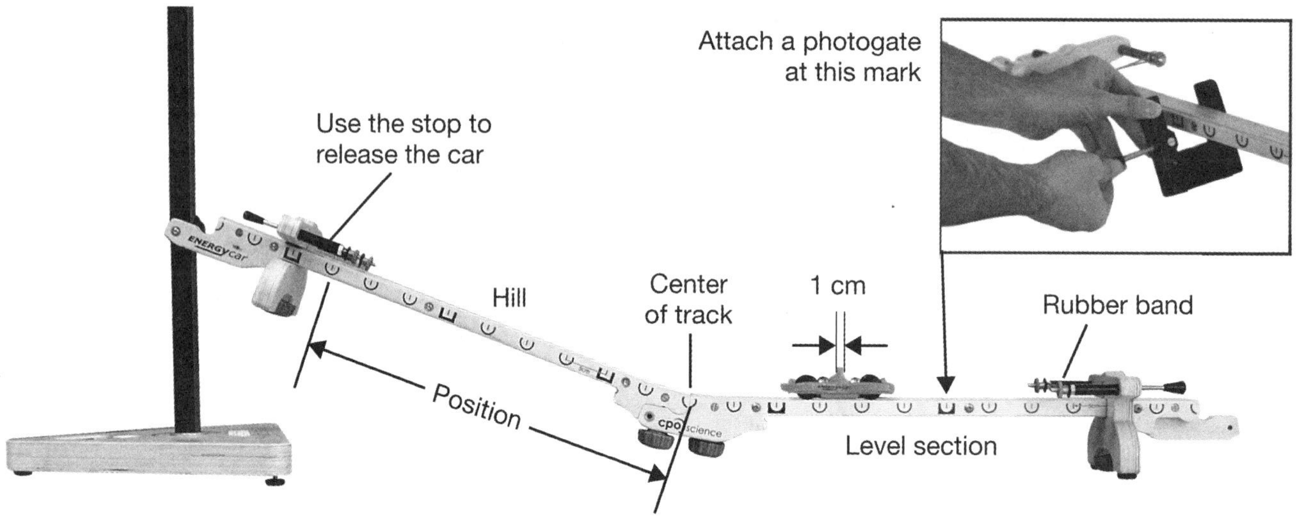

1. Set up the track with the steeper hill and a level section as shown above.
2. Place a rubber band on the thumb screws at the bottom of the track.
3. Attach a photogate near the middle of the level section at the spot marked with a square.

2 Collecting data

1. Measure the distance between the center of the track and the stopper at the top of the track as shown above. Record this distance (the drop position) in Table 1.
2. Hold the car against the stopper and release it without giving it a push.
3. Record the time through the photogate before and after the car bounces off the rubber band. You will have to use the memory button to get the time before the bounce.
4. Calculate the speed of the car before and after it bounces. The speed is the width of the flag (1 cm) divided by the time it takes the flag to pass through the beam of the photogate. Record the speeds in the table.
5. Move the wooden track stop part of the way down the hill. Measure the distance from the center of the track to the metal stopper.

6. Drop the car as you did before. Measure the times and calculate the speeds.
7. Repeat for several drop positions along the hill.

Table 1: Speed data

	Before bouncing		After bouncing	
Drop position (cm from center)	Time through photogate (s)	Speed (cm/s)	Time through photogate (s)	Speed (cm/s)

3 Thinking about what you observed

a. How high did the car climb up the hill after bouncing? Did it go higher, lower, or the same height as the drop position?

b. How is the drop position related to the speed of the car the first time it passes through the photogate (before bouncing)?

c. How do the speeds before bouncing compare to the speeds after bouncing? Is this the same for all five trials?

d. What could you do to make the car travel farther up the hill after bouncing?

e. In one paragraph, explain how the answers to a, b, c, and d are explained using the idea of energy.

7B Work and Energy

How does a system get energy?

Energy comes from somewhere. When you lift a box off the floor the increase in energy of the box comes from the work you do on the box. The box has a net increase in energy due to the additional potential energy you gave the box by doing work when you lifted the box. This stored potential energy does not always have to be in the form of gravitational potential energy. It can be stored through pressure, compression, or even elastic potential energy. In this Investigation you will examine the conversion of work into stored elastic energy, and then predict its conversion into kinetic energy.

Materials

- Energy Car
- Data Collector and photogates
- Electronic scale (or triple-beam balance)
- Metric ruler
- String
- 10 N Force scale

1 Measuring the work done

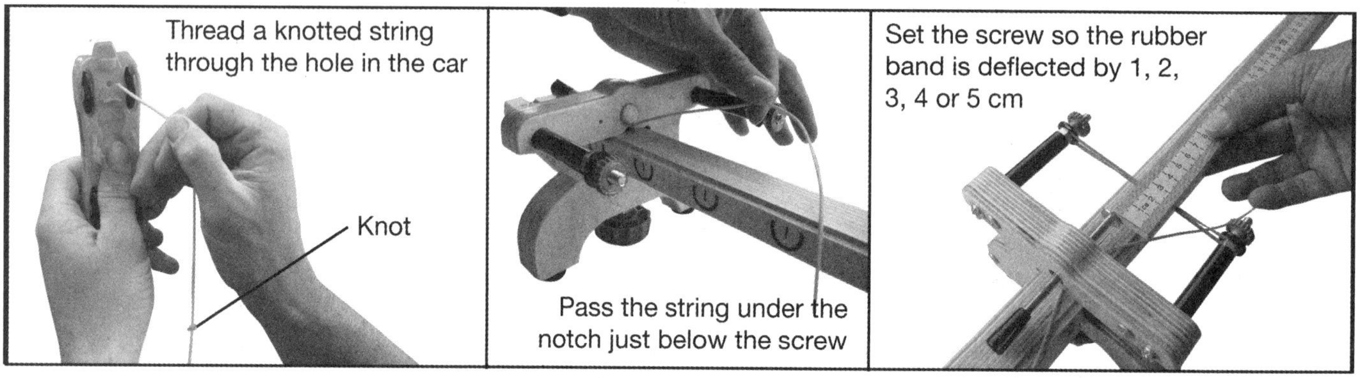

1. Set up the long straight track with a rubber band at one end and a clay ball at the other end.
2. Adjust the threaded screw until the distance between the screw and the front of the rubber band is one centimeter (see diagram).
3. Tie a knot in the string and pass the string through the hole in the car and under the notch just below the screw. Tie a loop in the other end of the string.
4. Use a spring scale to measure the force when the car is just touching the screw.
5. Adjust the screw for distances of 2, 3, 4, and 5 centimeters. Measure the force for each distance. Record your measurements in Table 1.

Table 1: Force vs. distance data

Distance rubber band is stretched (cm)	Force (N)

2 Thinking about what you observed

a. Graph the force from the rubber band vs. the distance.

b. Write down a one sentence definition of work in physics.

c. You would like to know how much work the rubber band does on the car during a launch. Since the force changes with distance, use the graph to do some averaging. Divide your graph up into bars each representing one centimeter of distance. Make the height of each bar the average force over the distance interval covered by the bar. The area of each bar is the work done over that interval of distance. Your graph will have data out to 5 cm. The sample graph below shows data from 0 to 3 cm as a demonstration that you can follow for your entire graph.

Step 1:
Graph force vs. distance

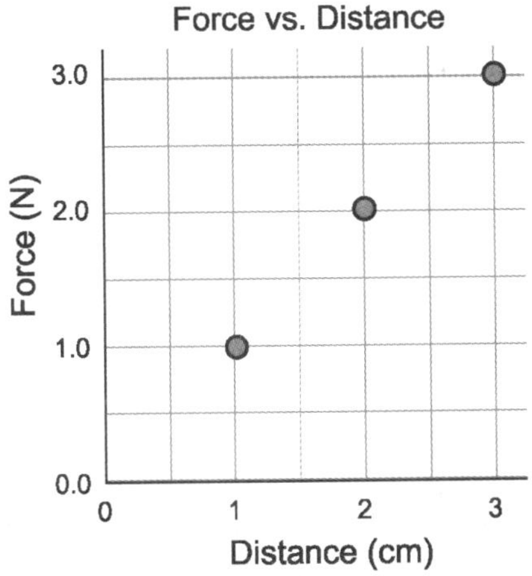

Step 2:
Divide graph into 1 cm bars

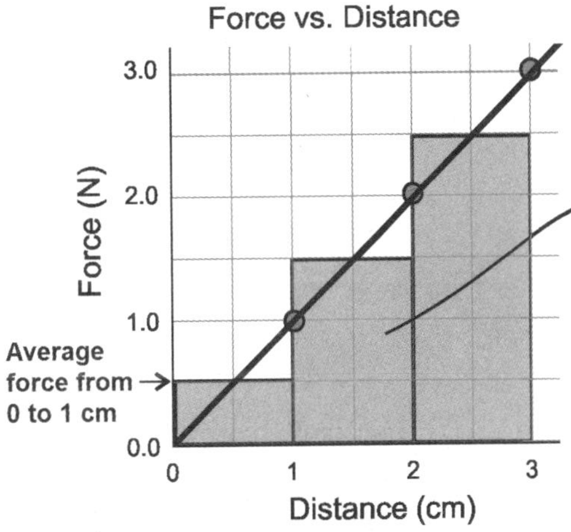

Step 3:
Calculate work for each bar

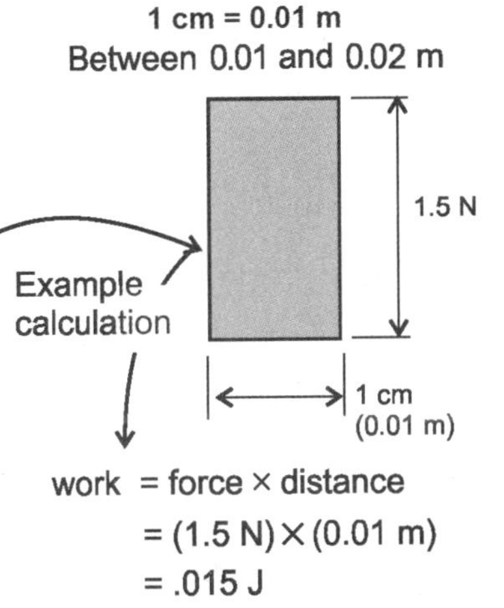

d. To get the total work done on the car you need to add up all the work done as the rubber band straightens out and moves the car forward. Use the table below to calculate the work done.

Rubber band deflection	4 - 5 cm		3 - 4 cm		2 - 3 cm		1 - 2 cm		0 - 1 cm		Total work done
1 cm										=	
2 cm								+		=	
3 cm						+		+		=	
4 cm				+		+		+		=	
5 cm		+		+		+		+		=	

e. Make a graph showing the work done on the *y*-axis and the deflection of the rubber band on the *x*-axis.

f. Assume all the work done becomes kinetic energy of the car. Derive a formula for the speed of the car that depends only on the car's mass and the work done by the rubber band.

3 Testing the theory

Set the photogate just ahead of the flag when the rubber band is straight.

Launch the car at the same measured deflections for which you measured the force.

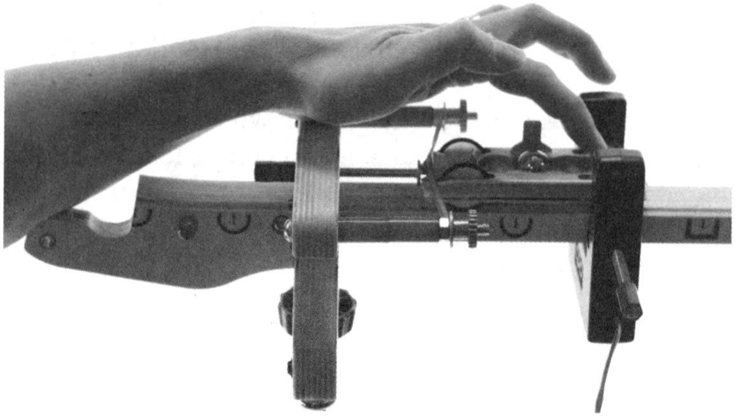

1. Put a photogate on the track so the flag on the car breaks the light beam about one centimeter after leaving the rubber band.
2. Use the adjustment screw to launch the car at the same measured deflections of the rubber band for which you measured the forces (1, 2, 3, 4, and 5 cm).
3. Measure the speed of the car for each deflection of the rubber band.
4. At each deflection, take data using cars with one steel ball of added mass. Record the speed in the column for "measured speed".

Table 2: Deflection, mass, and speed data

Deflection of rubber band (cm)	Mass of car (kg)	Photogate time (s)	Measured speed (m/s)	Predicted speed (m/s)

4 Thinking about what you observed

a. Use your formula to predict the speed the car should have at each combination of mass and deflection. Write the results in Table 2 in the column "predicted speed".

b. Graph the predicted speed of the car vs. the deflection of the rubber band. Draw a smooth curve through the plotted points. On the same graph, show the measured speeds.

c. Graph the measured speed of the car vs. the deflection of the rubber band on the same graph as part c.

d. Does your data support the theory that the energy of the car is equal to the work done by the rubber band? Your answer should provide evidence from your results and discuss possible sources of errors.

e. Use your theory to predict the speed if a car with 2 and 3 steel balls is launched at a deflection of 3 cm. Do the experiment and see if your prediction is accurate.

8A Determining Freezing/Melting Point

How do you determine the freezing/melting point of cetyl alcohol?

A cooling/heating curve is a plot of temperature vs. time. It illustrates the effect on a substance as the temperature decreases/increases through a phase change. In this experiment, you will study the effects of cooling and heating cetyl alcohol through a phase change. The data will be recorded using a data collector and then plotted on graphs. Based on your observations, measurements, and graphs you will be able to determine the freezing/melting point of cetyl alcohol.

Materials

- Data Collector
- Temperature probe
- Test tube, 22 × 150-mm
- Cetyl alcohol
- Hot water
- Safety goggles
- Lab apron
- Heat resistant gloves or hot mitts
- 250 mL beaker
- 400–600 mL beaker

WARNING — This lab contains chemicals that may be harmful if misused. Read cautions on individual containers carefully. Not to be used by children except under adult supervision.

1 Thinking about what you will do

a. The cetyl alcohol you will be using is a pure substance. What do you expect to observe in terms of the temperature range for the freezing/melting point?

b. Suppose your cetyl alcohol sample became contaminated with another substance. How would this affect the freezing/melting point?

2 Melting cetyl alcohol

1. Fill a 250 mL beaker with cold water.
2. Fill a 400 to 600 mL beaker with approximately 250 mL of hot water (75–80 °C).
3. Measure out approximately 4–5 grams of cetyl alcohol in a test tube.
4. Measure the temperature of the hot water to make sure it is within the range in step 2.
5. Quickly immerse the temperature probe into the cold water. Then remove and dry it making sure it is cooled to room temperature.
6. Place the temperature probe into the test tube and put the test tube containing the solid into the water bath and press START on the Data Collector. View the experiment in graph mode.
7. Stir the solid vigorously with the temperature probe.

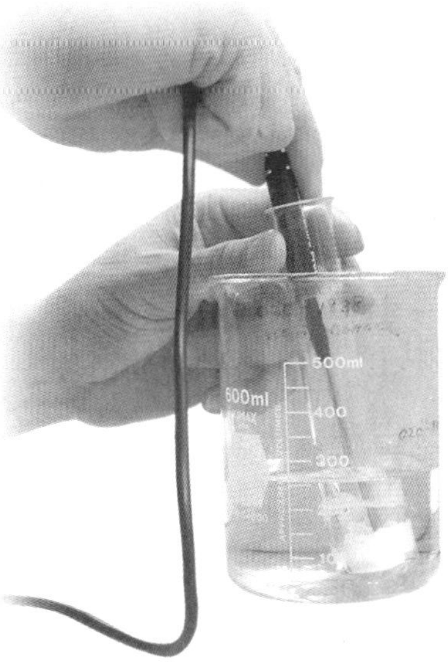

8. Continue to stir as long as some solid remains.
9. Once all of the material is liquid, stop and store the experiment. Record the experiment name assigned by the Data Collector.
10. Leave the sample in the warm water for use in Part 4.

3 Stop and think

a. Predict the effect of sample size on the melting/freezing point.

b. Describe and carefully sketch the shape of the graph from Part 2.

4 Freezing cetyl alcohol

1. Observe the cetyl alcohol in the test tube to make sure it is still all liquid. If some solid has formed, pour out the water in the warm beaker and add more warm water.
2. Remove the test tube from the warm beaker and start a new experiment on the data collector.
3. Stir the liquid vigorously with the temperature probe as the solid begins to form.
4. Continue to stir until all of the liquid turns into a solid. (NOTE: If you need to repeat any part of the experiment, you may re-use your sample.)
5. When you are finished recording data return your sample of cetyl alcohol to your teacher.
6. Record the name of the experiment from the Data Collector in your notebook.

5 Thinking about what you observed

a. Describe and carefully sketch the graph from Part 4.

b. Referring to both of your graphs, determine the melting and freezing point of cetyl alcohol. Are they the same? Should they be?

c. Based on the shapes of your curves, which data do you think is more reliable—the heating or cooling data? Why do you think this is so?

d. What is happening to the molecules of cetyl alcohol during the diagonal portions of the heating curve? What about the plateau?

e. Your graph from Part 2 shows that during a change from solid to liquid, the temperature stays the same. Explain why the temperature does not increase, even though energy is being added.

8B Freezing Point of a Stable Mixture

What is the freezing point of a unique ice cream topping?

Have you ever put "shell" topping on a frozen desert like ice cream? The topping is a liquid in the plastic bottle, but when you put it on ice cream, it hardens very quickly to form a shell or crust over the desert. The ice cream topping contains lecithin, a substance that stabilizes the mixture, so it acts like a pure substance when heated or cooled. In this investigation, you will compare the freezing points of different shell topping brands.

Materials

- Shell ice cream topping
- Foam cup (large)
- 1 large glass or plastic test tube
- Data Collector
- Temperature probe
- Ice water

1 Doing the experiment

1. Fill a foam cup about 3/4 full with ice water.
2. Shake the bottle of ice cream topping well, and pour about 15 mL into the test tube. This step might already have been completed for you.
3. Place the temperature probe in the test tube.
4. Connect the temperature probe to the Data Collector and set up a new experiment. Choose a sample rate of 5 seconds per sample (5 sec/sample). This means temperature data will be taken and placed in the table every 5 seconds.
5. Start the Data Collector. Place the test tube into the ice water bath. Use the temperature probe to gently stir the mixture while the temperature data is collected. Continue until the topping is a solid, and you can't move the thermometer any more. When the topping is a solid, continue collecting data until the temperature reaches 4 or 5°C (the temperature of the ice water bath).
6. Save the Data Collector experiment and make a note of the file name.
7. Follow your teacher's instructions for cleanup.

2 Analyzing the data

Go to the graph screen on your Data Collector. The graph should show temperature on the *y* axis and time on the *x*-axis. If this is not the case, select the correct variable for each axis. Study the graph and answer the questions below.

a. As you cooled the mixture, what happened to the temperature at first?

b. At some point in the cooling process, you will see that the line on the graph is relatively flat (this is called a plateau). What does this tell you about the temperature during this time period?

c. What is happening to the ice cream topping during this point in the cooling process?

d. Did the temperature drop again after the plateau? Discuss.

e. On a piece of graph paper, make a sketch of the graph you have on the Data Collector. You do not have to use numbers, but you should label each axis with the correct variable. On your sketch, label the area of the graph where freezing occurred, and make a note of the average temperature (freezing point). Or, you can download the data to a spreadsheet and create a graph to print and place in your lab notebook.

3 Thinking about what you observed

Different brands of shell ice cream topping have similar ingredients, but they are not all identical mixtures. Are their compositions different enough to affect the freezing point? You will compare results with other groups to find out.

a. Based on your data analysis, at what temperature did your brand or type of ice cream shell topping freeze?

b. Compare your freezing point with other groups that used the same brand. Is the temperature similar? Discuss.

c. Compare your freezing point with other groups that used a different brand or type of topping. Is the temperature similar? Discuss.

d. Based on your observations, if your teacher gave you an unknown brand of shell topping, would you be able to identify the brand or type by finding its freezing point? Explain your answer.

4 Applying your knowledge

a. When the topping was in the ice water bath, thermal energy was transferred away from the topping. At first, the temperature dropped rapidly, and then the temperature stalled. Why?

b. How would the experiment be different if the freezing point of the shell were 20°C?

c. How would the experiment be different if the freezing point of the shell were 0°C?

d. Pure substances have unique freezing/melting points, but mixtures generally do not. Shell ice cream topping is a mixture that *does* freeze and melt at a certain temperature range. Why does the ice cream topping act like a pure substance?

9A Temperature and Heat

How are temperature and heat related?

Hot and cold are familiar sensations. What happens when something hot comes in contact with something cold? Think about putting some ice cubes in a drink. Things don't remain the same. Changes occur and these changes have to do with the movement of energy from one material to the other. This investigation will explore the difference between temperature and thermal energy.

Materials

- Temperature probe
- Data Collector
- Electronic scale (or triple-beam balance)
- Aluminum cube from density kit
- 3 × 12-oz. foam cups
- Ice
- Hot water
- Tongs
- 100 mL graduated cylinder
- Safety goggles
- Lab apron

1 Making a prediction

Suppose you mix equal masses of water. One sample is at 0°C and the other is at 50°C. What do you think the final temperature of the mixture will be? Why?

2 Mixing hot and cold water

1. You will need three 12 ounce foam cups for this experiment. Label two of the cups as follows: HOT and COLD.
2. Prepare an ice bath by placing approximately 250mL of water in the unlabeled cup and add 4 or 5 ice cubes.
3. Connect the temperature probe to the Data Collector and select Meter mode for this experiment.
4. Place the temperature probe in the ice water bath.
5. Use a graduated cylinder to measure 100 mL of very hot tap water. Pour the hot water into the HOT cup.
6. Use a graduated cylinder to measure 100 mL of the ice water. Pour the cold water into the COLD cup.
7. Place the temperature probe into the COLD cup, wait until the reading stabilizes and record the temperature in Table 1.
8. Place the temperature probe into the HOT cup. Wait until the reading stabilizes and record the temperature in Table 1.
9. Immediately pour the hot water into the cold water. This is the mixture.
10. Stir well using the temperature probe and measure the final temperature (when it has stabilized). Record your data in Table 1.
11. Do not throw away your ice water. You will be using it again in the second part of the investigation.

Table 1: Temperature data for mixing equal masses of water

Cold water temp. (°C)	Hot water temp. (°C)	Mixture temp. (°C)

3 Thinking about what you observed

a. Did the result of your experiment agree with your prediction (within 3 degrees)? Discuss why the temperature value may not have exactly matched your predicted value.

b. How do you think your results would have been different if you had used more hot water than cold water, instead of equal masses?

4 Making another prediction

Suppose you mix equal masses of cold water and hot metal. Will the final temperature follow the same pattern as the experiment you did in Part 2? Explain your answer.

5 Combining hot metal and cold water

1. Find the mass of the aluminum cube and record in Table 2.
2. Connect the temperature probe to the Data Collector and select Meter mode for this experiment.
3. Place the temperature probe in the ice water bath that was left from part 2 of the experiment.
4. Place the aluminum cube in a foam cup and cover it with very hot tap water. Let the aluminum cube stay in the hot water for several minutes so it gets warm.
5. Measure out a mass of water from the ice water bath that is equal to the mass of the aluminum cube and pour it into a foam cup. Record the mass of the cold water in Table 2. Make sure there is no ice in the foam cup of cold water.
6. Move the temperature probe from the ice water bath to the cup of cold water and record the temperature of the cold water in Table 2.
7. Record the temperature of the hot water and aluminum cube just before putting the cube in the cup of cold water. Record the value in Table 2.
8. Use the tongs to put the aluminum cube in the foam cup containing the cold water.
9. Stir the mixture with the temperature probe.
10. Record the temperature of the cold water and metal cube when the temperature has stabilized. Do not wait longer than one minute to measure the temperature. Record this value in Table 2.

Table 2: Temperature data for combining water and metal

Metal mass (g)	Metal temp. before mix (°C)	Cold water mass (g)	Cold water temp. before mix (°C)	Mixture temp. (°C)

6 Thinking about what you observed

a. Why didn't the temperature of the water and aluminum mixture come out halfway between the temperature of the cold water and hot aluminum cube, even though you mixed equal masses?

b. Explain what is happening between the aluminum cube and water in terms of temperature and energy.

c. How much energy does it take to raise the temperature of one gram of aluminum by 1°C compared to raising the temperature of one gram of water by 1°C? (Look up this value.) Relate your answer back to your experimental results.

d. Heat and temperature are related, but they are not the same thing. According to your results, what does the concept of heat energy (or thermal energy) take into account that temperature does not?

9B Convection in Earth's Atmosphere

How is convection responsible for the movement of air through the atmosphere?

If you have ever seen a hot air balloon in the sky, you may have asked yourself a few questions: "How does the balloon rise up?" or "How does it sink back to the ground?" In a hot air balloon, a flame is ignited inside the balloon to warm the air, causing it to become less dense. The less dense air in the balloon causes it to take flight. When the flame is allowed to go out, the air inside the balloon cools off and becomes denser so the balloon sinks back toward the ground. In this investigation, you will explore how *convection* moves air from one place to another, forming winds.

Materials

- GeoBox
- Candle
- Long fireplace matches
- Safety goggles
- Incense
- Aluminum weighing dish
- A small piece of cardboard (6 cm × 6 cm)

1 Setting up and making observations

Safety tip: Use caution when working with an open flame. Long hair should be tied back. Loose or bulky outer layers of clothing should be removed.

WARNING — This lab contains chemicals that may be harmful if misused. Read cautions on individual containers carefully. Not to be used by children except under adult supervision.

1. Gather the materials listed above. Remove the lid of the GeoBox.
2. Carefully place the candle into the GeoBox so that it will be under one of the two chimneys that are on the lid. Put the incense in the weighing tray and place it in the GeoBox so that it will be under the other chimney. See the diagram below.
3. Carefully light the candle. Place the lid back onto the box.
4. Place your hand over the chimney above the candle. What do you feel with your hand?
5. Remove the lid from the GeoBox, and light the incense. Place the lid back on, and observe what you see inside the box.

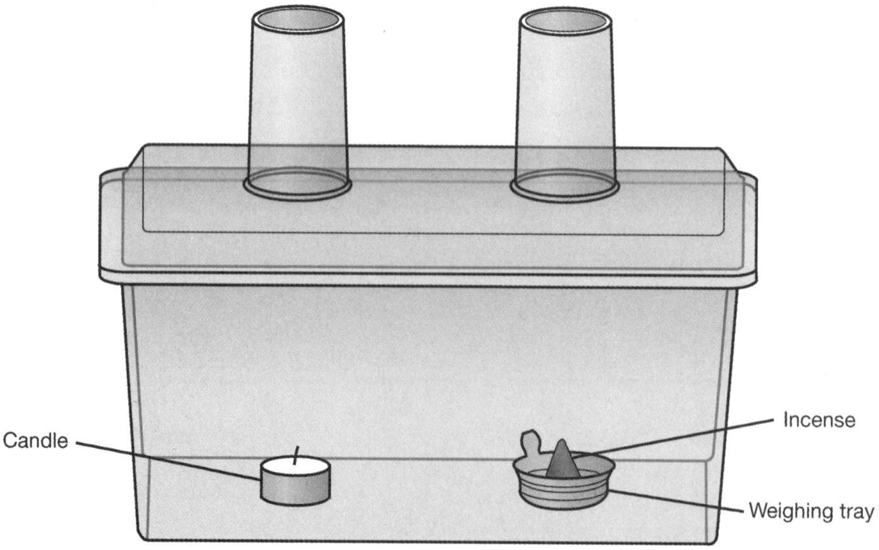

6. Observe what is happening inside the box and chimneys for several minutes. Sketch your observations.

2 Stop and think

a. Define convection. Where is the air rising and sinking in the box/convection chamber?

b. Describe the movement of the smoke inside the box. Be sure to describe the movement of smoke in the two chimneys as well.

c. What is the purpose of the smoke from the burning incense?

d. Based on your sketch in part 1 and your answers to these questions (a–c), make a diagram that illustrates how convection is taking place in the GeoBox. Draw and label where air is rising, sinking, becoming less dense, and becoming more dense.

3 Further exploration

1. Carefully place your fingers inside the two chimneys. Take note of what you feel. Is the air rising or sinking in each of the chimneys?
   Safety note: Your fingers will be at a safe distance from the candle when you place your fingers in the chimneys. However, it is always a good idea to work carefully around heat sources, especially candles!
2. Cover the chimney above the candle with the piece of cardboard. Observe what happens to the movement of the air inside the GeoBox.
3. Now, cover the chimney above the burning incense with the piece of cardboard. Observe what happens to the movement of the air inside the GeoBox.

4 Thinking about what you observed

a. Describe the motion of the air in the GeoBox above the candle. Be sure to describe the change in density of the air and whether air is rising or sinking.

b. Describe the motion of the air in the box above the incense. Be sure to describe the change in density of the air and whether air is rising or sinking.

c. Why is the air rising in the chimney above the candle? Why is the air sinking in the chimney above the incense?

d. Describe how the air is moving between the candle and the incense. How can you explain this movement in terms of convection?

e. Think about what you observed inside the GeoBox. Explain how convection currents in the atmosphere form and how they are related to wind.

5 Exploring on your own

a. A sea breeze is a type of wind that is commonly found along coastal areas during daytime hours. During the day, the sand of a beach absorbs more heat than the ocean, causing it to become warmer. This causes a wind known as a sea breeze to flow from the ocean to the land. Using arrows, draw how convection currents would form in the situation in the picture. Label the sea breeze, the location where air becomes less dense, and the location where the air becomes denser.

b. A land breeze is a type of wind that is commonly found along coastal areas during the night time. During the night the ocean tends to retain more heat than the sand on the beach. This causes the ocean temperature to typically become warmer than the sand on the beach. This temperature difference causes convection currents to form and winds to blow from the beach out to the ocean. Using arrows, draw how convection currents would form in this situation. Label the land breeze, the location where air becomes less dense, and the location where air becomes denser.

10A Density

How is an object's density related to its volume, mass, and tendency to sink or float?

You may be familiar with the trick question "Which is heavier: a pound of feathers or a pound of bricks?" The answer, of course, is that they have the same weight. However, the pound of feathers has a much greater volume because feathers have a much lower density than bricks. The brick material is squeezed together tightly, while the feathers contain a large amount of empty space. In this investigation you will study the relationship between mass, volume, and density. You will also determine how an object's density affects whether it sinks or floats in water.

Materials

- Balance
- Displacement tank
- Disposable cup
- 250 mL beaker
- Set of six identical objects
- 100 mL graduated cylinder
- Graph paper
- Density cubes
- Metric ruler
- Water
- Paper towels

1 Measuring mass and volume

1. Each lab group has a unique set of six objects. Find the mass and volume of one of your objects. Add a second object and find the total mass and volume of both objects. Then find the total mass and volume of three, four, and five objects. Record your data in Table 1. Note: Although your objects look identical, there may be small differences. Do not obtain your data by multiplying the mass or volume of one object by the number of objects you have.

Table 1: Mass and volume data

	One object	Two objects	Three objects	Four objects	Five objects
mass (g)					
volume (mL)					

2. Plot your data on graph paper. Label the *x*-axis "volume" and the *y*-axis "mass." Be sure to use the entire space on your graph paper for making your graph.

2 Analyzing your results

a. Is there any pattern to the data points on your graph? For example, the points might form a smooth curve, a straight line, a random scattering, or a cluster in a certain region. Describe any pattern you see.

b. Line up your ruler along the points on your graph so it is as close as possible to all of the dots. The line may not touch all of the dots, but should have an equal number of dots on each side of it. This line is called the "line of best fit." Draw the line.

c. Find the slope of the line of best fit. To do this, choose any two points on the line. These will be represented as (X_1, Y_1) and (X_2, Y_2). Use the formula below to calculate the slope of the line:

$$\frac{(Y_2 - Y_1)}{(X_2 - X_1)} = \text{slope}$$

The slope tells how many grams of matter are contained in each milliliter of material. Some substances, like lead, have quite a few grams of matter packed into each milliliter. Other substances, like styrofoam, have less than a single gram of matter packed into each milliliter.

d. Compare your slope with the result obtained by other groups. Are your slopes similar or different?

e. The relationship between a substance's mass and volume is called its density. What is the density of the material you tested?

3 Using your knowledge

a. Your graph shows data for five objects. Use your graph to predict the mass of *six* objects.

b. Next, use the balance to find the total mass of all six objects.

c. How does your value from your graph compare to the mass obtained using the balance?

d. Use the mass that you found in step 3b. Find that number on the y-axis of your graph. Now find the point on the line with that y-value. What is the x-value of the point?

e. Now, find the volume of six objects experimentally.

f. How does the x-value from the graph compare with the measured volume?

4 Comparing class data

Collect data from each group in the class to fill in Table 2.

Table 2: Class data for density of objects

	Group1	Group 2	Group 3	Group 4	Group 5	Group 6
volume of one object (mL)						
type of material						
density (g/mL)						

Using the data above, answer the following questions:

a. Does density depend on the size of the material? Give evidence to support your answer.

b. Does density depend on the type of material? Give evidence to support your answer.

c. Using what you have observed in this lab, do you suppose that density depends on the shape of the material? Why or why not?

5 Using a different method to find volume

You used the displacement method to find the volume of your objects in the first part of the investigation. The displacement method works because an object's volume is equal to the volume of water it displaces, or pushes aside. This method is useful for objects with complicated shapes.

If an object has a simple shape, such as a cube, its volume can be found by measuring its dimensions. The volume of a cube is found using the formula:

Volume = length × width × height

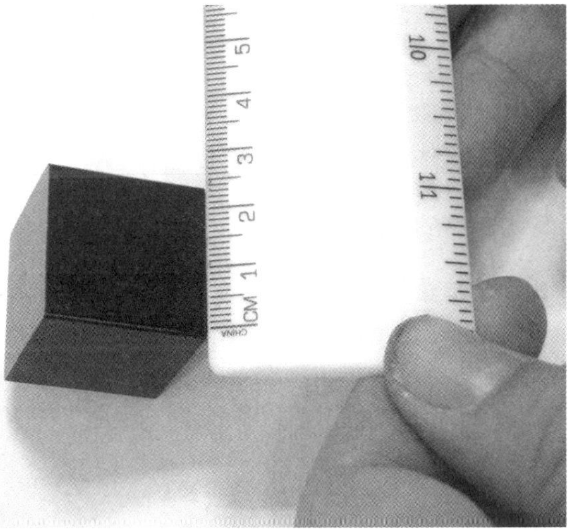

When length, width, and height are measured in centimeters, volume is in cubic centimeters or cm^3. A 1 cm × 1 cm × 1 cm cube displaces one milliliter of water, so 1 cm^3 = 1 milliliter.

1. Use the method demonstrated in the diagram on the right to measure the length, width, and height of the steel cube in centimeters.
2. Record your measurements in Table 3.
3. Calculate the volume of the steel cube. Record your volume calculations in cubic centimeters in Table 3.
4. Repeat steps 1–5 for the other 4 cubes.

Table 3: Cube volume table

Material of solid cube	Length (cm)	Width (cm)	Height (cm)	Volume from calculation (cm^3)
Steel				
Oak				
Aluminum				
Copper				
PVC				

6 Calculating the density

Each cube's volume is almost exactly the same, but their masses are different because they are all made of different materials. Use Table 4 to calculate the density of each cube.

1. Use a balance to determine the mass of the steel cube, and record it in Table 4.
2. Divide the mass by the volume of the steel cube to calculate its density: Density (g/cm^3) = mass (g) / volume (cm^3). Record the density value in Table 4.
3. Repeat steps 1 and 2 and calculate the density of each cube.

Table 4: Cube density data

Material of solid cube	Mass (g)	Volume (cm^3)	Density (g/cm^3)	Prediction (sink or float)	Result (sink or float)
Steel					
Oak					
Aluminum					
Copper					
PVC					

7 Stop and think

a. How do the volumes compare to each other? Why do you think they might be different?

b. Pick up and hold each cube. Predict whether it will sink or float in water. Record your predictions in Table 4. What did you base your predictions on?

c. What is the density of water in g/cm^3?

d. Compare the density of water to the density you calculated for each cube. Take another look at your sink/float predictions. Make any changes you need to based on density.

e. What rule did you use to make your prediction? Write the rule down in one sentence.

8 Testing the hypothesis

Your predictions from part 7d, and the rule from part 7e, represent a hypothesis. Test the hypothesis by dropping each cube in a beaker of water. Record your results in Table 4.

9 Thinking about what you learned

a. Describe two different ways you can find the density of a regularly-shaped object like a cube.

b. Explain why two different objects can have equal volumes but different masses.

c. Which method of prediction was better, testing the weight of the cube in your hand, or comparing the density of the cube to the density of water? Why?

10B Buoyancy

Can you make a clay boat?

A solid material will float if it is less dense than the liquid in which it is immersed, and sink if it is denser than the liquid. You may have noticed, however, that ships are often made of steel, which is obviously denser than water. So how does a steel boat float? In this investigation, you will experiment with modeling clay to discover how and why boats can be made of materials that are denser than water.

Materials

- Stick of clay (1/2 stick can be used as well)
- Centimeter ruler
- Displacement tank
- 250 mL beaker for overflow
- Water
- Paper towels, sponges
- 100 mL graduated cylinder
- Electronic scale (or triple-beam balance)
- Disposable cup to catch displaced water when refilling tank
- Dishpan or shallow bucket for testing clay boats
- Wax paper

1 Measuring your stick of clay

Take your stick of clay and find its density. Use a balance to measure its mass. Use the length × width × height method to calculate its volume. Record your measurements in Table 1.

Volume = length × width × height

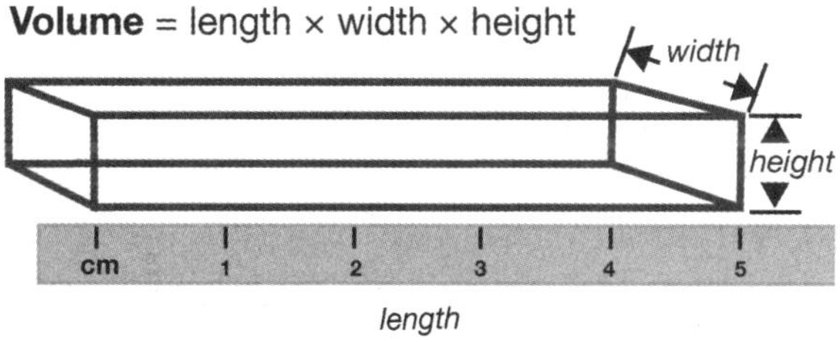

Table I: Mass, volume, and density of clay stick

length (cm)		width (cm)		height (cm)		Volume (cm^3)
	×		×		=	

Mass (g)		Volume (cm^3)		Density (g/cm^3)
	÷		=	

a. The density of water is $1 g/cm^3$. How does the density of your clay compare to water?

b. Make a prediction: Will your stick of clay sink or float? Why?

2 Doing the Experiment

A. Testing your prediction

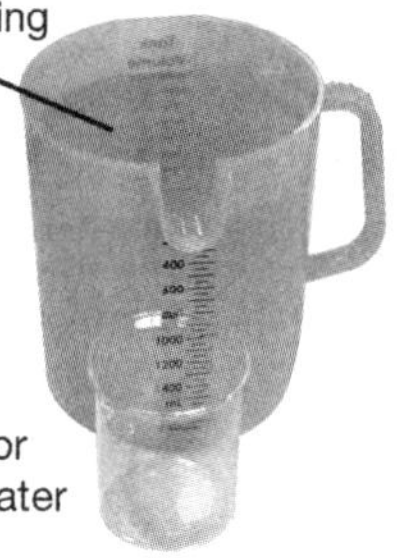

1. Prepare the displacement tank for use. Place a cup or beaker under the spout to catch the overflow water. Fill the displacement tank all the way up until it just starts to overflow into the cup or beaker. Once it stops remove the cup or beaker that was catching the overflow water.
2. Place a dry beaker under the spout to catch the overflow water displaced by your clay.
3. Gently put your stick of clay in the water. Did your stick of clay sink or float?

B. Finding the mass and volume of the displaced water

1. Pour the displaced water from the beaker into a 100-mL graduated cylinder. Record the volume in Table 2.
2. Now, calculate the mass of the displaced water using water's density and the volume of displaced water you measured. Record this value in Table 2.

C. Calculating the weight of your clay and the displaced water

Mass and weight measure two different properties of matter. Mass refers to how much matter the object contains. Weight measures the gravitational pull between the object and (in our case) Earth. The gravitational force between a 1-kilogram object and Earth is 9.8 newtons. So a 1-gram object's weight on Earth is 0.0098 newtons.

1. Use the mass of your clay to calculate its weight. Use the formula below and record the mass and weight in Table 2.

mass in grams x (0.0098 N/g) = weight in Newtons

2. Calculate the weight of the displaced water. Use the formula above and record your data in Table 2.

Table 2: Mass and volume of clay and displaced water

	Mass of clay (g)	Weight of clay (N)	Volume of displaced water (mL)	Mass of displaced water (g)	Weight of displaced water (N)
Stick of clay					
Clay boat					

3 Challenge: Can you mold your clay into a shape that floats?

For this part of the investigation, you must use ALL of your clay. Mold it into a shape that you believe will float. The clay boat must be able to fit in the displacement tank without touching the sides. Before you do, make a prediction;

How much water will the boat displace compared to the stick of clay?

1. When you are ready to test a shape, lower it into a container of water approximately three-quarters full. If the clay sinks, retrieve it immediately and dry it with a paper towel. Avoid mixing water into your clay, or it will get very slimy. Keep trying until you get a boat that floats.
2. When you have successfully molded a boat that floats, take it out of the water and dry it with a paper towel. Then, prepare your displacement tank just as you did in step 2A.
3. Carefully place your boat into the displacement tank. Avoiding making waves. When the water stops flowing, move the beaker away from the displacement tank spout. Safely remove a little water from the displacement tank so it doesn't overflow. Retrieve your boat and set it aside to dry.
4. Pour the displaced water from the beaker into a graduated cylinder. Record its volume in Table 2.
5. Use the density of water to calculate the mass of the displaced water. Calculate its weight using the formula from step C1. Record both the water's mass and weight in Table 2.
6. When your boat is dry, first measure its mass, then calculate its weight using the formula from step C1. Record both the mass and weight in Table 2.

4 Analyzing your data

a. Did the weight of the clay change during the investigation? Give a reason for your answer.

b. Which displaced more water—the stick of clay or the clay boat, and how much more?

c. Which weighed more—the stick of clay or the water it displaced, and how much more?

d. Which weighed more—the clay boat or the water it displaced, and how much more?

5 Why the boat floats

Use your mass and volume data from Table 2 to calculate the apparent density of your clay boat. This value is called the *apparent* density because the total volume of the floating boat is not displaced. The part of your floating boat that is above the surface isn't displacing any water. To find out how much of the boat is below the surface and **is** displacing water, look at the total amount of water you measured that spilled out into the beaker when the boat was floating.

Table 3:Data for boat

Mass of boat (g)	Volume of water displaced by the boat (mL)	Apparent density of the boat (g/mL)

6 Thinking about what you observed

a. Which displaced more water, the stick of clay or the floating clay boat?

b. Assuming the mass of the clay did not change, how do you explain the difference in the volumes displaced by the stick of clay and the clay boat?

c. Look at the boat's apparent density. Why is it different than the density of the stick of clay? What other substance has a density very similar to the boat's apparent density?

d. Out of all the properties you measured and calculated in this investigation, which property tells you the most about whether an object or material will float in water, and how would you use that property to determine whether it will float?

e. Explain why a solid stick of clay sinks but a clay boat can be made to float.

f. What would happen if you added "cargo," like pennies, to your boat? Is there a limit to how much mass you can add before the boat sinks? Does the volume of displaced water increase or decrease when the boat gets heavier? Why? Try the experiment.

11A Heating Land and Water

How does solar radiation affect the heating and cooling of continents and oceans?

When solar radiation reaches Earth's surface, it is absorbed unevenly. An important term, which will be useful in explaining the results of this investigation, is specific heat. *Specific heat* is the amount of energy needed to raise the temperature of 1 gram of a substance by 1 degree Celsius.

Which has a higher specific heat—the oceans or the land? Having a higher specific heat means that it takes more energy to raise a substance's temperature, but once the substance is warm, it takes longer to cool off. In this investigation, you will find out whether water or land has the higher specific heat, and how that affects local temperatures.

Materials

- 2 GeoBoxes
- 2 Physics stands
- Bridge
- 2 Data Collectors
- 2 Temperature probes
- 2 light sources with clips
- Sand (to fill one GeoBox 5 cm deep)
- Water (to fill one GeoBox 5 cm deep)

1 Doing the experiment

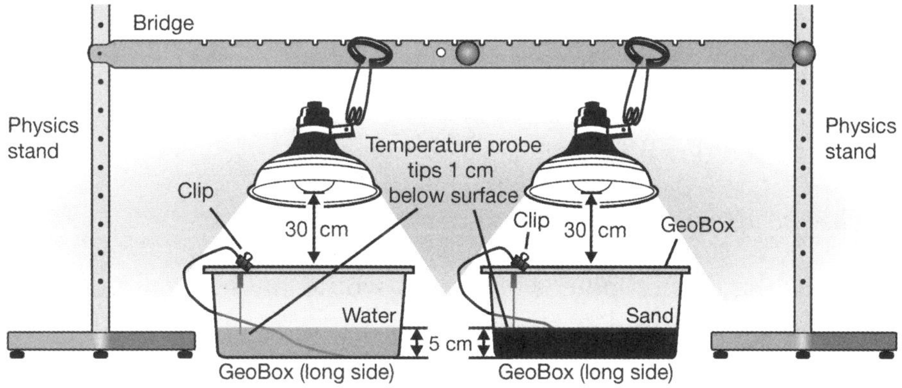

1. Set up your materials as shown above. You will need to team up with another lab group in your class to have two physics stands.
2. One GeoBox will contain water and the other will contain sand. Both the water and sand should be added to a depth of 5 centimeters. Each GeoBox will have its own light. Place the lights so that the bottom of the bulbs are about 30 cm directly above the top of their respective GeoBoxes.
3. Place one temperature probe in the sand so that the tip of the probe is about 1 cm below the surface. Place the other probe so that its tip is about 1 cm below the surface of the water. Secure both in place with a clip or clothespin. Clamp the probes to the rim of the GeoBox, securing them in place.
4. Wait three minutes. Record the initial temperature of each GeoBox in Table 1.
5. Turn the lights on at exactly the same time and have the Data Collectors record temperatures from both probes each minute for 10 minutes. Record your data in Table 1 to keep track of the temperature changes as they happen.

6. Turn the lights off and continue to record temperatures for an additional 10 minutes. Record your data in Table 1.

Table 1: Time and temperature

	Initial	Heating (light turned on)													Cooling (light turned off)							
Time (minutes)	0	1	2	3	4	5	6	7	8	9	10	11	12	13	14	15	16	17	18	19	20
Sand Temp. (°C)																					
Water Temp. (°C)																					

2 Graphing your results

Display and examine the data collected in Graph mode on both Data Collectors. Compare the two graphs and answer the questions in Part 3. If possible, download both sets of data into a spreadsheet program on a classroom computer and merge the data in order to get both sets to appear on the same graph.

3 Thinking about what you observed

a. Which received more heat (radiation) from its lamp, the sand or the water? Or did they receive the same?

b. Which material heated up faster?

c. Which material cooled down faster?

d. Using the term *specific heat*, explain your results.

e. Based on your results, which places on Earth would have the smallest/greatest temperature ranges—the land or the sea?

4 Exploring on your own

a. Find out which of the world's hemispheres has a higher proportion of water to land and describe how that might influence the temperature variations in the hemispheres.

b. Use the term specific heat to explain how sea and land breezes work. Also, refer to Section 5.3 of your text to answer this question in terms of convection in the atmosphere.

c. Give an explanation for why the temperature range (8 degrees) for Boston, MA (a coastal city) is smaller compared to the temperature range (11 degrees) for Las Vegas, NV (a landlocked city).

	Temperature range for January (°C)	Total degrees in range
Coastal city Boston, MA	–6 to 2	8
Landlocked city Las Vegas, NV	3 to 14	11

11B Observing the Weather

How can you use weather data to make predictions?

In this investigation, you will go outside to practice your observation skills. You will observe cloud types and use instruments to measure temperature, air pressure, and humidity. You will use the Beaufort wind scale to make observations about wind speed and direction.

In this investigation, you will keep a journal of weather conditions for four weeks. Then you will graph your data and look for patterns in the measurements you recorded.

Materials

- A thermometer
- A barometer
- A hygrometer
- Cloud chart
- Beaufort wind scale
- Compass
- Graph paper
- Blue and pink high lighter pens

1 Setting up

1. Gather the instruments and charts you'll need to collect your weather data.
2. Take a look at the data chart on the next page. Decide as a group which instrument or chart is needed to take each measurement.
3. In your group, pass around each instrument. Discuss the unit you'll use to report each of your measurements. Everyone should practice reading each instrument and chart.
4. Decide on an outdoor location and a time of day to make your measurements.

2 Stop and think

a. Why is it important to take measurements in the same location every day?

b. Why is it important to take measurements at the same time every day?

c. What season are you currently experiencing? Will you have more hours of daylight at the end of this four-week period, or fewer? Do you think the temperatures you measure will be getting warmer, cooler, or staying about the same over the next four weeks?

3 Doing the experiment

1. Assign each person in your group an instrument or chart and one type of data to collect. Rotate after four or five days so that each person has a chance to collect each type of data. Note: Be sure to use the correct units for your measurements.
2. Spend five minutes at the beginning or end of class each day collecting and recording your weather data in the chart on the following page. For percent cloud cover, estimate whether 0%, 25%, 50%, 75%, or 100% of the sky is cloudy. Remember to write your units and be consistent
3. In the notes section, record any precipitation (rain, snow, etc.) that has fallen since last time you took measurements, and whether any precipitation is currently falling.

Table 1: Weather data

Date:______ Time:______	Date:______ Time:______	Date:______ Time:______	Date:______ Time:______	Date:______ Time:______
Temp:________ Pressure:______ Humidity:_____ Wind speed and direction:______ __________ Cloud type and % sky covered: __________ __________ Notes:________ __________	Temp:________ Pressure:______ Humidity:_____ Wind speed and direction:______ __________ Cloud type and % sky covered: __________ __________ Notes:________ __________	Temp:________ Pressure:______ Humidity:_____ Wind speed and direction:______ __________ Cloud type and % sky covered: __________ __________ Notes:________ __________	Temp:________ Pressure:______ Humidity:_____ Wind speed and direction:______ __________ Cloud type and % sky covered: __________ __________ Notes:________ __________	Temp:________ Pressure:______ Humidity:_____ Wind speed and direction:______ __________ Cloud type and % sky covered: __________ __________ Notes:________ __________
Date:______ Time:______	**Date:______ Time:______**	**Date:______ Time:______**	**Date:______ Time:______**	**Date:______ Time:______**
Temp:________ Pressure:______ Humidity:_____ Wind speed and direction:______ __________ Cloud type and % sky covered: __________ __________ Notes:________ __________	Temp:________ Pressure:______ Humidity:_____ Wind speed and direction:______ __________ Cloud type and % sky covered: __________ __________ Notes:________ __________	Temp:________ Pressure:______ Humidity:_____ Wind speed and direction:______ __________ Cloud type and % sky covered: __________ __________ Notes:________ __________	Temp:________ Pressure:______ Humidity:_____ Wind speed and direction:______ __________ Cloud type and % sky covered: __________ __________ Notes:________ __________	Temp:________ Pressure:______ Humidity:_____ Wind speed and direction:______ __________ Cloud type and % sky covered: __________ __________ Notes:________ __________

Table 1: Weather data

Date:______ Time:______	Date:______ Time:______	Date:______ Time:______	Date:______ Time:______	Date:______ Time:______
Temp:______ Pressure:______ Humidity:______ Wind speed and direction:______ Cloud type and % sky covered: Notes:______	Temp:______ Pressure:______ Humidity:______ Wind speed and direction:______ Cloud type and % sky covered: Notes:______	Temp:______ Pressure:______ Humidity:______ Wind speed and direction:______ Cloud type and % sky covered: Notes:______	Temp:______ Pressure:______ Humidity:______ Wind speed and direction:______ Cloud type and % sky covered: Notes:______	Temp:______ Pressure:______ Humidity:______ Wind speed and direction:______ Cloud type and % sky covered: Notes:______
Date:______ Time:______	Date:______ Time:______	Date:______ Time:______	Date:______ Time:______	Date:______ Time:______
Temp:______ Pressure:______ Humidity:______ Wind speed and direction:______ Cloud type and % sky covered: Notes:______	Temp:______ Pressure:______ Humidity:______ Wind speed and direction:______ Cloud type and % sky covered: Notes:______	Temp:______ Pressure:______ Humidity:______ Wind speed and direction:______ Cloud type and % sky covered: Notes:______	Temp:______ Pressure:______ Humidity:______ Wind speed and direction:______ Cloud type and % sky covered: Notes:______	Temp:______ Pressure:______ Humidity:______ Wind speed and direction:______ Cloud type and % sky covered: Notes:______

4 Graphing your data

You have collected weather data for four weeks. Making graphs of this data can help you to find patterns that influence the weather in your region. Each member of your team will be responsible for a different graph. Make separate line graphs for temperature, pressure, humidity, wind speed, and percent cloud cover.

1. For each graph, plot the data you collected on the *y*-axis against time in days on the *x*-axis. Be sure to include even the days no data was collected. Remember to draw a line or curve of best fit rather than simply connecting the dots on your graph.
2. Using a blue high lighter pen, color in each column of your graph that represents a day where precipitation fell.
3. Label each graph appropriately.

5 Thinking about what you observed

The following questions will help you think like a meteorologist! You will be looking for patterns in your data that could help you recognize when certain weather events are likely to occur.

a. Examine the blue colored-columns on your five graphs. Is there a pattern to the data on days with precipitation? For example, is the air pressure usually higher or lower on rainy days? Does it tend to snow on the coldest days, or warmer days? Of course, your answers will depend on your season and climate. Just note any data patterns you find in your graphs.

b. Meteorologists look for patterns to help them predict what will happen in the future. Take a second look at your five graphs. Is there anything that usually happens *before* you get precipitation? For example, does the temperature usually rise, or the wind speed up? Provide specific evidence from your data as you answer these questions.

c. Look at your cloud cover data. Which type of cloud did you see most often? Mark those days with a pink high lighter. Since the column may already be colored blue, just highlight the number on the *x*-axis for those days. Do you see any patterns here? Does this type of cloud usually occur with rising or falling pressure, temperature, humidity, or wind speed? Or do the clouds seem to appear randomly?

d. Go back and look at the prediction you made in part 2c. Was there a relationship between hours of daylight and the temperatures you recorded? Why or why not?

6 Exploring on your own

Watch a local weather forecast. What sorts of evidence does the forecaster use to predict the weather? Which types of data—temperature, pressure, humidity, wind speed, cloud type, or cloud cover—were mentioned in the predictions? Were other types of measurements used as well?

12A Atomic Structure

What is inside an atom?

We once believed that atoms were the smallest units of matter. Then it was discovered that there are even smaller particles inside atoms! The structure of the atom explains why nearly all the properties of matter we experience are what they are. This investigation will lead you through some challenging and fun games that illustrate how atoms are built from protons, neutrons, and electrons.

Materials

- Atom Building Game

1 Modeling an atom

In the atom game, colored marbles represent the three kinds of particles. Red or green marbles are protons, blue marbles are neutrons, and yellow marbles are electrons.

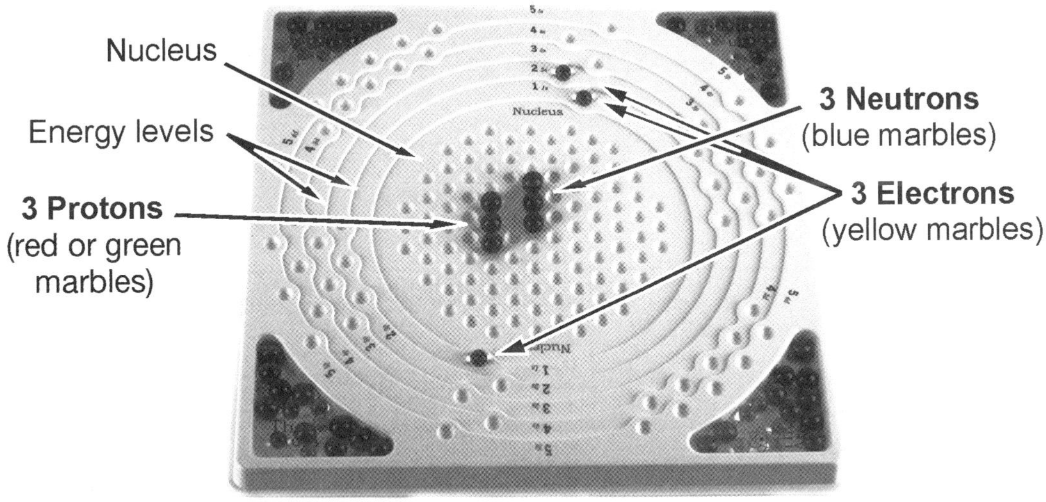

6,7
Li
3

1. Build the atom above using three red or green, three blue, and three yellow marbles.
2. Fill in the blanks in the empty periodic table box for the atom you constructed.

2 Thinking about the atom

a. What is the number below the element symbol? What does this number tell you about the the atom?

b. What is the number(s) above the element symbol called? What does this number tell you about the atom?

c. Why do some elements have more than one number above the symbol? What are the variations in this number called?

3 Making atoms

Build the 6 atoms shown on the chart and fill in the missing information Protons and neutrons go in the middle of the board. Electrons go in the outside and fill up the holes from the lowest row to the highest.

	Element	Atomic number	Mass number	Protons	Neutrons	Electrons
3a				●●●●	●●●●●	
3b				●●●●●●	●●●●●●	
3c					●●●●●●●	○○○○○○
3d		8			●●●●●●●●	
3e				●●●●●●●●●●	●●●●●●●●●●	○○○○○○○○○○
3f			27	●●●●●●●●●●●●●		

4 Stop and think

a. Two of the atoms you made were the same element. What was different about them?

b. One of the atoms had just enough electrons to completely fill the first two rows. Which atom was this? Where on the periodic table is it found?

c. Which atom had an atomic number of 8?

d. Which atom had a mass number of 13?

e. One atom is found in a lightweight, silvery metal used in airplanes. Which atom was it?

f. One atom represents an element that makes up about 21% of the air you breathe. You could not live without this element.

5 The game of atomic challenge

Each player takes a turn adding protons, neutrons, and electrons to the atom to build heavier and heavier elements. The winner is the first player to run completely out of marbles.

1. Each player should start with 6 blue (neutrons), 5 red or green (protons), and 5 yellow marbles (electrons).
2. Each player takes turns adding 1 - 5 marbles, but not more than 5. The marbles may include any mixture of electrons, protons, and neutrons.
   FOR EXAMPLE: you can add one blue, one red or green, and one yellow marble in a turn. That makes three total marbles, which is less than 5.
3. Marbles played in a turn are added to the marbles already in the atom.
4. Only atoms where the electrons, protons, and neutrons match one of the naturally occurring elements on the periodic table are allowed. If you add marbles that make an atom NOT on the periodic table you have to take your marbles back and try again.

Example of a good move

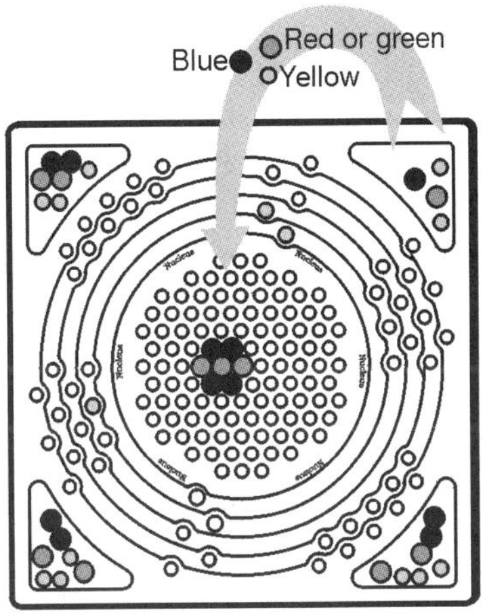

$Li^7 + p + n + e = Be^9$

The Three Rules

Rule #1: The number of protons matches the atomic number

Rule #2: The total number of protons and neutrons equals a stable mass number

Rule #3: The number of electrons matches the number of protons

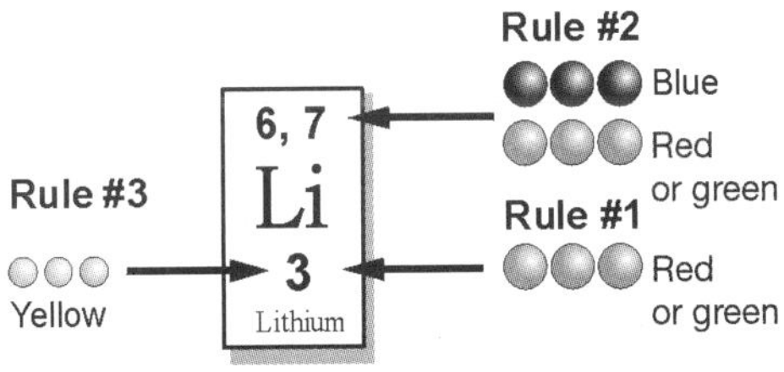

5. A player can trade marbles with the bank INSTEAD of taking a turn. The player can take as many marbles, and of as many colors as they need but must take at least as many total marbles as they put in. For example, a player can trade 2 yellows for 1 yellow, 1 blue, and 1 red.

6 Stop and think

Atoms which are not on the periodic table shown may exist in nature but they are radioactive and unstable. For example, carbon-14 is unstable and is not listed although carbon-12 and carbon-13 are stable.

a. What four elements make up almost all of the mass in your body?

b. How many stable isotopes does oxygen have?

c. Find one element on the chart that has no stable isotopes.

d. What element has atoms with 26 protons in the nucleus?

e. On most periodic tables, a single atomic mass is listed instead of the mass numbers for all the stable isotopes. How is this mass related to the different isotopes?

Periodic Table of the Elements 1- 54

(Stable isotopes)

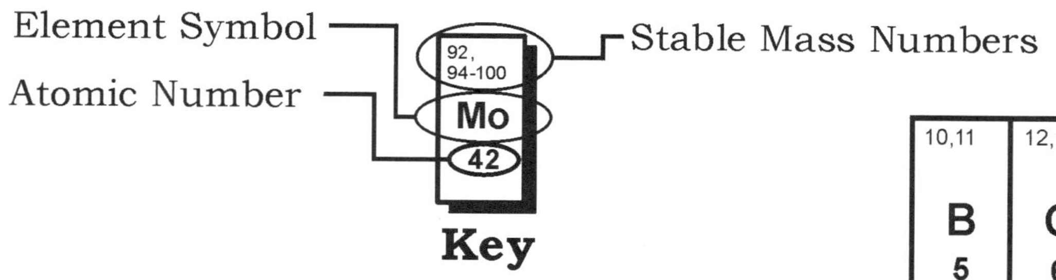

1,2 **H** 1																	3,4 **He** 2
6,7 **Li** 3	9 **Be** 4											10,11 **B** 5	12,13 **C** 6	14,15 **N** 7	16-18 **O** 8	19 **F** 9	20-22 **Ne** 10
23 **Na** 11	24-26 **Mg** 12											27 **Al** 13	28-30 **Si** 14	31 **P** 15	32-34, 36 **S** 16	35,37 **Cl** 17	36,38, 40 **Ar** 18
39,41 **K** 19	40,42-44,46, 48 **Ca** 20	45 **Sc** 21	46-50 **Ti** 22	51 **V** 23	50, 52-54 **Cr** 24	55 **Mn** 25	54,56-58 **Fe** 26	59 **Co** 27	58,60-62,64 **Ni** 28	63,65 **Cu** 29	64,66-68,70 **Zn** 30	69,71 **Ga** 31	70,72-74,76 **Ge** 32	75 **As** 33	74,76-78,80, 82 **Se** 34	79,81 **Br** 35	78,80, 82-84, 86 **Kr** 36
85 **Rb** 37	84, 86-88 **Sr** 38	89 **Y** 39	90-92, 94,96 **Zr** 40	93 **Nb** 41	92, 94-100 **Mo** 42	none **Tc** 43	96,98-103,104 **Ru** 44	103 **Rh** 45	102,104-106,108, 110 **Pd** 46	107,109 **Ag** 47	106,108, 110-112, 114,116 **Cd** 48	113 **In** 49	112,114-120,122, 124 **Sn** 50	121 **Sb** 51	120,122, 124-126, 128,130 **Te** 52	127 **I** 53	124,126, 128-132, 134,136 **Xe** 54

12B The Periodic Table

How is the periodic table organized?

Virtually all the matter you see is made up of combinations of elements. Scientists know of 118 different elements, of which about 90 occur naturally. Each element has its own unique kind of atom. The periodic table is a chart that shows all of the elements in order of increasing atomic number.

Each box on the periodic table tells you the element symbol, atomic number, and atomic mass for all the known elements. The arrangement of the elements on the periodic table gives you valuable information. Each major column of elements represents a group of elements with similar chemical behavior. In this investigation, you'll build the periodic table, and play a challenging game that will help you find out how and why the arrangement of elements on the periodic table is so important.

Materials

- Periodic Table Tiles
- Blank periodic table bingo card sheets; 1 per player
- Caller's clues and checklist; 1 per team
- Copy of the periodic table of elements; 1 per player
- Highlighter or other marker; 1 per player

1 Building the periodic table

Every element is given a symbol of one or two letters. For example, the symbol for hydrogen is a capital letter H. The symbol for lithium is two letters, Li. Each element also has a unique number called the atomic number. The *periodic table* is a chart that shows the elements in an arrangement that helps us recognize groups of elements with similar properties. The picture below shows the shape of the periodic table and the first few elements in sequence from left to right. The elements are arranged from lowest to highest atomic number.

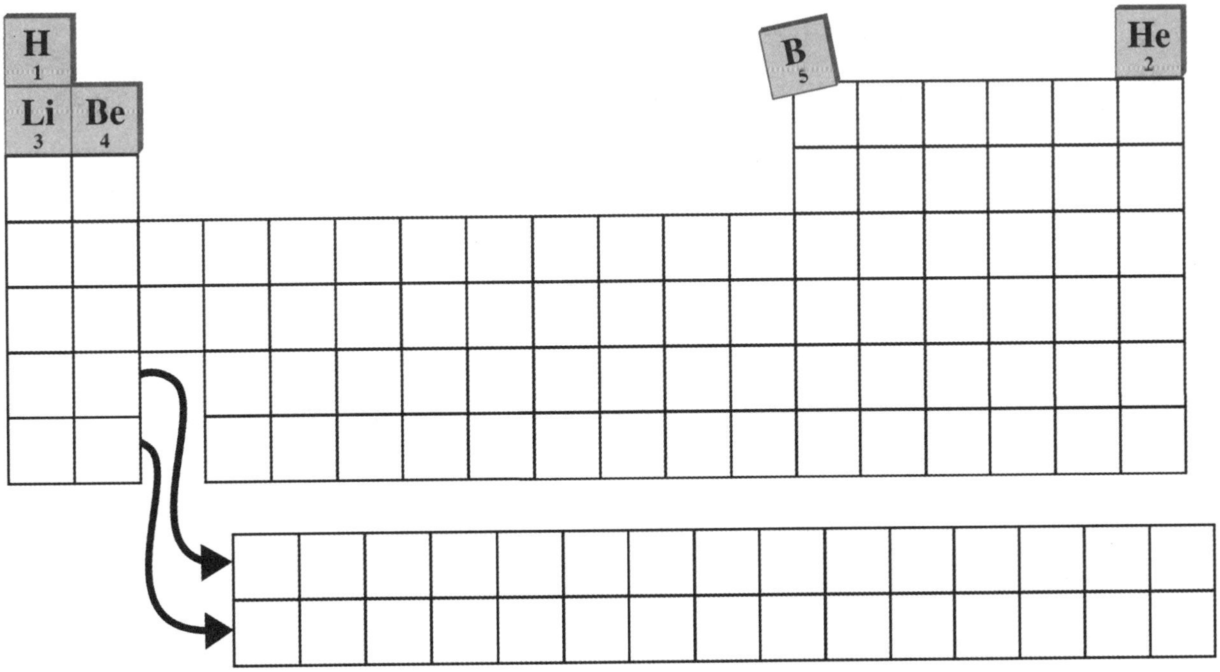

Using the chart as a guide, build the periodic table out of the periodic table tiles. The only tile you should use yellow-side up is the hydrogen tile. The colors will show a pattern when you are finished. There is a tricky part near the bottom of the table. The table breaks off between element 56 (Ba) and element 71 (Lu), and fills in the first of two long rows underneath the main part of the chart.

2 Organization of the periodic table

Look at the periodic table you put together out of the tiles. Use the diagram below, and your model of the periodic table, to answer the questions below.

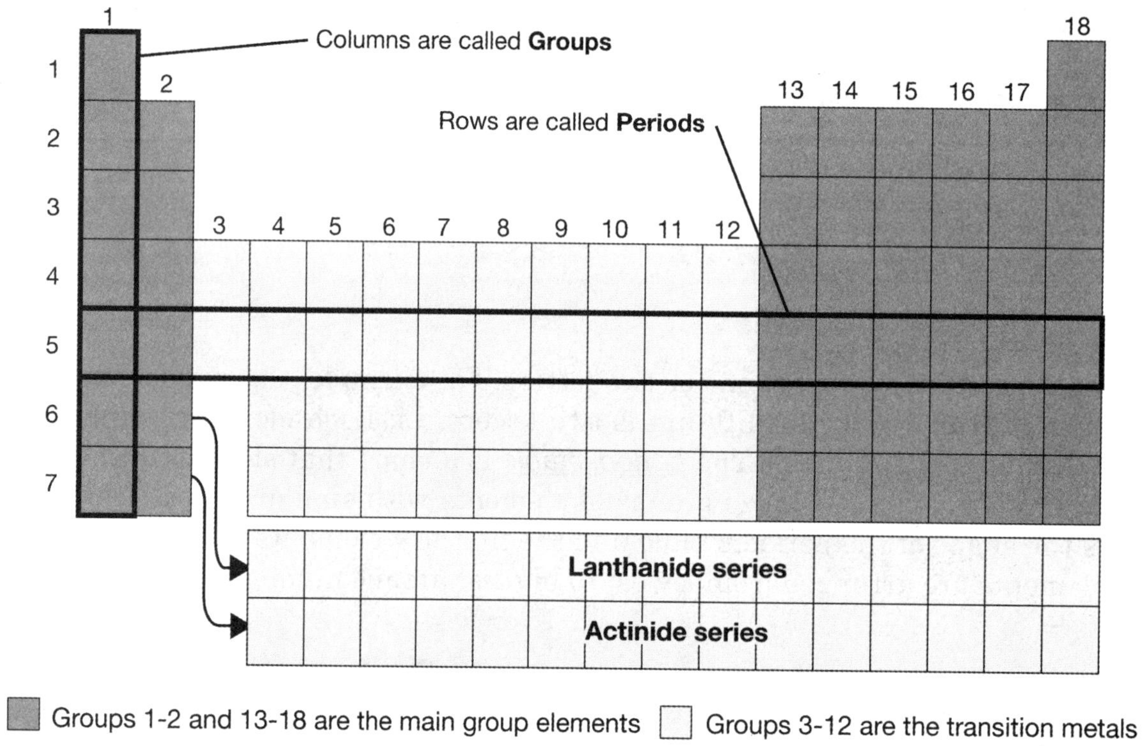

a. Which elements are in group 1?

b. Which elements are in group 8?

c. Name three transition metals.

d. To which group does chlorine belong? What other elements are in that group?

e. Which elements are in the 2nd period?

3 Applying what you have learned

a. Each row (period) of the periodic table contains only a certain number of elements. What does this have to do with the structure of the atom? Research this question in your textbook.

b. Which group of the periodic table above contains the element argon? What characteristic do the elements in this group share?

c. Which group contains the element carbon? What characteristic do the elements in this group share?

4 Periodic Table Challenge

Periodic Table Challenge is a bingo-like game that helps you understand how the elements on the periodic table are arranged. Each player will fill out their own five-by-five grid with element symbols, and then the caller will read element clues. The players must interpret the clues and highlight any boxes on the grid that fit the clue. The first player that correctly highlights five boxes across, up-and-down, or diagonally in a row is the winner.

5 Rules of play

1. Designate one member of your group to be the caller. The caller will call out clues and keep track of them on the checklist.
2. Each player will have a sheet of 4 blank grids. Fill out one of the grids with random element symbols (other grids can be used for additional games). You may choose elements in the atomic number range of 1–54 (hydrogen through xenon). Do not repeat symbols on the card. You will only be able to fit 25 of the possible 54 symbols on the card. You choose which ones to use, and where to place them on the grid.
3. The caller will randomly pick clues from the list, and as a clue is called out, the caller will check off the clue. The answers are only for the caller to check the winner's card!
4. When a clue is called, players check the grid to see if any of the elements fit the clue. Any elements that fit the clue must be highlighted. If no elements fit the clue, then no boxes are highlighted on that turn.
5. When a player has five boxes highlighted in a row up and down, across, or diagonally, play stops. The caller will double check the clue list and answers to see if the clues indeed match the elements. Play continues until a true winner is determined.

6 Caller's clues

Clues can be called in any order. Check off each clue as you use it.

Clue	Possible Answers
A member of the carbon family	C, Si, Ge, Sn
Chemical properties similar to calcium, but not calcium	Be, Mg, Sr
A transition metal that has a "C" in the symbol	Sc, Cr, Co, Cu, Tc, Cd,
A member of the oxygen family	O, S, Se, Te
Chemical properties similar to cesium, but not cesium	H, Li, Na, K, Rb
A member of the noble gas family	He, Ne, Ar, Kr, Xe
A nonmetal in the nitrogen family	N, P
A metal in the boron family	Al, Ga, In
A gas in the oxygen family	O
A solid in the halogen family	I
An element that is liquid at room temperature	Hg, Br
A transition metal with less than 25 protons	Cr, V, Ti, Sc
A transition metal commonly found in jewelry	Ni, Cu, Ag
A metalloid in the carbon family	Si, Ge
Chemical properties similar to aluminum, but not aluminum	B, Ga, In
A transition metal with 39–43 protons	Y, Zr, Nb, Mo, Tc
An element symbol with a first letter that is different from the first letter of the name	Na, K, Fe, Ag, Sn, Sb
A member of the nitrogen family with a one-letter symbol	N, P
A transition metal from period 5	Y, Zr, Nb, Mo, Tc, Ru, Rh, Pd, Ag, Cd

7 Periodic table challenge card

Fill in each of the squares with the symbol of an element with atomic number between 1 and 54 (hydrogen–xenon).

You may not use the same element symbol twice.

13A Chemical Bonds and Electrons

Why do atoms form chemical bonds?

Most of the matter on Earth is in the form of compounds. Even when a substance exists as a pure element, it tends eventually to combine with other elements. For example, if you leave an iron nail outside in the rain, it will quickly combine with the oxygen in the air to form iron oxide, better known as rust. In this investigation, you will build models of atoms and discover one of the fundamental ideas in chemistry: how electrons are involved in the formation of chemical bonds.

Materials

- Atom Building Games (2)

1 Reviewing atomic structure

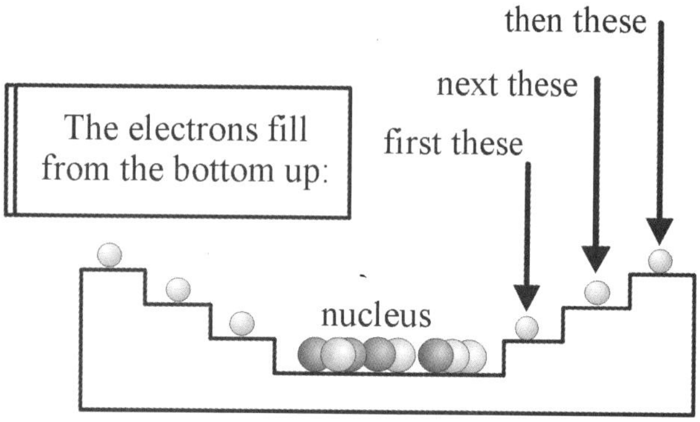

Let's review what you already know about atoms:

- A neutral atom has the same number of electrons and protons.
- The electrons occupy energy levels surrounding the nucleus.
- Since electrons are attracted to the nucleus, they fill the lower energy levels first.

Once a given level is full, electrons start filling the next level.

2 How many electrons are in the outermost level?

Using the atom building game, build each element in the table. For each element, record the number of electrons in the outermost energy level and the number of unoccupied spaces in the outermost energy level.

Table 1: Element data

Element	Atomic number	Electrons in outermost level	Unoccupied spaces in outermost level
hydrogen			
helium			
lithium			
fluorine			
neon			
sodium			
chlorine			
argon			
potassium			

3 What are valence electrons?

Examine the table you just completed and record the answers to the following questions:

a. Use your textbook to find out about *valence electrons*.

b. What do lithium, sodium, and potassium have in common?

c. What do fluorine and chlorine have in common?

d. What do neon and argon have in common?

4 Modeling a chemical bond

Atoms that have a complete outermost energy level are stable. If there are empty holes, an atom will either gain, lose, or share electrons with another atom in order to complete its outermost level and become stable. When atoms gain, lose, or share electrons with another atom, they form *chemical bonds*.

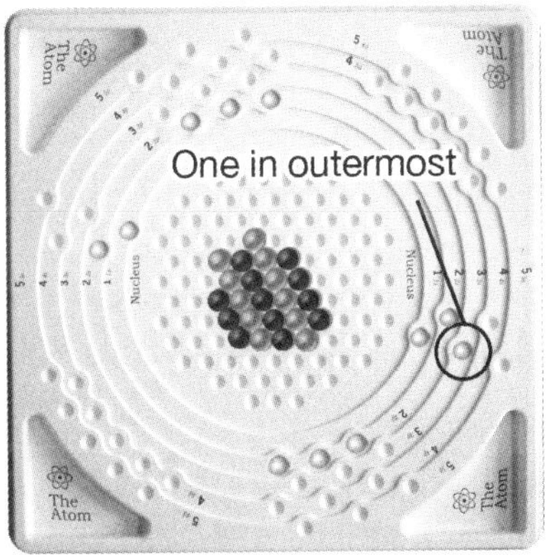

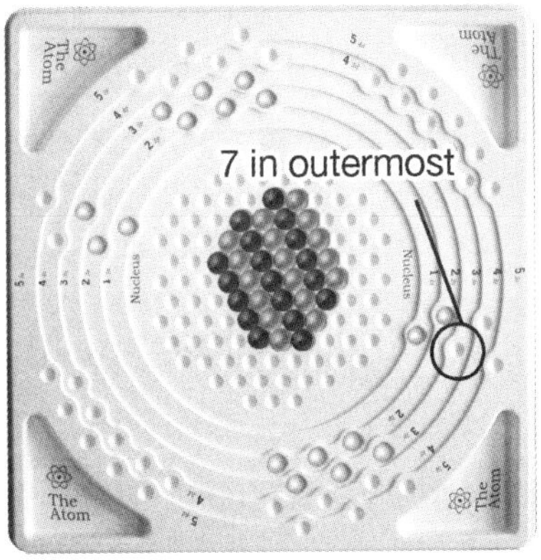

Using two atom building games, build a sodium atom and a chlorine atom. Put them next to each other and answer the questions below.

a. In order to complete its outermost energy level, do you think sodium will tend to lose its only valence electron, or gain seven? Explain your answer.

b. In order to complete its outermost energy level, do you think chlorine will tend to lose all of its valence electrons or gain one electron? Explain your answer.

c. Why might these two atoms bond together to form a molecule? In your answer, describe what you think might happen when sodium and chlorine form a chemical bond.

5 Determining oxidation numbers

An element's *oxidation number* is equal to the charge an atom has when it **ionizes**, that is, gains or loses electrons.

Use your models of sodium and chlorine to answer the questions below.

- **a.** Remove the valence electron from sodium. What has happened to the balance of positive and negative charges? What is sodium's oxidation number?
- **b.** Move the electron you took from sodium into the chlorine. What happens to chlorine's charge when it gains the electron from the sodium atom? What is chlorine's oxidation number?
- **c.** When sodium and chlorine form a chemical bond, what is the overall charge of the molecule? Why do you think sodium and chlorine combine in a 1:1 ratio?

6 Modeling chemical bonds

Use multiple atom boards for this part of the investigation. First, build each atom. Next, move the boards together to model the formation of a chemical bond.

- **a.** One carbon atom and four hydrogen atoms
- **b.** Two lithium atoms and one oxygen atom
- **c.** One beryllium atom and two fluorine atoms

7 Applying your knowledge

Read in your textbook about Lewis dot diagrams. Draw a dot diagram for each compound in Part 6.

13B Chemical Formulas

Why do atoms combine in certain ratios?

Chemists have long noticed that groups of elements behave similarly. The periodic table is an arrangement of the elements grouped according to similar behavior. In this investigation, you will discover how the arrangement of electrons in atoms is related to groups on the periodic table. You will also learn why atoms form chemical bonds with other atoms in certain ratios.

Materials

- Periodic Table Tiles
- Periodic table with oxidation numbers
- Special Bonds card

1 Oxidation numbers and ions

An element's <u>*oxidation number*</u> indicates how many electrons are lost or gained when chemical bonding occurs. The oxidation number is equal to the charge an atom has when it *ionizes*, that is, gains or loses electrons to become an <u>*ion*</u>. The partial periodic table below shows the most common oxidation numbers of the elements. The oxidation numbers are written above the group number above each column on the table. The most common oxidation numbers for the main group elements are shown.

Oxidation Numbers from the Periodic Table

1+	2+											3+	4+	3-	2-	1-	
H 1 hydrogen																	He 2 helium
Li 3 lithium	Be 4 beryllium											B 5 boron	C 6 carbon	N 7 nitrogen	O 8 oxygen	F 9 fluorine	Ne 10 neon
Na 11 sodium	Mg 12 magnesium											Al 13 aluminum	Si 14 silicon	P 15 phosphorus	S 16 sulfur	Cl 17 chlorine	Ar 18 argon
K 19 potassium	Ca 20 calcium	Sc 21 scandium	Ti 22 titanium	V 23 vanadium	Cr 24 chromium	Mn 25 manganese	Fe 26 iron	Co 27 cobalt	Ni 28 nickel	Cu 29 copper	Zn 30 zinc	Ga 31 gallium	Ge 32 germanium	As 33 arsenic	Se 34 selenium	Br 35 bromine	Kr 36 krypton
Rb 37 rubidium	Sr 38 strontium	Y 39 yttrium	Zr 40 zirconium	Nb 41 niobium	Mo 42 molybdenum	Tc 43 technetium	Ru 44 ruthenium	Rh 45 rhodium	Pd 46 palladium	Ag 47 silver	Cd 48 cadmium	In 49 indium	Sn 50 tin	Sb 51 antimony	Te 52 tellurium	I 53 iodine	Xe 54 xenon
Cs 55 cesium	Ba 56 barium		Hf 72 hafnium	Ta 73 tantalum	W 74 tungsten	Re 75 rhenium	Os 76 osmium	Ir 77 iridium	Pt 78 platinum	Au 79 gold	Hg 80 mercury	Tl 81 thallium	Pb 82 lead	Bi 83 bismuth	Po 84 polonium	At 85 astatine	Rn 86 radon
Fr 87 francium	Ra 88 radium		Rf 104 rutherfordium	Db 105 dubnium	Sg 106 seaborgium	Bh 107 bohrium	Hs 108 hassium	Mt 109 meitnerium	Uun 110 ununnilium	Uuu 111 unununium	Uub 112 ununbium	113	Uuq 114 ununquadium	115	116	117	118

NOTE: Many elements have more than one possible oxidation number.

La 57 lanthanum	Ce 58 cerium	Pr 59 praseodymium	Nd 60 neodymium	Pm 61 promethium	Sm 62 samarium	Eu 63 europium	Gd 64 gadolinium	Tb 65 terbium	Dy 66 dysprosium	Ho 67 holmium	Er 68 erbium	Tm 69 thulium	Yb 70 ytterbium	Lu 71 lutetium
Ac 89 actinium	Th 90 thorium	Pa 91 protactinium	U 92 uranium	Np 93 neptunium	Pu 94 plutonium	Am 95 americium	Cm 96 curium	Bk 97 berkelium	Cf 98 californium	Es 99 einsteinium	Fm 100 fermium	Md 101 mendelvium	No 102 nobelium	Lr 103 lawrencium

2 Stop and think

a. Describe the groups on the periodic table according to their valence electrons

b. Why do elements in group 2 have an oxidation number of 2+?

c. Why do elements in group 17 have an oxidation number of 1–?

d. Why do the oxidation numbers in the first two groups tend to be positive?

3 Predicting chemical formulas

A *binary compound* is composed of two different elements. Predict the chemical formulas for the binary compounds that are made up of the pairs of elements in the table below. Use the following steps:

1. Using the periodic table on the previous page, determine the ion formed by each element.
2. Figure out how many periodic table tiles of each element will be needed to make the compound electrically neutral.
3. Form the compound with your tiles and write the chemical formula for each compound based on the number of tiles of each element.

Table 1: Writing chemical formulas for binary compounds

Element 1	Element 2	Oxidation number 1	Oxidation number 2	Number of tiles of element 1	Number of tiles of element 2	Chemical formula
hydrogen	fluorine					
magnesium	sulfur					
calcium	bromine					
aluminum	oxygen					
potassium	chlorine					
lithium	argon					
rubidium	sulfur					

4 Naming compounds

Naming binary ionic compounds is very simple if you follow these rules:

1. Write the name of the element with a positive oxidation number first.
2. Write the root name of the element with a negative oxidation number second. For example, *chlor-* is the root name of chlorine. Subtract the *-ine* ending.
3. Add the ending *-ide* to the root name. *Chlor-* becomes chloride.

Using these rules, write the name of each of the compounds in Table 1.

5 Playing Compound Crossword

Now that you understand how elements combine to form compounds, you are ready to play Compound Crossword. In this game, you will score points by forming stable compounds, crossword style. Players use the oxidation numbers of the elements to form correct compounds. Points are determined by adding up the atomic numbers of each atom in the compound.

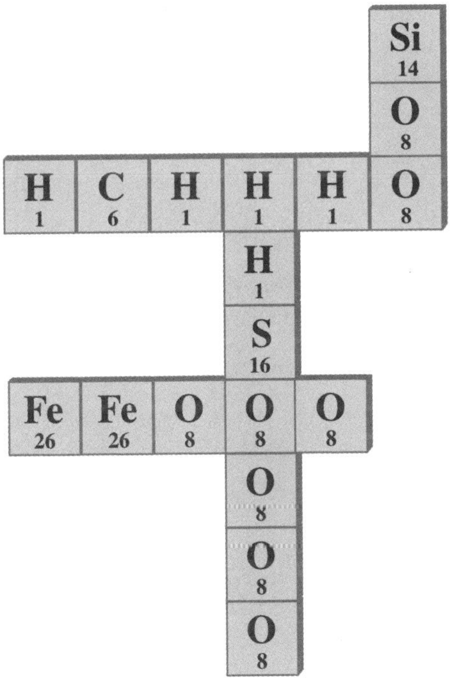

Sample game after four turns:

CH_3OH scores 18 points

Fe_2O_3 scores 76 points

SiO_2 scores 30 points

H_2SO_4 scores 50 points

Starting the game

Each player starts with ten randomly selected tiles. The remaining tiles should be placed in a box or paper bag so that additional tiles may be drawn without being seen. The playing surface can be any flat table (or floor) with minimum dimensions of 75-by-75 centimeters.

Each player draws a tile from the bag; the highest atomic number goes first. Once the starting player has been determined, those tiles are returned to the bag. The play continues to the starting player's right.

Play

1. Players take turns adding a compound to the crossword by using tiles from their set of ten. Either side of a tile may be used.
2. The elements in a compound may be arranged in any order, as long as they are in a single horizontal or vertical row.
3. Each new compound must be shown by the player to have oxidation numbers that add up to zero. Otherwise the player must take back the compound and wait until the next turn.

4. If the compound is correct, the player adds up the atomic numbers of all the atoms in the compound to determine the points and then draws new tiles to restore a set of ten.
5. Play continues until all the tiles in the bag are used and one player is out of tiles, or until all players are unable to make a compound with their remaining tiles.
6. The winner is the player with the highest score at the end of the game.

Determining correct compounds

The oxidation numbers found directly above each element on the periodic table provided are used to determine whether a molecule is correct. The oxidation numbers of all the elements in the compound must add up to zero. In some cases there is more than one oxidation number for an element. Iron (Fe), for example, has oxidation numbers of +2 and +3. The player may choose either oxidation number to add to the total.

The Special Bonds card included with the periodic table tiles gives some additional possibilities for forming compounds.

Examples of correct and incorrect compounds:

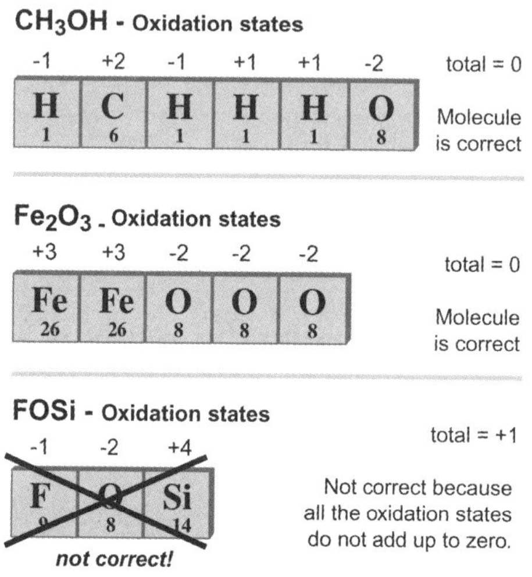

The oxidation number for each element in a compound is counted once for every atom present. The two iron atoms contribute a total of +6, which is balanced by the -6 from the three oxygen atoms.

Note: The Special Bonds card has some extra ways compounds can form when two atoms of the same element bond together, and when hydrogen is involved in unusual ways. The example of CH_3OH (methanol) above uses one of the special bonds that hydrogen can make with carbon.

Miscellaneous rules and strategies:

The noble gases do not form bonds; therefore they can't be played. Players may decide at the start of the game to remove these tiles. As another option, players can choose to turn the noble gas tiles over, revealing the + symbol. This symbol could be used as a wildcard in which the player chooses which element the + should represent. The symbol scores no points but allows the player to complete a molecule. Players must agree on the use of the noble gas tiles before the game begins.

14A Conservation of Mass

How do scientists describe what happens in a chemical reaction?

A French chemist named Antoine Lavoisier was the first to prove the law of conservation of mass. This law says that the total mass of the reactants in a chemical reaction is always equal to the total mass of the products. This is not as easy to see as you might think! As you do this investigation, you will discover how tricky it is to show the law of conservation of mass.

Materials

- Periodic Table Tiles
- Effervescent tablet
- Electronic scale (or triple-beam balance)
- Water
- Small paper cup

WARNING — This lab contains chemicals that may be harmful if misused. Read cautions on individual containers carefully. Not to be used by children except under adult supervision.

1 Testing the reaction

Step 1:
Tare the balance to zero with the empty cup

Step 2:
Measure the mass of water and tablet

Step 3:
Add tablet to water and observe the reaction

Table 1: Conservation of mass data

Step	Data and observations
1. Find the mass of the effervescent tablet.	
2. Put the paper cup on the balance and tare it to zero. Fill the cup about halfway with water. Record the mass.	
3. Put the tablet on the balance beside the cup, but don't put it in the water yet. Record the total starting mass.	
4. Drop the tablet into the cup of water. You can do this while the cup is still on the balance. Record your observations.	
5. Wait for the reaction to stop. Then, tap the cup gently to release as many bubbles as you can. Measure the mass.	
6. Subtract the final mass (5) from the starting mass (3). This is the mass difference between the products and reactants.	

2 Stop and think

a. Does this experiment agree with the law of conservation of mass? Look at the data that you just recorded. Use it to help you to explain why or why not.

b. Explain why you observed a difference in mass. Where did the missing mass go? Did it really disappear?

3 Modeling the reaction

Scientists write chemical reactions like mathematical formulas. The reactants are to the left of the arrow and the products are to the right of the arrow.

Reactants → Products

The effervescent tablet contains a chemical called sodium bicarbonate. This chemical reacts with water according to the following reaction.

$$H_2O + NaHCO_3 \rightarrow NaOH + CO_2 + H_2O$$

1. Build the reactants side (H_2O + $NaHCO_3$) of the chemical reaction above using the periodic table tiles.
2. Build the products side ($NaOH$ + CO_2 + H_2O) of the chemical reaction using more periodic table tiles.

4 Stop and think

Table 2: Counting atoms of each element

Element	Reactants	Products
Hydrogen		
Carbon		
Oxygen		
Sodium		

a. Fill in Table 2 with the numbers of each type of atom on the reactant side of the equation and on the product side of the equation.

b. How do the numbers of atoms of each element compare on the reactant and product side of the equation? What does this imply for the law of conservation of mass?

c. In what phase are each of the reactants (solid, liquid, or gas)? In what phase are each of the three products (solid, liquid, or gas)?

5 Proving that mass is conserved in a reaction

According to the law of conservation of mass, the mass of the products of the reaction should be exactly equal to the mass of the reactants. Can you design an experiment to prove this is true for the reaction you just observed?

Examine the materials your teacher has given you. These include:

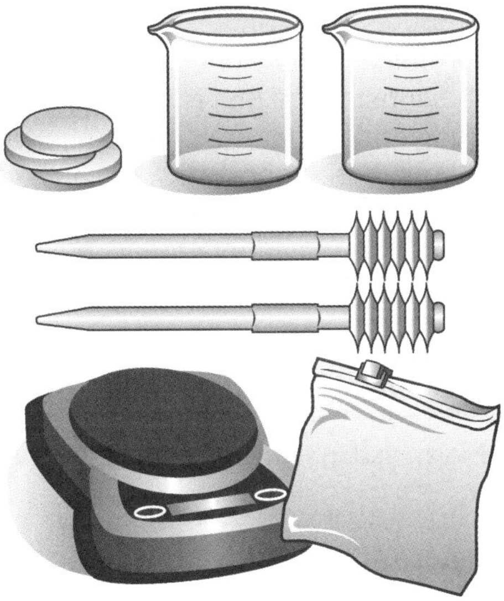

- effervescent tablet
- 2 beakers
- beaker of water
- 2 plastic pipettes
- 2 baggies with zippers
- electronic balance or mass scale

1. Working with your lab partner, devise an experiment that will prove that mass is conserved in the reaction of the tablet and water. You may request additional materials if your teacher has them available.
2. List the materials you will need and their use in the experiment.
3. List the steps you will follow in the experiment.
4. Before you try out your experiment, request approval from your teacher.
5. If your experiment does not work, adjust your procedures and/or materials and try it again.
6. Record your procedures, data, and results.

6 Presenting your results to the class

Prepare a brief presentation for the class about your experiment. Use the following format for your presentation:

a.	**Purpose**	What questions were you trying to answer?
b.	**Materials**	What materials and equipment did you choose and why?
c.	**Procedures**	What were the steps you followed? You may demonstrate your procedures if time and materials allow.
d.	**Data**	What was the data you collected?
e.	**Conclusions**	What does your data prove? If your experiment did not yield satisfactory results, what would you change in your procedures or materials and why?

14B Balancing Chemical Equations

How are atoms conserved in a chemical reaction?

A chemical reaction involves changes in substances that react to form new products. This process involves the breaking of chemical bonds and the formation of new ones. A chemical equation shows the chemical formulas of the substances that react, called reactants, and the chemical formulas of the substances that are produced, called products. The number and type of atoms in the reactants must be exactly equal to the number and type of atoms in the products. How do you write a chemical equation so that the number and type of atoms on the reactants and products sides are balanced?

Materials

- Periodic Table Tiles

1 Writing chemical equations

Magnesium metal reacts with water to produce magnesium hydroxide and hydrogen gas.

The statement above is the word form of a chemical reaction. It tells you the names of the reactants and the products. To write it as a chemical equation, you need to determine the chemical formulas of each of the substances in the reaction:

1. Magnesium metal is an element and exists as an atom. Its chemical formula is Mg.
2. The chemical formula for water is H_2O.
3. Magnesium hydroxide is an ionic compound. To write its chemical formula, you need to find out the charges of each ion it is made out of. The magnesium ion is Mg^{2+}. The hydroxide ion is OH^-. You need 1 Mg^{2+} and 2 OH^- to make a neutral compound so the formula is $Mg(OH)_2$.
4. Pure hydrogen gas always exists as a diatomic molecule so its chemical formula is H_2.

The chemical equation is written as:

magnesium metal	reacts with	water	to produce	magnesium hydroxide	and	hydrogen gas
Mg	+	H_2O	→	$Mg(OH)_2$	+	H_2

2 Trying out the reaction with periodic table tiles

Use periodic table tiles to make the reactants above.

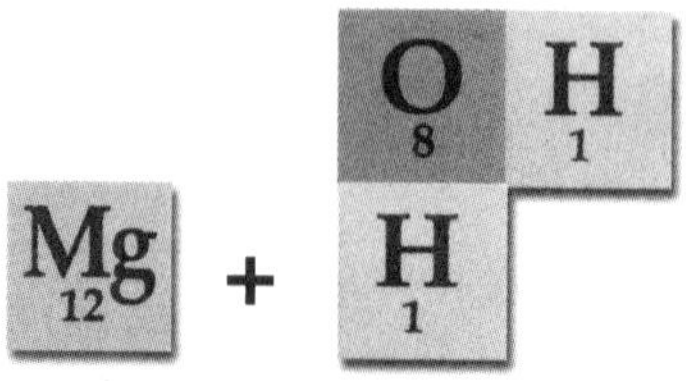

Rearrange the reactants to make the products. Is there any problem? What are you missing?

3 Balancing the reaction

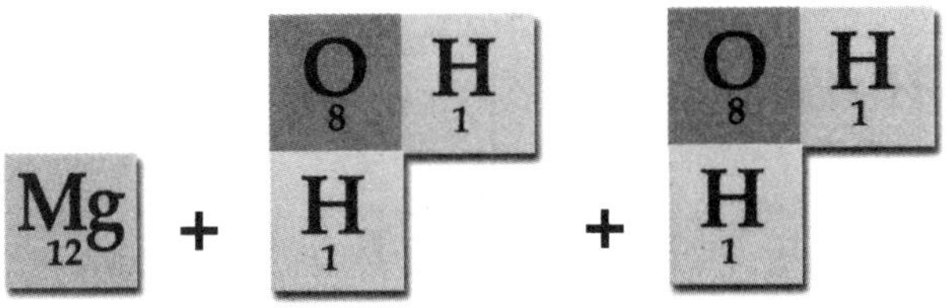

Chemical equations must always balance. This means that you must use all of the atoms you start with and you cannot have any leftover atoms when you are finished. If you need more atoms to make the products, you can only add them in the form of a chemical formula. You cannot simply add the extra atoms that you need, unless the chemical formula is a single atom–like Mg. Which atoms did you need more of for the reaction you tried? Since you needed more oxygen and hydrogen atoms, you can only add them in the form of another water molecule. Try adding another water molecule to the reactants and rearrange them to form the products again. Did the reaction work this time?

4 Writing balanced chemical equations

To balance the equation for this reaction, you needed to add another water molecule to the reactants side. You ended up with the correct amount of products. Since one magnesium atom reacted with two water molecules to form one magnesium hydroxide molecule and one hydrogen gas, the proper way to write the balanced chemical equation is:

$$Mg + 2H_2O \longrightarrow Mg(OH)_2 + H_2$$

The **2** in front of water is called a coefficient. This number tells you how many water molecules are needed in the reaction. The rest of the reactants and products in the reactants show no coefficients. This is because when the coefficient is **1**, there is no need to write it.

5 Try balancing these chemical equations

The following chemical equations have the proper reactants and products. Try to balance each using the following steps:

1. Assemble the reactants out of the appropriate tiles.
2. Rearrange the reactants to form the products.
3. Figure out the number of each reactant and product required to make the equation balance and write the numbers (the coefficients) in the boxes.

	Fe	+		O_2	$\longrightarrow$		Fe_2O_3			
	CH_4	+		O_2	$\longrightarrow$		CO_2	+		H_2O
	HCl	+		NaOH	$\longrightarrow$		NaCl	+		H_2O
	Na	+		Cl_2	$\longrightarrow$		NaCl			
	CO_2	+		H_2O	$\longrightarrow$		$C_6H_{12}O_6$	+		O_2

6 Challenge! Balancing difficult equations

Use the Periodic Table Tiles to help you balance these harder equations.

$\square\ CaCO_3 + \square\ HCl \rightarrow \square\ CaCl_2 + \square\ CO_2 + \square\ H_2O$

$\square\ GaBr_3 + \square\ Cl_2 \rightarrow \square\ GaCl_3 + \square\ Br_2$

$\square\ Al_4C_3 + \square\ HCl \rightarrow \square\ AlCl_3 + \square\ CH_4$

$\square\ CuSO_4 + \square\ Fe \rightarrow \square\ Fe(SO_4)_3 + \square\ Cu$

$\square\ Mg_3N_2 + \square\ HCl \rightarrow \square\ MgCl_2 + \square\ NH_3$

Each of these equations has three reactants:

$\square\ Li + \square\ CO_2 + \square\ H_2O \rightarrow \square\ LiHCO_3 + \square\ H_2$

$\square\ Al + \square\ NaOH + \square\ H_2O \rightarrow \square\ NaAl(OH)_4 + \square\ H_2$

Can you balance this equation without using the tiles?

$\square\ C_{30}H_{62} + \square\ O_2 \rightarrow \square\ CO_2 + \square\ H_2O$

15A Carbon Dioxide and Living Things

How do living things exchange carbon dioxide?

The *carbon cycle* is a series of different pathways through which carbon atoms are exchanged and stored. The air you breath contains carbon in the form of molecules of carbon dioxide. Plants use a process called *photosynthesis* to convert carbon dioxide into high-energy carbon molecules. Plants and animals break down those carbon molecules and release energy, carbon dioxide, and water in a process called *respiration*. Both processes also involve oxygen. In this investigation, you will explore how living things exchange carbon dioxide.

Materials

- Safety goggles and apron
- 4 40 mL clear plastic containers with screw caps
- Masking tape
- Marker
- 2 5 cm sprigs of an aquatic plant
- Bromothymol blue (BTB) solution
- Straw
- Beakers (500 mL)
- Funnel
- Aluminum foil
- Tweezers
- Blank white paper
- Lamp
- Plant seeds
- Jars (2) with covers
- Strainer
- Container used for filtering
- Ruler

Safety tip: Wear safety goggles and an apron during this investigation.

WARNING — This lab contains chemicals that may be harmful if misused. Read cautions on individual containers carefully. Not to be used by children except under adult supervision.

1 Thinking about respiration

In respiration, glucose (a carbohydrate) reacts with oxygen to produce carbon dioxide and water. Every cell in your body uses this reaction to release energy stored in carbohydrates. When you breathe, you take in oxygen, a reactant in respiration. You exhale carbon dioxide, a product of respiration. All types of cells including animal and plant cells undergo this process.

Respiration:

$$\underset{\text{Glucose}}{C_6H_{12}O_6} + \underset{\text{Oxygen}}{6O_2} \longrightarrow \underset{\text{Carbon dioxide}}{6CO_2} + \underset{\text{Water}}{6H_2O} + \text{Energy}$$

1. Obtain 100 mL of bromothymol blue (BTB) solution in a large beaker from your teacher. Bromothymol blue is a chemical that turns color when carbon dioxide is present.
2. Using a straw, **gently** blow bubbles into the solution until it completely changes color. **Do not swallow the BTB**! Your teacher may have you practice the technique using a beaker of plain water first.

a. What was the initial color of the BTB solution?

b. What was the color of the solution after you blew bubbles into the beaker?

c. What color does BTB change to when carbon dioxide is present?

d. What do you think would happen to the color of the solution if the carbon dioxide molecules were removed from the solution?

2 Thinking about photosynthesis

Plant cells use a chemical reaction called photosynthesis to store energy from the Sun in carbohydrate molecules. In the reaction, carbon dioxide reacts with water to make glucose (a high-energy molecule) and oxygen. The word "light" above the arrow means that light is required to make the reaction happen:

Photosynthesis:

$$6CO_2 + 6H_2O \xrightarrow{\text{Light}} C_6H_{12}O_6 + 6O_2$$

Carbon dioxide, Water, Glucose, Oxygen

a. Would the presence of plants in water cause the carbon dioxide to increase, decrease, or stay the same? Explain your answer.

b. When would you expect photosynthesis to happen, in the day or at night? Explain your answer.

3 Investigating photosynthesis—Experiment A

1. Obtain four clear plastic containers with screw caps. Label each container with a marker and masking tape as follows.
   #1 Plant: light
   #2 Plant: dark,
   #3 No plant: light
   #4 No plant: dark.
2. Use the beaker of BTB solution you obtained in Part 1 for this experiment. Gently blow bubbles into the solution again to make sure the color has completely changed to yellow. *The initial color of the solution should be yellow for this experiment.*
3. Record the initial color of the solution for all four containers in Table 1.
4. Pour 25 mL of the solution into each of the four test containers using a funnel. Cap tubes #3 and #4.
5. Place a 5 cm piece of aquatic plant in containers #1 and #2. Make sure each plant is completely submerged in the solution. Cap these containers.
6. Wrap tubes #2 and #4 with aluminum foil to block out any light.
7. Place the four capped tubes upright in front of a lamp. Make sure that the tubes are at least eight inches away and that light is hitting all tubes equally on the side. Your setup should look similar to the one shown below.

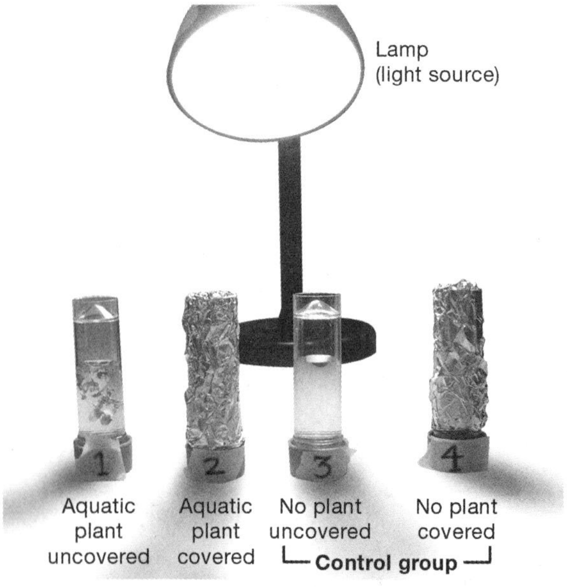

8. Let the plants sit for 50 minutes. Set up the experiment in Part 5 while you wait.

4 Stop and think

a. Why did you need to make sure the BTB solution had completely changed color before starting the experiment?

b. Why are tubes 3 and 4 called the *control group*? Why is a control group necessary in this experiment?

c. Predict what you think will happen to the color of the BTB solution in each of the four containers. *State your prediction in the form of a hypothesis.*

5 Investigating respiration—Experiment B

1. Obtain two plastic jars with screw caps and label each with a marker and tape: #1 With Seeds and #2 No Seeds.
2. Cover the bottom of jar #1 with plant seeds.
3. To each jar, add equal amounts of BTB solution. Be sure to use enough solution to cover the seeds. *The initial color of the solution should be blue for this experiment.*
4. Record the initial color of the solution for both jars in Table 2.
5. Cover both jars and place them in similar conditions for at least 45 minutes.
6. While you are waiting, *write a hypothesis about what you think will happen to the color of the solution in each container.*

6 Recording your data

1. After 45 minutes, carefully unwrap each container from the photosynthesis experiment (Experiment A).
2. Using tweezers, carefully remove the plants from containers #1 and #2.
3. Record the final color of the solutions in Table 1.

Table 1: Data from photosynthesis experiment (Experiment A)

Container	Initial color	Final color
#1 Plant: light		
#2 Plant: no light		
#3 No plant: light		
#4 No plant: no light		

Now you are ready to collect data from Experiment B.

4. Carefully pour the solution from container #1 into another clear container. Try not to let any seeds come out of the original container.
5. Compare the colors and record your data in Table 2.
6. Follow all clean up instructions from your teacher.

Table 2: Data from respiration experiment (Experiment B)

Container	Initial color	Final color
#1 With seeds		
#2 No seeds		

7 Analyzing the results

a. From Experiment A, in which container(s) did photosynthesis occur? Use your data and knowledge of photosynthesis to explain your answer.

b. Did the results from Experiment A support your hypothesis from Part 4c?

c. From Experiment B, in which container did respiration occur? Use your data and knowledge of respiration to explain your answer.

d. Did the results from Experiment B support your hypothesis from Part 5, step 6?

e. Why was light a variable in Experiment A, but not a variable in Experiment B?

f. Explain the importance of bromothymol blue in both experiments.

8 Applying your knowledge

a. Is there a relationship between photosynthesis and respiration? Explain how these two processes are related, and how they are important pathways of the carbon cycle.

b. How is energy involved in the processes of photosynthesis and respiration? Which process stores energy? Which process releases energy? Where does the energy come from?

15B Oceans and the Carbon Cycle

What role do the oceans play in the carbon cycle?

CO_2 is important to living things, but it is also a major greenhouse gas that traps energy from the Sun. Together with water vapor and methane, CO_2 helps keep Earth's surface at an average temperature of about 15°C. In the last investigation, you learned that CO_2 is produced during cellular respiration. It is also being produced in larger and larger quantities by the burning of fossil fuels and deforestation. The oceans absorb CO_2 from the atmosphere, accumulating and storing it for an indefinite period.

In this investigation, you will explore how the oceans absorb CO_2 and the chemistry involved in that process.

Materials

- 2 500mL beakers
- Red cabbage juice solution
- BTB (bromothymol blue) solution
- Simulated seawater
- Distilled water
- pH paper (range 4.5 to 9)
- 250 mL graduated cylinder
- 2 straws

Safety tip: Wear safety goggles and an apron during this investigation.

WARNING — This lab contains chemicals that may be harmful if misused. Read cautions on individual containers carefully. Not to be used by children except under adult supervision.

1 Thinking about the carbon cycle

The carbon cycle has carbon sources and carbon sinks. Anything that releases CO_2 into the atmosphere is a *carbon source*. Anything that absorbs CO_2 and holds it is a *carbon sink*. Study the diagram and then answer the questions below.

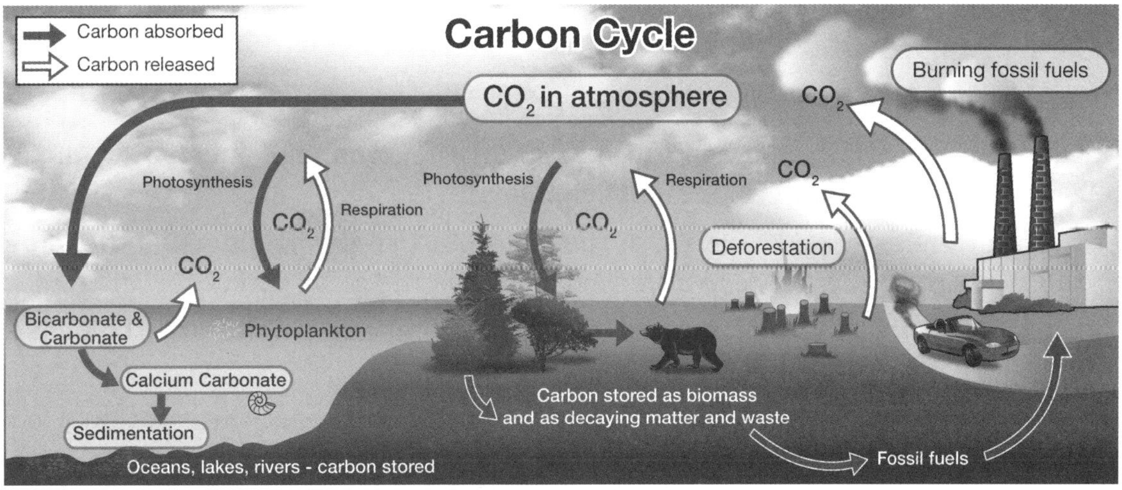

a. List the carbon sources in the diagram.

b. List the carbon sinks in the diagram.

c. Over time, carbon sources and sinks tend to stay in balance. Today, CO_2 levels are on the rise. What are the sources for this extra CO_2?

d. Based on the diagram, do you think the oceans are a net carbon producer or sink? Explain your answer.

2 CO_2 absorption and pH

Water and other solutions like seawater absorb CO_2 through a chemical reaction. When CO_2 diffuses into water, carbonic acid forms and then ionizes to form hydrogen, bicarbonate, and carbonate ions.

Step 1: $CO_{2\,(aq)} + H_2O_{(l)} \rightleftharpoons H_2CO_{3\,(aq)}$ (Carbonic acid)

Step 2: $H_2CO_{3\,(aq)} \rightleftharpoons H^+_{(aq)} + HCO_3^-{}_{(aq)}$ (Hydrogen ion; Bicarbonate ion)

Step 3: $HCO_3^-{}_{(aq)} \rightleftharpoons H^+_{(aq)} + CO_3^{2-}{}_{(aq)}$ (Carbonate ion)

We measure the *acidity* of a solution using a value called *pH*. The pH scale is shown below. As you can see, the lower a solution's pH number, the greater its acidity. In this experiment, you will determine the effects of CO_2 absorption on the pH of a solution.

pH Scale

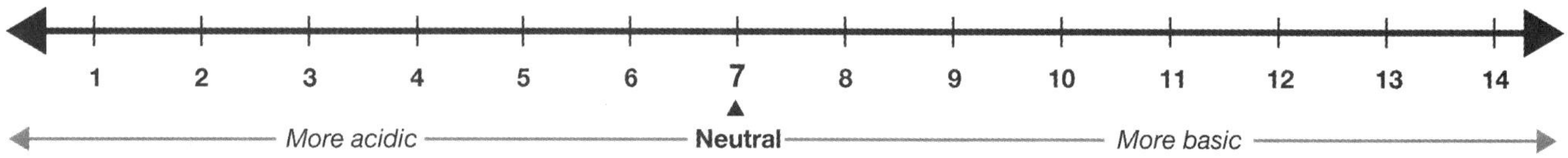

a. What do you think will happen to the acidity of a solution as it absorbs more CO_2. Will it become more acidic, less acidic, or will the acidity stay the same? State your answer in the form of a hypothesis.

1. Put on your safety goggles and apron.
2. Obtain a 500 mL beaker containing 100 mL of red cabbage juice solution. Red cabbage juice changes color as pH changes.
3. Record the initial color and pH of the solution. Use the pH chart on the left side of the table below to find the pH value that matches the color you observe.
4. Choose a "bubbler" from your group. Other members will observe the beaker and record observations.
5. Using the straw, the bubbler will gently blow bubbles into the solution. Do not inhale the solution. Observe and record each color you observe in the table. Use the chart provided by your teacher to record the pH. Keep recording until the color appears to stop changing.

Table I: Color and pH

Red cabbage juice pH scale	
color	**pH**
blue	7-8
violet	5-6
dark red	3-4
pink	1-2

Data from the experiment		
	Color	**pH**
Initial		

3 Analyzing the results

a. Was the initial pH of your solution acidic, neutral, or basic?

b. Was your hypothesis correct? Explain why or why not based on your results.

c. What happened to the pH of the solution as CO_2 was absorbed?

4 Comparing CO_2 absorption in freshwater vs. seawater

BTB turns from dark blue to green to yellow depending on the concentration of ionized carbonic acid in the solution. As the solution absorbs more CO_2, more carbonic acid forms and ionizes which lowers the pH. The color of the solution gets lighter as the pH gets lower.

In this experiment, you will bubble CO_2 into distilled water (representing fresh water) and then simulated seawater using your breath and a straw. You will then observe and compare the color change in both solutions.

1. Put on your safety goggles and apron.
2. Fill one 500 mL beaker with 100 mL of distilled water and label it "distilled water."
3. Fill another 500 mL beaker with 100 mL of seawater and label it "seawater."
4. Add 20 drops of BTB solution to each beaker.
5. Using pH test strips, measure the pH of each beaker and record it in the first row of Table 2.
6. Place each beaker on a blank white sheet of paper.
7. Choose a "bubbler", "timer", and "observer" from your group.
8. Using a straw, the bubbler will gently blow bubbles into the distilled water beaker while the timer measures the time each minute for 6 minutes and the observer watches for color changes in the beaker.
9. Record your observations in the data table below.
10. Use a pH test strip to measure the final pH and record in the last row of Table 2.
11. Repeat steps 7 and 8 for the seawater beaker using a clean straw. If the seawater does not change to a pale yellow, record the lightest color change you observed within 6 minutes of gentle bubbling.

Table 2: Time vs. color change

Time (min)	Distilled water	Seawater
Initial color and pH		
1		
2		
3		
4		
5		
6 min (final color and pH)		

5 Analyzing the data

a. Which was the control in this experiment, distilled water or seawater? Why did you need a control?

b. Based on your results, which do you think has the ability to absorb more CO_2 before becoming acidic, distilled water or seawater? Explain your answer.

c. Might the results of this experiment explain part of the reason why oceans are a carbon sink? Justify your answer.

d. What other processes in the oceans store CO_2? What processes release CO_2?

e. Based on your results, what effect might continued CO_2 absorption have on the pH of seawater?

f. Based on the results of the experiment, could the release of excess CO_2 from the burning of fossil fuels and deforestation have an effect on the pH of the oceans? Explain your reasoning.

6 Impact of ocean acidification

Although the natural absorption of CO_2 by the oceans helps absorb excess CO_2, many scientists believe the decrease in pH will have a negative effect on marine *calcifying organisms*. These are organisms that construct shells out of calcium carbonate and include some phytoplankton, coccolithophores, corals, crustaceans, and some molluscs. Calcifying organisms help store carbon in their calcium carbonate shells and many absorb CO_2 for photosynthesis. When these organisms die, their shells sink to the bottom of the ocean and store carbon in calcium carbonate sediments. In fact, England's famous white cliffs of Dover are made mostly of the calcium carbonate shells of coccolithophores.

a. Conduct research on ocean acidification and its effect on calcifying organisms. Try to find answers to these questions: (1) What is the pH of seawater in the oceans and has it changed since the Industrial Revolution? (2) How could rising CO_2 levels impact calcifying organisms? (3) What are some steps humans need to take in order to lower our impact on the pH of oceans?

b. Suppose you are a marine biologist who is designing an experiment to test the effects of ocean acidification on a species of coral. Answer these questions: (1) What is your research question and hypothesis? (2) What are the control variables in your experiment? What is your experimental variable? (3) What materials would you need for your experiment? (4) Briefly describe your procedures and the data you would need to collect.

16A Electricity

How do you measure voltage and current in electric circuits?

We use electricity every day, nearly every minute! In this investigation you will build circuits and learn about voltage (volts) and current (amps) which are fundamental quantities that describe the electricity we use.

Materials

- Electric Circuits kit
- 1 "D" battery
- Multimeter
- Metal paper clip, plastic straw, piece of string, rubber band, pen cap or other similar objects

1 Building a circuit

Single bulb circuit

1. Build the circuit shown in the diagram with one battery, a switch, and a bulb.
2. Open and close the switch and see what happens.

2 Thinking about what you observed

a. How can you tell electric current is flowing in the circuit? Can you see the current?

b. Current flows from positive to negative. Trace the flow of current around the circuit with your finger.

c. How does the switch cause the current to stop flowing?

d. Why does the bulb go out when you open the switch?

3 Conductors and insulators

Materials through which electric current flows easily are called *conductors*.

Materials through which current does not flow easily are called *insulators*.

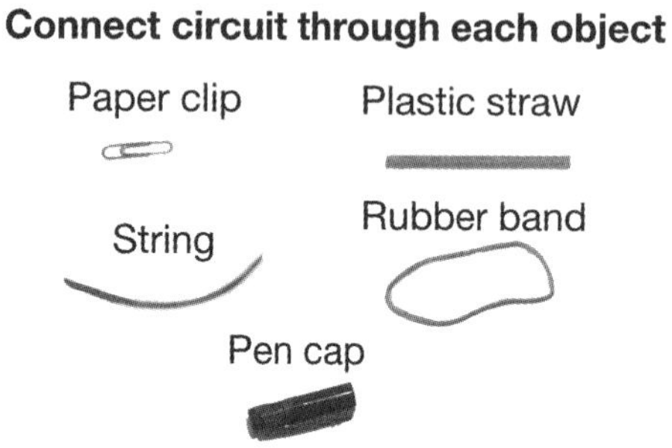

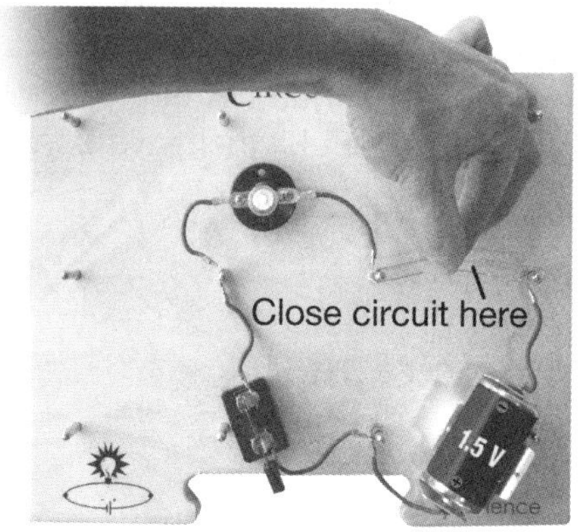

1. Break one connection in your one-bulb circuit.
2. Complete the circuit by touching different materials between the wire and the post.
3. Which materials allow the bulb to light and which do not?

4 Thinking about what you observed

a. Make a table listing the materials as either conductors or insulators.

b. What characteristics are shared by the conductors you found?

c. What characteristics are shared by the insulators you found?

5 Circuit diagrams

For describing electric circuits we use the language of *circuit diagrams*. In a circuit diagram wires are represented by solid lines. Electrical devices like switches, batteries, and bulbs are represented by symbols.

a. Using these symbols, draw a picture of the circuit you built with one battery, switch, and light bulb.

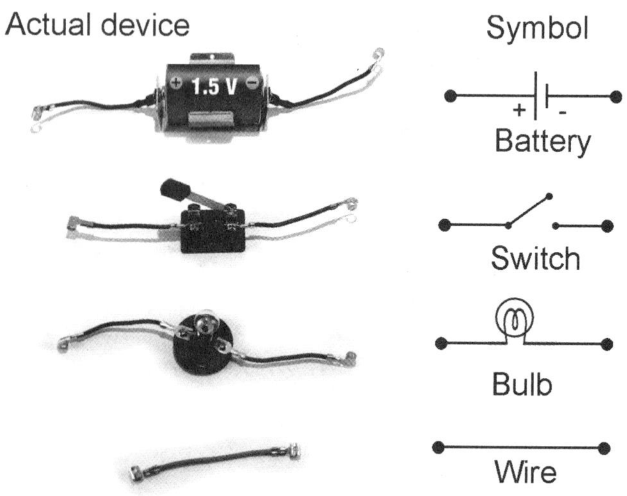

6 Measuring the voltage of a battery

Turn the dial of your multimeter to DC volts. Red goes to the positive terminal and black to the negative terminal. When you touch two points in a circuit with the leads, the meter reads the voltage between the two points.

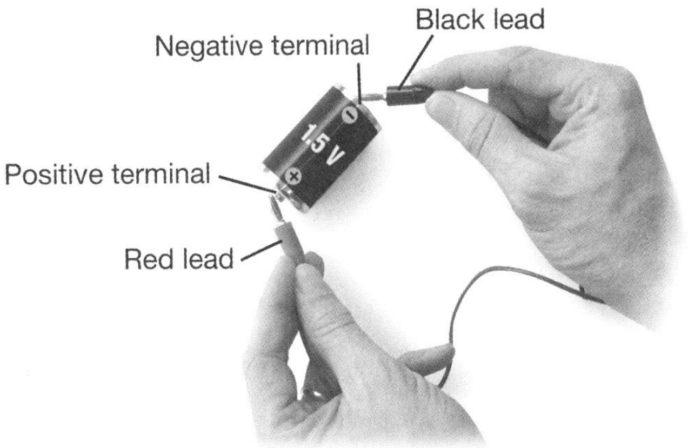

a. Measure the voltage of the battery and record your reading.

b. Take a second battery and connect it to the first by touching the ends together.

c. Measure the voltage for the four possible ways to connect two batteries (+ to –, + to +, – to –, – to +). How do your readings compare to the voltage of just one battery?

7 Measuring current

To measure current, the meter must be connected so the current has to flow through it. This is different from voltage measurement. To measure current you must force the current to flow through the meter by eliminating all other paths the current could go. Follow the instructions below carefully. Too much current can damage the meter.

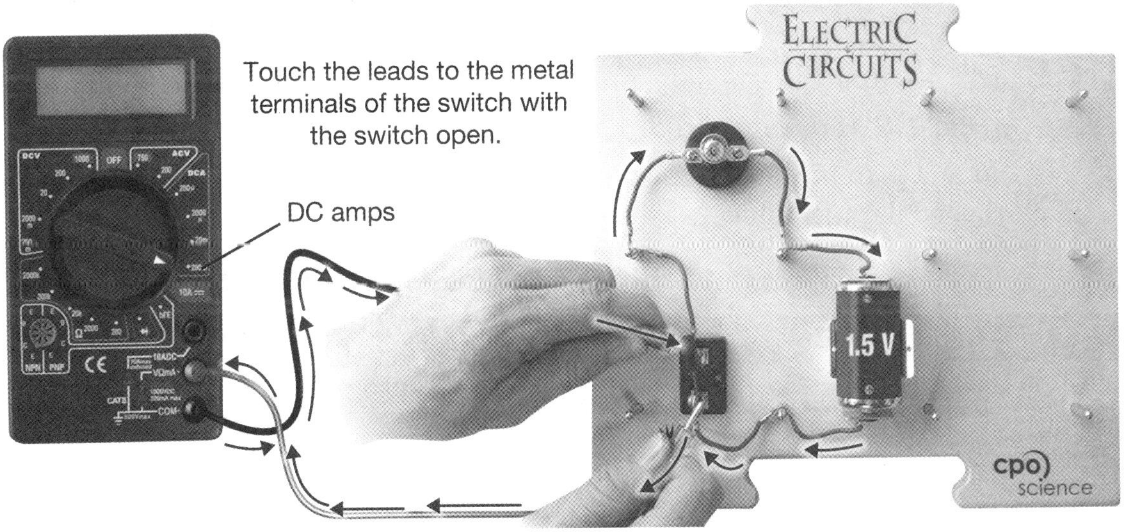

1. Set the multimeter to measure DC amps (current).
2. Open the switch and touch the red lead of the meter to the metal part of the switch closest to the battery's positive terminal (+).
3. Touch the black lead of the meter to the metal part on the other side of the switch.
4. The bulb should light, showing you that current is flowing through the meter. The meter should display the current in amps. This is the total current flowing around the circuit carrying power from the battery to the bulb. Remove the meter.

a. How much current is flowing in the circuit when the bulb is making light?

8 A circuit with a dimmer switch

The potentiometer (or *pot*) is an electrical device that can be used to make a dimmer switch. When the dial on the pot is turned one way the pot acts like a closed switch and current flows freely through it. When the dial is turned the other way the pot resists the flow of current.

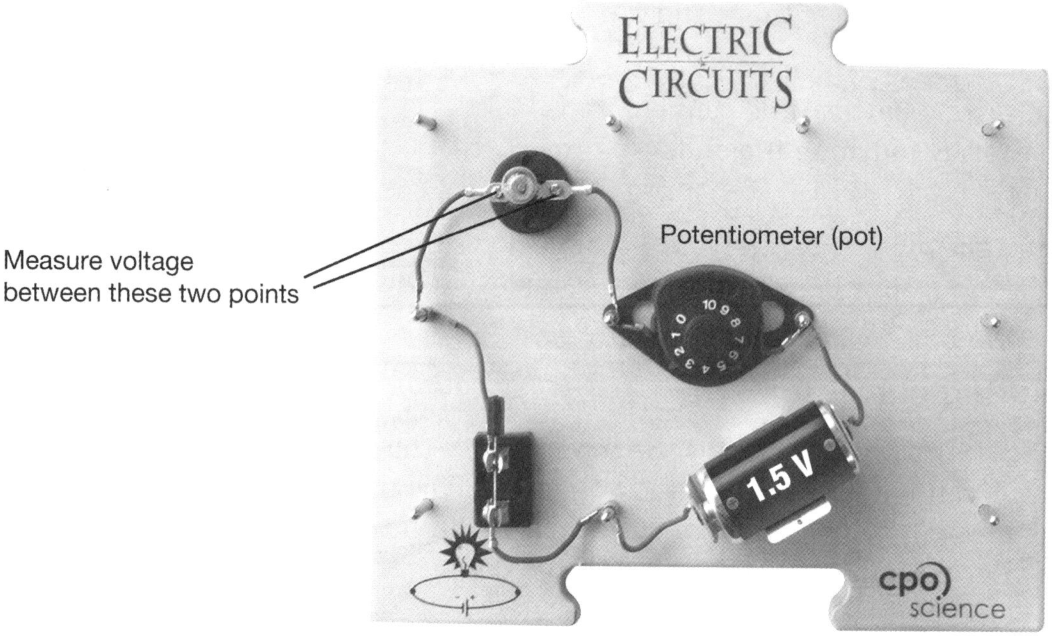

1. Connect the circuit in the diagram using the pot, a battery, wire, and a bulb.
2. Adjust the dial and watch what happens to the bulb.
3. Use the meter to measure the voltage across the bulb for different settings of the pot. Record your data in table 1.

Table I: Pot settings and voltage across bulb

Pot dial position	Voltage across bulb (V)	Observed light output of bulb

9 Thinking about what you observed

a. As you changed the settings of the pot, what happened to the voltage across the bulb?

b. Did you observe a relationship between the voltage across the bulb and the light output?

c. Propose a relationship between power and voltage that would explain the light output of the bulb.

16B Resistance and Ohm's Law

What is the relationship between current and voltage in a circuit?

Electrical devices get the energy they use from the current that flows through them. When designing an electrical device or a circuit, it is important for the proper amount of current to flow for the voltage that is available. Resistance is the property of electricity that helps regulate the current in the circuit. You will explore resistance and Ohm's law, the equation that relates voltage, current, and resistance.

Materials

- Electric Circuits kit
- 2 "D" batteries
- Multimeter

1 Mystery resistors

A *resistor* is used in a circuit to provide resistance. You have green, blue, and red resistors in your kit with values of 5 Ω, 10 Ω, and 20 Ω, but you don't know which is which!

1. Make a circuit with a switch, battery, and resistor, as pictured above.
2. Set the meter to measure current (DC amps). Open the switch. Measure the current by connecting the meter across the open terminals of the switch.
3. Set the meter to measure voltage (DC volts). Close the switch. Measure the voltage across the mystery resistor as pictured above.
4. Repeat the current and voltage measurements for each of the mystery resistors.

Table 1: Resistor Currents

Resistor color	Voltage across resistor (V)	Current (A)

a. Use your knowledge of Ohm's law to determine which resistor is which. The resistance you calculate from Ohm's law will not come out exactly to 5, 10, or 20 because the meter itself has a small resistance.

2 Resistance and potentiometers (pots)

The potentiometer (pot) you used in the previous investigation is really a *variable resistor.* A variable resistor allows you to change its resistance by turning a dial. Many dials you use every day, like dimmer switches, are actually potentiometers.

1. Use the meter to measure the resistance of the pot for different positions of the dial. The pot does not have to be in the circuit; you can just touch the leads across the pot.
2. Take your first reading with the pot turned all the way to the left and take 4 or 5 readings until the pot is turned all the way to the right.

Table 2: Pot settings and resistance

Pot dial position	Resistance (Ω)

3 The bulb dimmer circuit

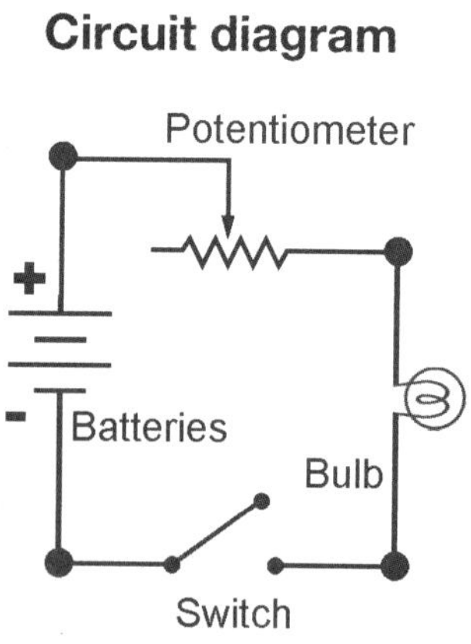

1. Build the dimmer circuit with two batteries, the pot, a switch, and a bulb.
2. Close the switch and observe how the brightness changes as you change the dial on the pot.

a. Use the concept of resistance to explain how the pot controls the brightness of a bulb.

4 The voltage drop

The voltage in a circuit is reduced whenever current flows through a device that has a resistance greater than zero. The reduction of voltage is called the *voltage drop*.

DC volts

To measure the voltage drop across the pot, touch the leads to (A) and (B).

Use points (B) and (C) to measure the voltage drop across the bulb.

Circuit diagram

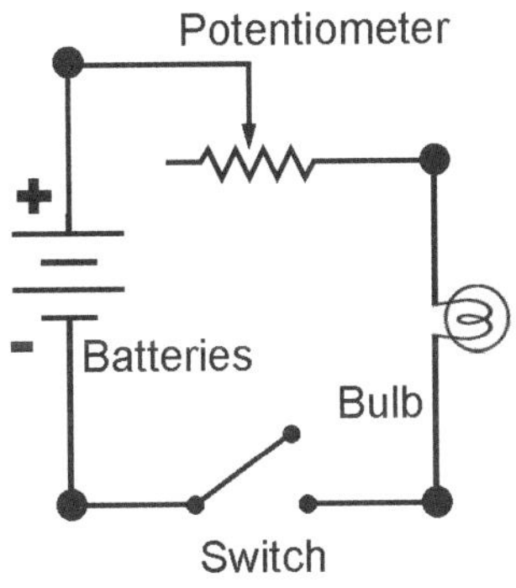

1. Connect the dimmer circuit with the pot, switch, two batteries, and a light bulb.
2. The voltage drop is measured by touching the meter leads to the terminals of each device.
3. Measure the voltage drop across the pot and across the bulb for different dial settings of the pot. Record the observed light output at each dial position measured.

Table 3: Pot settings and voltage drops

Pot dial position	Voltage drop across pot (V)	Voltage drop across bulb (V)	Observed light output

a. What is the relationship between the voltage drop across the pot and the voltage drop across the bulb?

b. What does the voltage drop tell you about the electrical energy carried by the current?

c. What relationship do you observe between the measured voltage drops and the battery voltage?

17A Magnetism

How do magnets and compasses work?

Magnets are used in almost all electrical and electronic machines from motors to computers. How far does magnetic force reach? How can you use a compass to detect magnetic forces? In this investigation, you will use magnets and a compass to answer these and other questions about magnetism.

Materials

- 2 magnets (from kit)
- Compass (from kit)
- Metric ruler

1 How far does magnetic force reach?

How far does the magnetic force of a magnet reach? This is an important question concerning machines such as motors and generators that use magnets.

How far does the magnetic force reach?

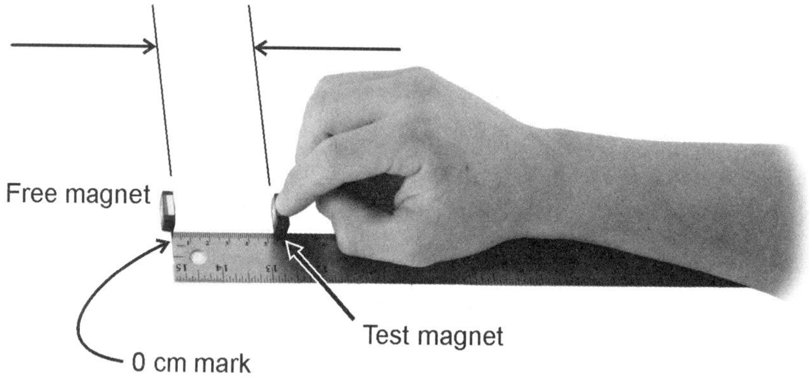

1. Place one magnet at the 0 cm mark of the ruler and slide a second magnet closer and closer until the first magnet moves. Practice the technique several times before recording data.
2. Record the distance between the magnets when you first see movement.
3. Try each of the combinations of poles—north–north, south–south, and north–south.
4. For each combination, complete three trials, and average your three distances.

Table 1: Magnetic forces between two magnets

	North-South	South-South	North-North
Distance 1 (mm)			
Distance 2 (mm)			
Distance 3 (mm)			
Average distance (mm)			
Average estimated error (mm)			

2 Thinking about what you observed

a. What is the average estimated error for each magnet combination? Subtract each individual distance for the North-South magnet combination from the average of the three distances. Drop any negative signs. Once you have found these 3 differences, average them and record in Table 1. Repeat for the other 2 magnet combinations.

b. Are the attract and repel distances *significantly* different? Your answer should include a comparison between average estimated errors and the differences between magnet combination average distances.

3 Using a compass to detect magnetic forces

The needle of a compass is a permanent magnet. Earth is magnetic, so a compass needle is attracted to north in the absence of other (stronger) magnets.

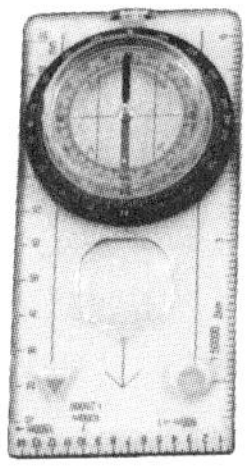

Start:
Align the ring, arrow, and needle with north.

Measure the distance to deflect the needle 20 degrees East.

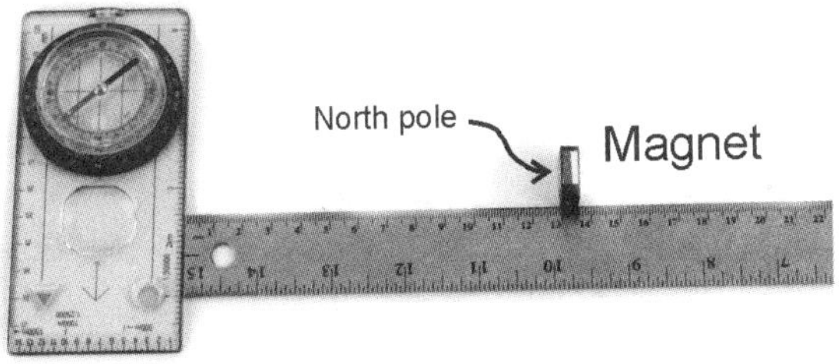

Measure the distance to deflect the needle 20 degrees West.

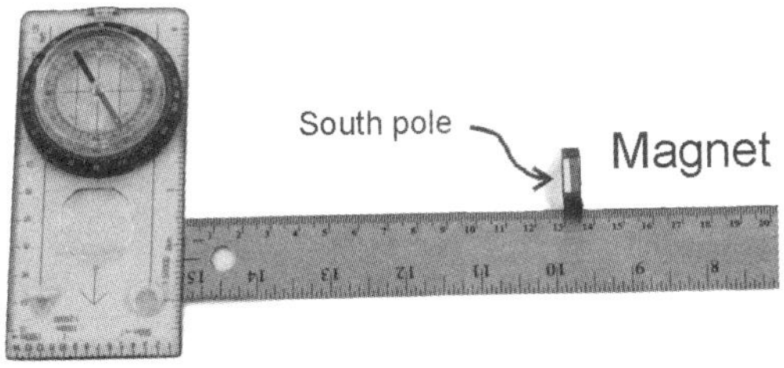

1. Set a compass on your table far from any magnets. Rotate the compass so the needle, dial, and arrow are all aligned with north.
2. Place a metric ruler to the side of the compass and line it up perpendicularly with the north pole of the compass. Move a small magnet near the compass and note the distance at which the needle moves 20 degrees from north.
3. Reverse the pole of the small magnet and note the distance at which the needle moves 20 degrees in the opposite direction.

4 Thinking about what you observed

a. At a distance of 10 cm, which is stronger: the magnetic force from Earth or the magnetic force from the small magnet? How is your answer supported by your observations?

b. Is the end of the compass needle a magnetic north or a magnetic south pole? How is your answer supported by your observations?

c. Is the geographic north pole of the planet Earth a magnetic north or a magnetic south pole? How is your answer supported by your observations?

17B Electromagnets

How are electricity and magnetism related?

Almost every electrical device that creates motion, such as a motor, uses magnets. Permanent magnets are not the only type of magnets used in these devices. Often, electromagnets are used. Electromagnets create magnetic forces through electric currents. This investigation will explore the properties of electromagnets.

Materials

- Electric Circuits kit
- Electromagnet coil (in kit)
- 1 permanent magnet (in kit)
- Compass (in kit)
- 1 "D" battery

1 Electromagnets

1. Attach the coil, battery, and switch in the circuit shown above. Leave the switch open so no current flows.
2. Place a permanent magnet about 1 centimeter away from the coil. Stand the magnet up on its end.
3. Close the switch and watch what happens to the magnet. DON'T leave current running or the coil will overheat. Open the switch after each trial.
4. Turn the permanent magnet around so its other pole faces the coil. Close the switch and see what happens now.
5. Reverse the wires connecting the battery to the circuit. This makes the electric current flow the other way. Repeat steps 3 and 4 of experiment with the magnet.

2 Thinking about what you observed

a. Write 2–3 sentences that explain what you saw when the switch was closed.

b. Propose an explanation for why the magnet moved.

c. When the magnet was reversed, did the force between it and the coil change direction? How did the force change?

d. When the coil wires were switched, did the force from the coil change direction? How do you know?

e. How is a current-carrying coil like a magnet? How is it different? Explain how this shows that electricity and magnetism are related.

3 Comparing the electromagnet to a permanent magnet

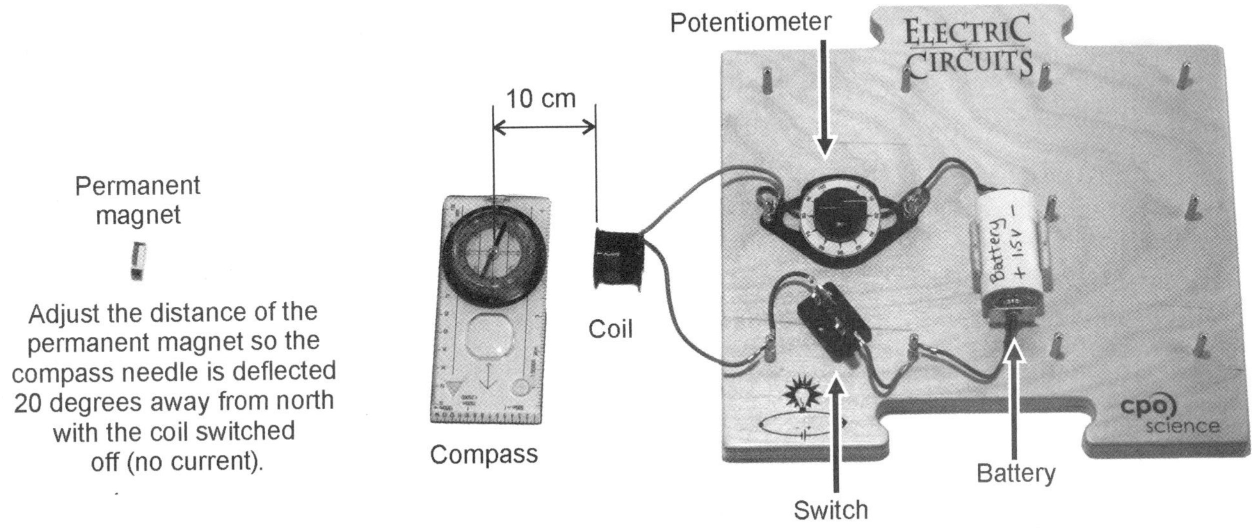

1. Attach the potentiometer, coil, battery, and switch in the circuit shown in the diagram. Leave the switch open so no current flows.
2. Set the compass so the needle, ring, and arrow are all aligned with north. Put the coil about 10 cm from the center of the compass.
3. Place a permanent magnet on the side of the compass opposite the coil. Bring the magnet close enough to deflect the needle 20 degrees away from north.
4. Close the switch and adjust the potentiometer so the needle returns to north. The coil should deflect the compass needle back toward north. Reverse the permanent magnet if the needle moves the wrong way. DON'T leave current running or the coil will overheat. Open the switch after each trial.
5. Try moving the permanent magnet to different distances and using the potentiometer to return the compass needle to north with force from the electromagnet.

4 Stop and think

a. The permanent magnet is pulling the compass needle to the left. The electromagnet is pulling the needle in the opposite direction, to the right. When the needle returns to north what can you say about the magnetic forces from the permanent magnet and electromagnet?

5 Iron and electromagnets

Trial # 1: Coil only

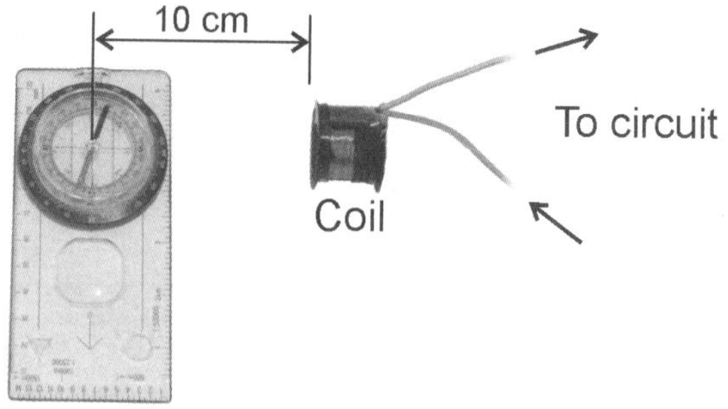

Measure the current it takes to bring the compass needle back to north

Trial # 2: Coil and steel pin

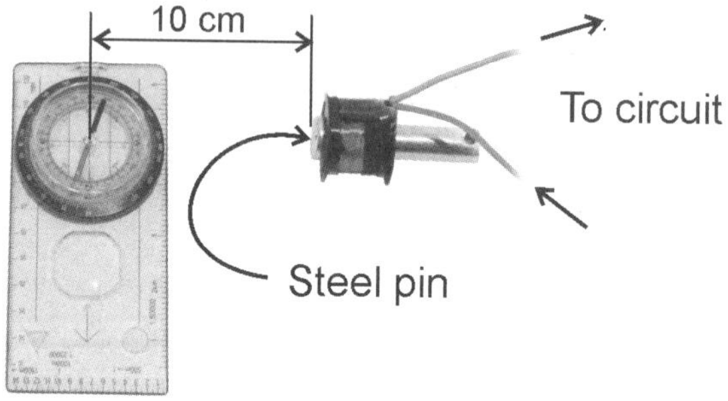

1. Use the same circuit as for part 3 with one battery, switch, coil and potentiometer.
2. Rotate the compass until the needle and dial are aligned with north. *There should be no magnets nearby, and no current in the coil for this step.*
3. Move a permanent magnet close enough to deflect the needle 20 degrees from north.
4. The coil should be 10 cm from the center of the compass (see diagram above).Close the switch, then use a multimeter to measure and record how much current it takes for the coil to bring the needle back to north. Adjust the current with the potentiometer. Once you have recorded the measurement in Table 1, open the switch to stop the current.
5. Put the steel pin in the coil so its head is against the coil and 10 cm from the center of the compass.
6. Adjust the distance of the permanent magnet so the compass needle is deflected 20 degrees from north, like you did in step 3. *There should be no current in the coil for this step.*
7. Close the switch, then use a multimeter to measure and record the current it takes to return the needle to north with the steel pin in the coil.

Table 1: Electromagnet current with and without the steel pin

Current with bare coil	Current with steel pin	Difference in current	Percent difference

6 Thinking about what you observed

a. How did the steel pin affect the magnetic force created by the coil? Was the magnetic force reduced, increased, or did it stay about the same? Use your observations to support your answer.

18A Time and Tree Rings

Do tree rings tell a story?

If you look at the cross-section of a tree, you will see tree rings. Tree rings tell a story about the growing conditions of the tree. The number of rings that a tree has equals its age.

Careful examination of tree rings can give detailed information about the age of a piece of wood and the growing conditions a tree experienced during its life time. Scientists compare tree rings to a catalog of tree ring history to figure out the age of certain wooden objects like old ships, log cabins, and archeological artifacts. In this investigation, you will determine the age and growing conditions of several samples of wood by examining tree rings and tree core samples yourself.

Materials

- Microscope slide of a tree cross-section
- Tree cross section
- 7 paper strips representing tree cores, a set of 4 and a set of 3 (included as a graphic in the lab)
- Two blank pieces of paper
- Graph paper
- Scissors
- Tape
- Metric ruler (with millimeter ruling)
- Colored pencils
- Magnifying lens

1 How do trees grow?

Each year, a tree gets larger by one ring. It is easy to count tree rings because each ring has a light and a dark band.

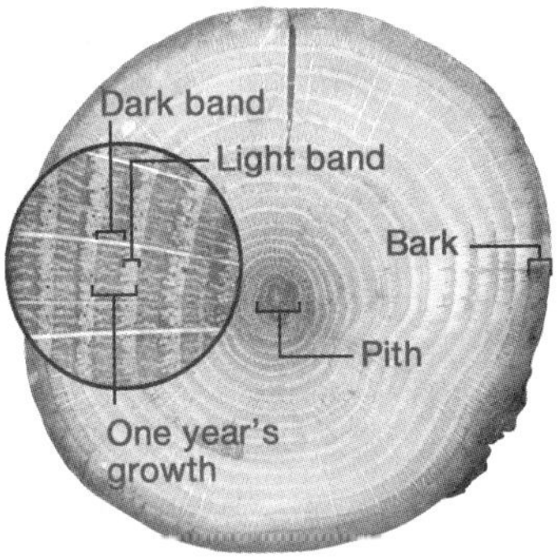

Tree rings (and the individual light and dark bands) vary in width depending on the growing climate. Wet, warm years allow for more growth than cool, dry years. A wide ring means the tree grew during a wet year. A narrow ring band means the tree grew during a cool, dry year.

a. Look at the microscope slide of the tree cross-section. Use the magnifying lens to help you see the details. Write a short paragraph describing what you see.

b. You will see small holes in the tree cross-section. What are these holes? Come up with a hypothesis that answers this question.

c. Looking at the slide with your eyes, how many tree rings do you see?

2 Investigating tree rings

1. **Tree cross-section:**
Now, look at the tree cross-section. Use a magnifying lens to help you see the rings. Notice the width of the bands. Look at the bark and the pith. See if you can figure out the age of the tree. Remember that one ring (one year's growth) includes a light band and a dark band. Discuss what you see with your group. *NOTE: The pith and bark are not counted in determining the age of a sample. The youngest wood is under the bark.*

2. **Tree cores:**
   Obtain copies of the four paper strips below and cut them out. The strips represent tree cores from four different trees. The trees are the same species and grew in the same woodland.

3. Determine the age of each tree by counting the rings (one ring includes a wide light band and dark line). Record your answers in Table 1.

Table 1: Tree core data

Sample	Age of Tree	Year Cut or Cored	Year Growth Began
1			
2			
3			
4			

4. Look for patterns in the rings. Line up all the samples by matching the patterns before taping them onto a blank piece of paper. Color the overlapping sections so you don't lose track. Figure out the age of each tree and record this information, and the years that the trees began growing and were cut in Table 1.
   Note: These matching patterns indicate the trees grew in the same woodland and help prove what the climate was like in the past for that area.
5. **More tree cores:**
   Now look at the three new tree cores (see next page). Write down the age of each core. Then, cut out the paper strips and line them up according to the corresponding dates that overlap on each sample.
6. Measure the width in millimeters of each tree ring, and record your measurements for each year in Table 2. Two years have already been done for you: 2006 and 2005.

Table 2: Tree core data

Year	Tree ring width (mm)	Year	Tree ring width (mm)	Year	Tree ring width (mm)
2006	10 mm	1993		1980	
2005	8 mm	1992		1979	
2004		1991		1978	
2003		1990		1977	
2002		1989		1976	
2001		1988		1975	
2000		1987		1974	
1999		1986		1973	
1998		1985		1972	
1997		1984		1971	
1996		1983		1970	
1995		1982			
1994		1981			

7. Make a graph of your tree core data. Plot the ring-width in millimeters on the y-axis against the year on the x-axis. One ring includes a wide light band and dark line. Label the graph appropriately.

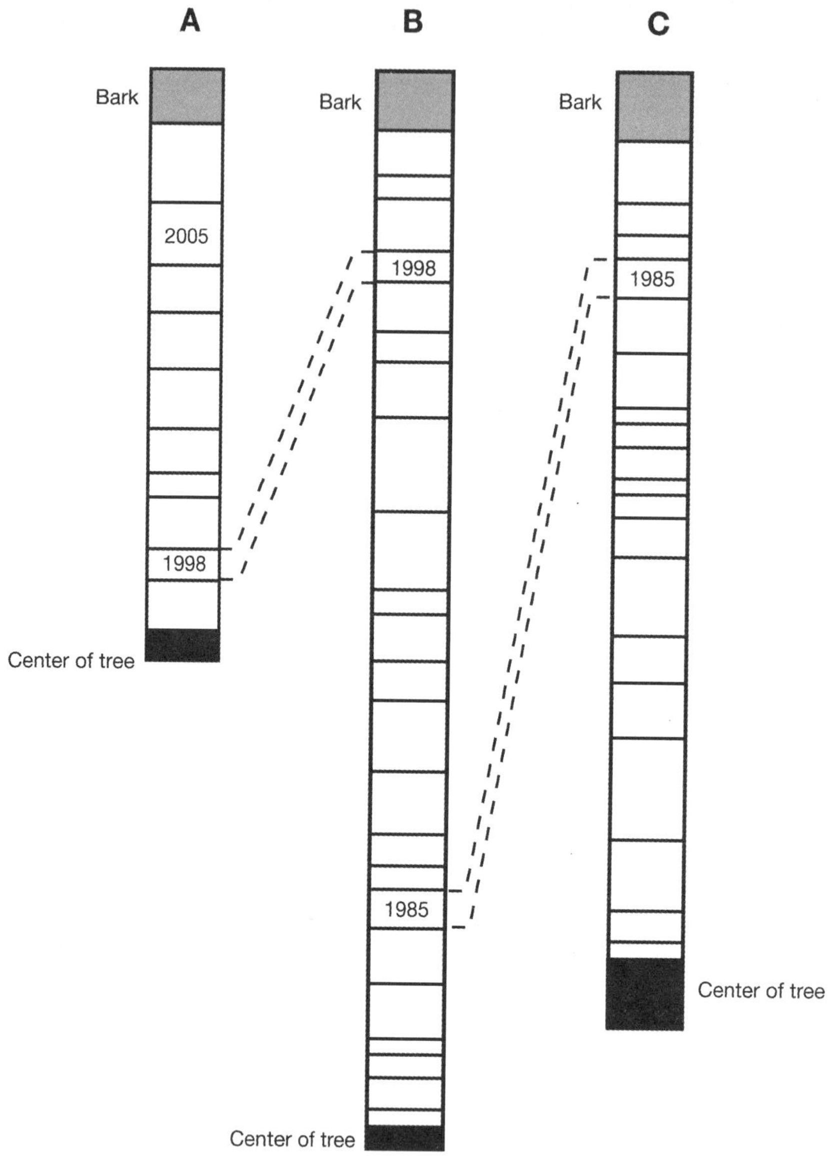

3 Thinking about what you observed

The set of four ring cores:

a. What kind of growing season existed for the first four cores in 1967? How can you tell?

b. If poor tree growth is mainly caused by drought, which years were probably drought years? How can you tell?

c. Did this woodland have more years of drought or plentiful rainfall?

d. Why might a climatologist be interested in tree ring data from this woodland?

The set of three ring cores:

a. What does your graph show about the three cores (A, B, and C) and the climate?

4 Exploring on your own

a. Global climate change (caused by an increase of carbon dioxide in the atmosphere) is a current environmental topic that concerns scientists. The result of global climate change is an increase in Earth's average temperature. Due to global climate change, what differences would you expect to see between tree rings of today and tree rings in the future (100 years from now)?

b. What factors can influence the growth of trees? List all the factors you can think of.

c. Trees are like the history books of a forest. But, what other objects in nature can we observe and measure to tell us something about certain environments? You may want to go to an outdoor location to brainstorm answers to this question.

d. What is a dendrochronologist and what does one do?

18B Relative Dating

How does relative dating tell a story?

Earth is very old and many of its features were formed before people came along to study them. For that reason, studying Earth now is like detective work—using clues to uncover fascinating stories. The work of Earth science researchers is very much like the work of forensic scientists at a crime scene. In both fields, the ability to put events in their proper order is the key to unraveling the hidden story.

Materials

- None required

1 Sequencing events after a thunderstorm

Carefully examine this illustration. It contains evidence of following events:

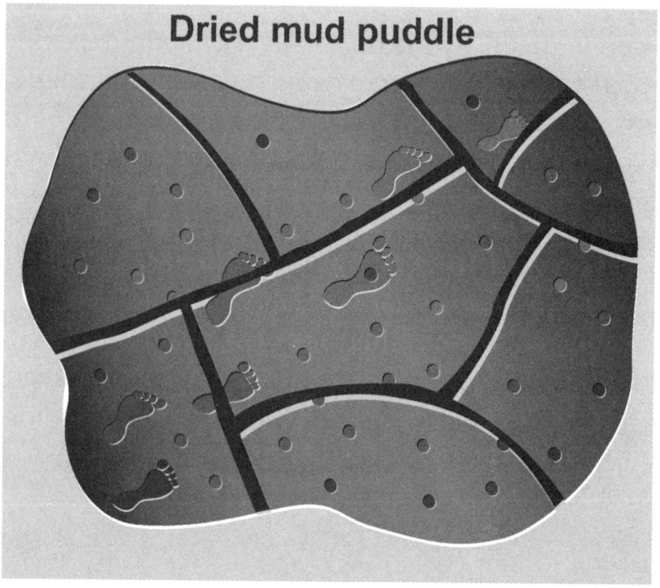

- The baking heat of the sun caused cracks to formed in the dried mud puddle.
- A thunderstorm began.
- The mud puddle dried.
- A child ran through the mud puddle.
- Hailstones fell during the thunderstorm.

a. From the clues in the illustration, sequence the events listed above in the order in which they happened.

b. Write a brief story that explains the appearance of the dried mud puddle and includes all the events. In your story, justify the order of the events.

2 Determining the relative ages of rock formations

Relative dating is an Earth science term that describes the set of principles and techniques used to sequence geologic events and determine the relative age of rock formations. Below are graphics that illustrate some of these basic principles used by geologists. You will find that these concepts are easy to understand.

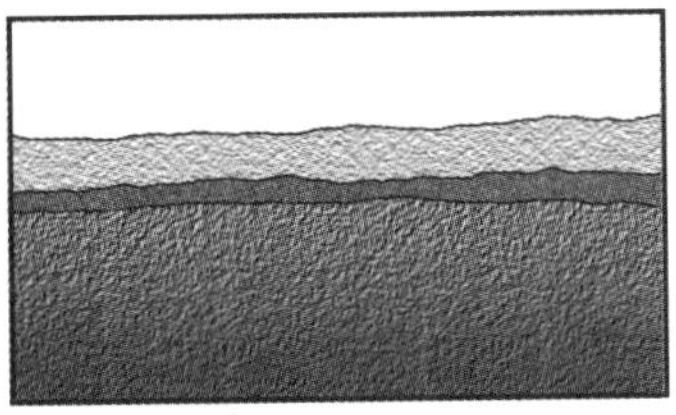

_____ 1. Original Horizontality

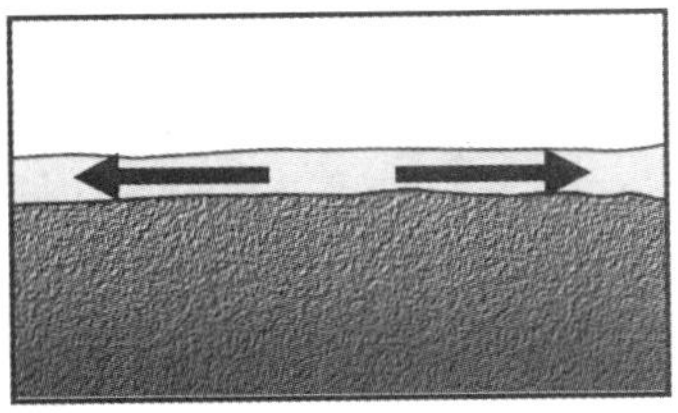

_____ 2. Lateral Continuity

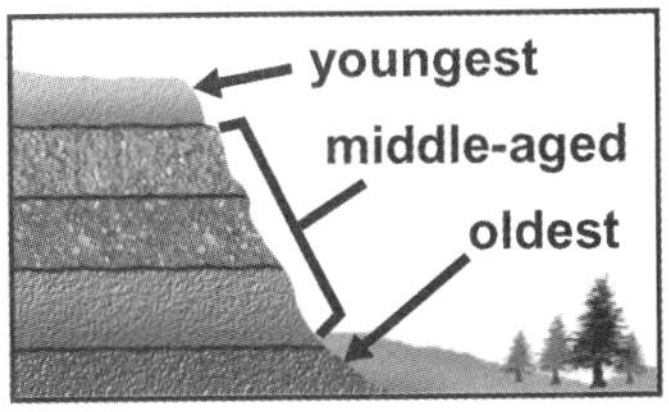

_____ 3. Superposition

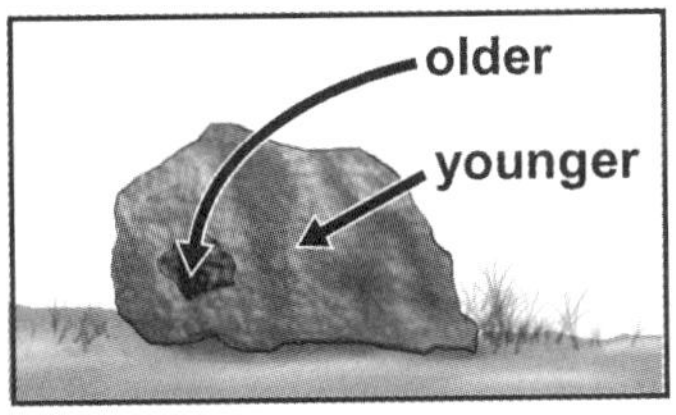

_____ 4. Inclusions

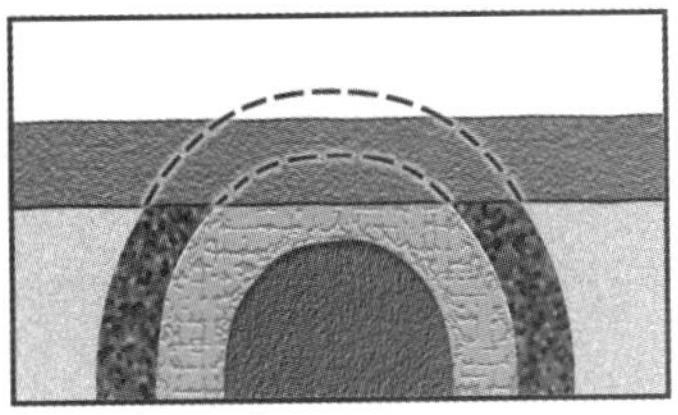

_____ 5. Unconformities

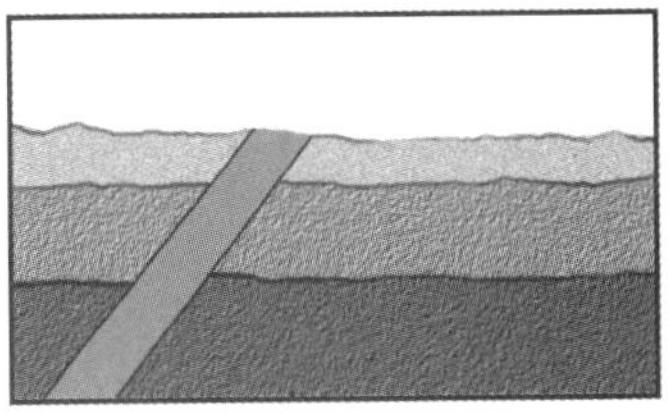

_____ 6. Cross-Cutting Relationships

Match each principle to its explanation. Write the letter of the explanation in the space provided under each graphic.

Explanations:

A. In undisturbed rock layers, the oldest layer is at the bottom and the youngest layer is at the top.

B. In some rock formations, layers or parts of layers may be missing. This is often due to **erosion**. Erosion by water or wind removes sediment from exposed surfaces. Erosion often leaves a new flat surface with some of the original material missing.

C. Sediments are originally deposited in horizontal layers.

D. Any feature that cuts across rock layers is younger than the layers.

E. Sedimentary layers or lava flows extend sideways in all directions until they thin out or reach a barrier.

F. Any part of a previous rock layer, like a piece of stone, is older than the layer containing it.

3 Sequencing events in a geologic cross-section

Understanding how a land formation was created with its many layers of soil begins with the same time-ordering process you used in Part 1. Geologists use logical thinking and geology principles like the ones described in Part 2 to determine the order of events for a geologic formation. Cross-sections of Earth, like the one shown below, are our best records of what has happened in the past.

Rock bodies in this cross-section are labeled A through H. One of these rock bodies is an **intrusion**. Intrusions occur when molten rock called **magma** penetrates into layers from below. The magma is always younger than the layers that it penetrates. Likewise, a fault is always younger than the layers that have faulted. A **fault** is a crack or break occurs across rock layers, and the term **faulting** is used to describe the occurrence of a fault. The broken layers may move so that one side of the fault is higher than the other. Faulted layers may also tilt.

a. Put the rock bodies illustrated below in order based on when they formed.

b. Relative to the other rock bodies, when did the fault occur?

c. Compared with the formation of the rock bodies, when did the stream form? Justify your answer.

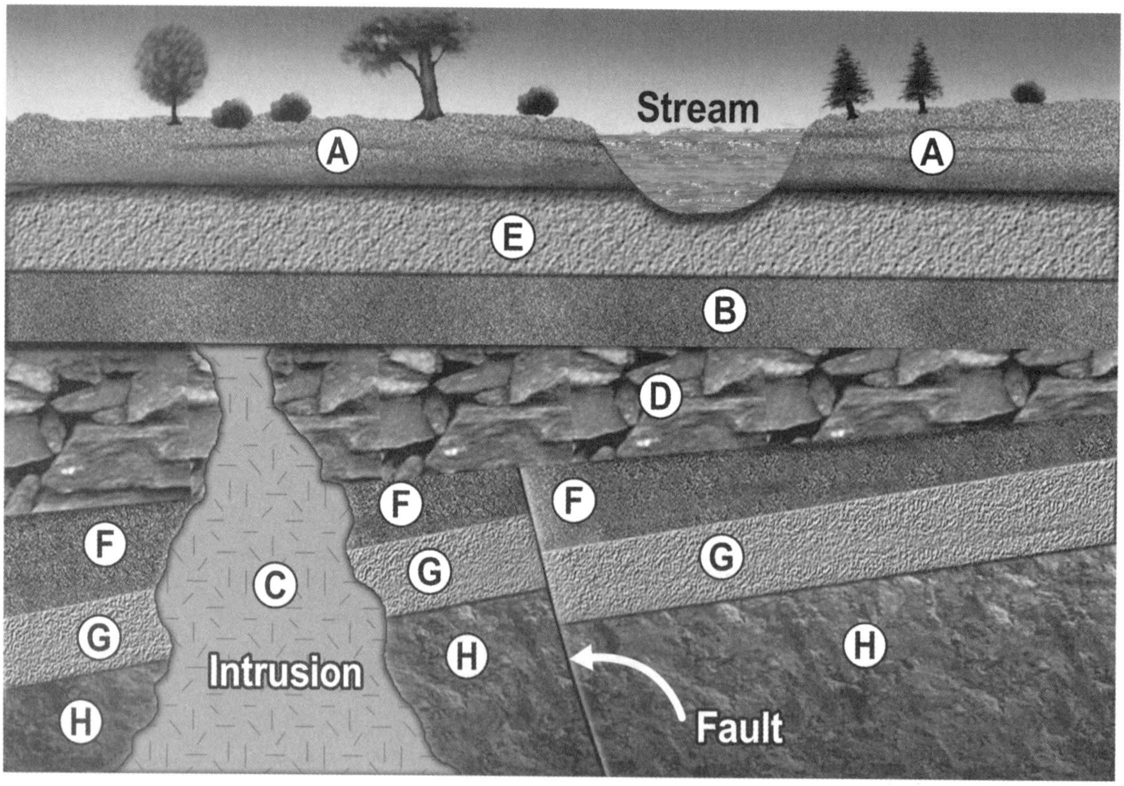

4 How do they know what is down there?

The cross section diagram from part 3 shows various geologic features present underground very clearly. But how do geologists know what is down there? To get a clear understanding of what is underground geologists extract core samples from the ground.

A core sample is a solid cylinder of the material that makes up the ground itself. Geologists obtain a core sample by drilling down into the ground with a large, hollow drill bit. The exposed layers in the core sample can be examined, identified, and labeled, showing geologists what is underground. By taking many core samples over a large area, geologists can get a fragmented "picture" of what is beneath the ground. The geologists fill in the missing spaces using principles of Relative Dating. Give it a try for yourself. Below are 15 core samples from an area (these are oversized; actual core samples would be much thinner). They are labeled A–O. Use the diagram to answer questions a-d below.

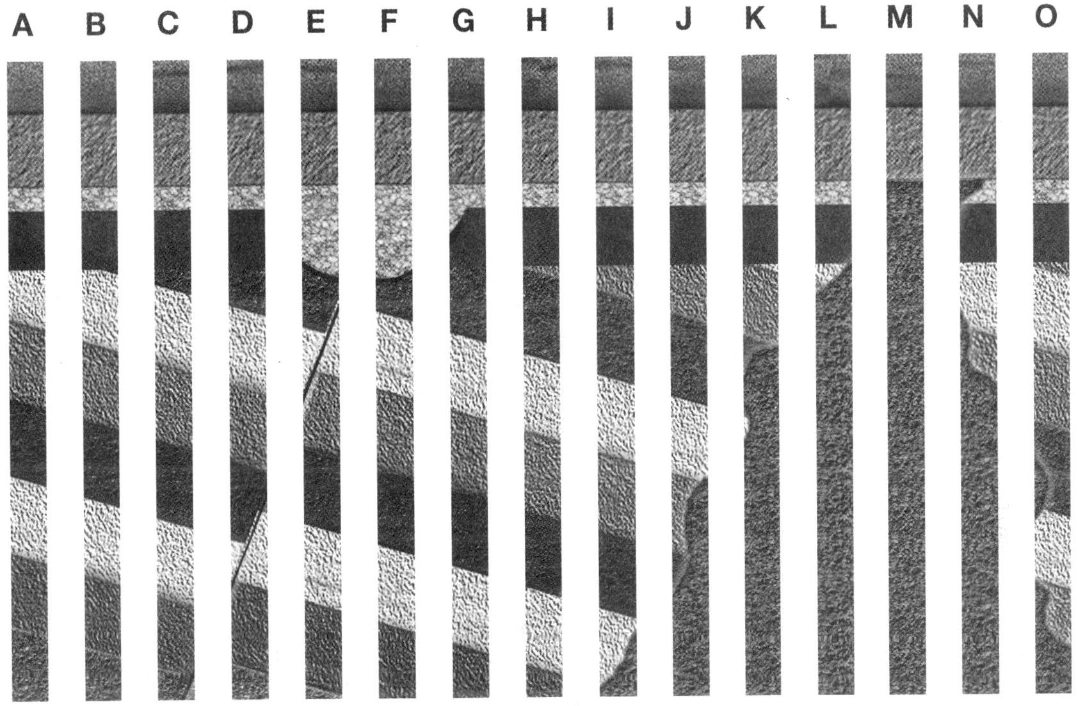

a. Do you see evidence of a fault in your core samples? Describe the evidence of any fault you see and list the letters of the core samples that you think show the fault.

b. Do you see evidence of an intrusion in your core samples? Describe the evidence of any intrusion you see and list the letters of the core samples that you think show the intrusion.

c. Do your core samples reveal any evidence for a stream to have existed at one time? If so, describe how you can tell and where it may have been.

d. Which happened first, the fault or the stream? How can you tell?

e. Place the stream, fault, and intrusion in chronological order.

19A All Cracked Up

What is a good way to model Earth?

Using *models* is an important practice when studying science. Many of the objects that a scientist would study, such as a planet or a molecule, are much too big or too small to work with by hand. Models can help a scientist solve this problem. Some examples of scale models a scientist might build include shrinking a planet down to the size of a basketball or blowing up a molecule to the size of a tennis ball. These models make objects easier for scientists to conduct their research. Think about a geologist studying plate tectonics and continental drift—two concepts you will learn more about in Chapter 11. The geologist would likely build small scale models of each of Earth's lithospheric plates in order to more easily understand how the movements of the plates affect one another.

Different kinds of scale models serve different purposes. In this activity, you will examine several geologic models of Earth to learn more about some of Earth's important features.

Materials

- Three objects that can serve as models of Earth. *Examples: a ping-pong ball, a baseball, a golf ball, an orange, a basketball, etc.*
- Blank sheet of paper
- Colored pencils for sketching

1 Procedure

1. You will be shown three different objects that can serve as models for Earth. Make a sketch of each model on a blank piece of paper or in your notebook.
2. Write the name of each object in the first column of Table 1. Then, based on your understanding of Earth, determine each model's strengths and weaknesses. Write your evaluations in Table 1.

Table 1: Evaluating your models

Model	Strengths	Weaknesses

3. Of the three models, which do you feel is the most accurate for modeling the layers of Earth? Which model do you feel is the least accurate? Explain your reasoning behind both answers.
4. Your teacher will now conduct a short demonstration using a hard-boiled egg as a model for the interior of Earth. Record and sketch your observations of the egg model. Be sure to make observations and sketches of both the interior and exterior of the egg.

5. Based on the demonstration, label the layers of Earth in the diagram.

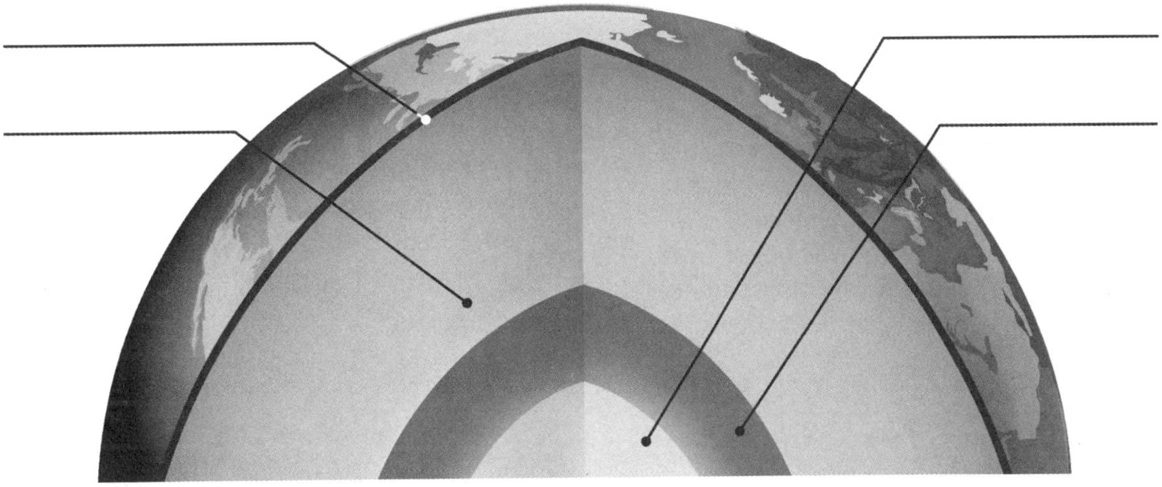

2 Discussion questions

a. Which layer of the interior of the egg corresponds to which layer of the interior of Earth? What part of Earth does the cracked egg shell model?

b. What are two strengths and two weaknesses of the cracked egg model?

c. How do Earth's layers compare in their composition and consistency with the egg layers?

d. Earth's lithospheric plates move relative to one another, unlike the eggshell sections which do not move at all. In some places plates move apart. In other places, plates come together, and in still others, plates rub sideways against each other.

- What geologic event might occur where plates rub sideways against each other?
- What geologic formation might result when two plates collide?
- What geologic formation might form where two plates move apart?

19B Plate Tectonics

What is plate tectonics?

Earth's crust plus the upper mantle forms the *lithosphere*. Earth's lithosphere is broken in a number of different pieces. How these pieces move and interact is what plate tectonics is all about. In this investigation, you will be an Earth detective and identify lithospheric plates using geologic evidence.

Materials

- Bathymetric map
- Colored pencils
- Pencil

1 Setting up

Part 1: Reading a bathymetric map

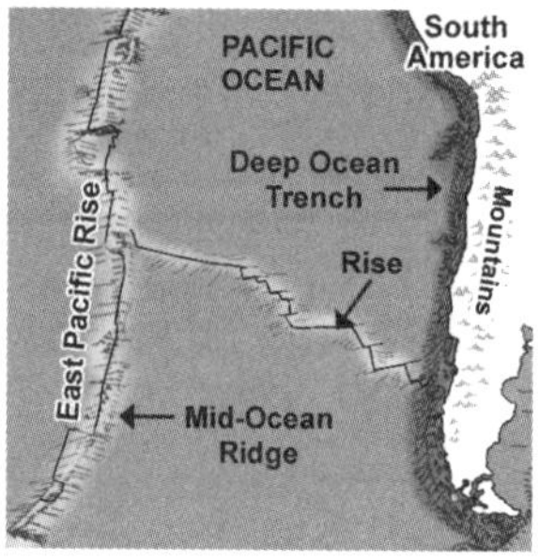

1. Examine your map. A bathymetric map shows what land looks like under a body of water like the ocean.
2. Find examples of the following features on your bathymetric map: mid-ocean ridges, rises, deep ocean trenches, and mountain ranges. List one example of each from your map in the second column of Table 1.
3. In the third column of Table 1, list which type of plate boundary, convergent or divergent, is associated with each feature. Note: A rise is like a ridge.
4. In the fourth column, there are small diagrams showing two plates and the boundary between these. Draw arrows showing how the plates move relative to each other at these boundaries.

Table 1: Features on a bathymetric map

Features	Examples from the map	Kind of plate boundary (convergent or divergent?)	How do the plates at this boundary move?
mid-ocean ridge			Plate 1 / Plate 2
rise			Plate 1 / Plate 2
deep ocean trench			Plate 1 / Plate 2

Part 2: Starting to find plate boundaries.

1. Earth has seven to ten large pieces and many small plates. To keep things simple on your map, you will identify seven large plates.

2. Keep in mind that mid-ocean ridges, rises, deep ocean trenches, and mountain ranges are all geologic features that are formed at tectonic plate boundaries.
3. With your pencil, draw a single line along the mid-ocean ridges, rises, deep ocean trenches, and mountain ranges. Draw your lines along the center of each of these features. Your lines represent the boundaries between tectonic plates.

Part 3: Using earthquake activity to find plate boundaries

1. Once you have identified all of the mid-ocean ridges, deep ocean trenches, and mountain ranges, you may notice that many of your lines may not connect. Complete the following steps to help fill in some of the missing parts of your plate boundaries.

Table 2: Earthquake data table

	Location	Magnitude
1	***Lat*** 36.3° N, ***Long*** 23.4° E	6.8
2	***Lat*** 32.5° N, ***Long*** 104.5° W	4.2
3	***Lat*** 5.7° S, ***Long*** 76.4° W	7.5
4	***Lat*** 18.8° N, ***Long*** 155.4° W	5.2
5	***Lat*** 34.5° N, ***Long*** 73.6° E	7.6
6	***Lat*** 10.9° N, ***Long*** 140.8° E	6.6
7	***Lat*** 45.1° N, ***Long*** 167.2° E	7.2
8	***Lat*** 2.1° N, ***Long*** 97.0° E	8.7
9	***Lat*** 55.5° N, ***Long*** 165.8° E	6.7
10	***Lat*** 39.8° N, ***Long*** 43.8° E	5.2
11	***Lat*** 4.5° S, ***Long*** 153.4° E	7.7
12	***Lat*** 35.0° N, ***Long*** 119.0° W	4.7
13	***Lat*** 36.4° N, ***Long*** 140.8° E	6.4
14	***Lat*** 7.9° N, ***Long*** 92.1° E	7.3
15	***Lat*** 9.6° N, ***Long*** 84.2° W	6.4
16	***Lat*** 51.2° N, ***Long*** 179.4° W	6.8
17	***Lat*** 41.3° N, ***Long*** 125.9° W	7.2
18	***Lat*** 36.3° N, ***Long*** 51.6° E	6.3
19	***Lat*** 49.3° N, ***Long*** 128.8° W	6.7
20	***Lat*** 19.9° S, ***Long*** 69.0° W	7.8

2. Earthquakes are common along tectonic plate boundaries. The earthquake data table (Table 2) shows the location and magnitude of recent earthquakes around the globe.
3. You will be plotting 20 earthquakes according to latitude and longitude. Plot the earthquakes using a colored pencil according to their magnitude. Use this key for your earthquakes, draw this key on your map.

Key for Earthquake Magnitude

- ● Magnitude less than 5.0
- ◉ Magnitude 5.0 to 6.9
- ✸ Magnitude 7.0 and above

4. When you have finished plotting the earthquake data, use your pencil and draw a single line along the earthquake dots. You may find that many of these lines will connect with the lines you drew for mountain ranges, deep-ocean trenches, and mid-ocean ridges. These lines represent the boundaries between tectonic plates.

Part 4: Using volcanic activity to find plate boundaries

1. Once you have plotted your earthquake data, you may find that some of the lines still do not connect. Follow these steps to help fill in some of the missing parts of your plate boundaries.
2. Volcanoes, like earthquake activity, are often found along plate boundaries. The volcano data table (Table 3 below) shows the location of recent volcanic eruptions around the globe.

Table 3: Volcano data table

	Location
1	***Lat*** 59.4° N, ***Long*** 153.4° W
2	***Lat*** 12.3° N, ***Long*** 93.9° E
3	***Lat*** 1.2° N, ***Long*** 77.4° W
4	***Lat*** 5.5° S, ***Long*** 150.0° E
5	***Lat*** 19.5° N, ***Long*** 155.3° W
6	***Lat*** 16.5° S, ***Long*** 168.4° E

	Location
7	***Lat*** 35.2° S, ***Long*** 70.6° W
8	***Lat*** 19.0° N, ***Long*** 98.6° W
9	***Lat*** 16.7° N, ***Long*** 62.2° W
10	***Lat*** 46.2° N, ***Long*** 122.2° W
11	***Lat*** 54.0° N, ***Long*** 159.5° E
12	***Lat*** 14.5° N, ***Long*** 90.9° W

3. You will be plotting volcano activity according to latitude and longitude. Plot the volcanoes using a colored pencil. Use a triangle to represent volcanoes. Include this in your key on your map.
4. When you have finished plotting your volcano data, use your pencil and draw a single line along the triangles. You may find that the lines you draw for the volcanoes will connect with the previous lines you have drawn.

Part 5: Using your evidence to locate Earth's major tectonic plates

1. Using your pencil, continue working on connecting the lines that mark tectonic plate boundaries. Remember you are trying to locate seven major tectonic plates: the American Plate, Eurasian Plate, African Plate, Pacific Plate, Antarctic Plate, Indo-Australian Plate, and the Nazca Plate. Label each of the plates you found.

2 Stop and think

a. How well do you think you did in accurately locating the seven major tectonic plates? What further evidence would have been helpful in making your map more accurate?

b. In what areas of your map could you have used more geologic evidence to help locate plate boundaries?

c. Why is a bathymetric map more useful in this investigation than a regular world map that does not show the ocean floor?

3 Doing the experiment

1. Now that you have come up with a hypothesis of where the seven major tectonic plates are located, your teacher will help you distinguish the actual location of the seven plates.
2. When you have the actual locations of the seven major tectonic plates, use your colored pencils to shade in each of the plates a different color. Be sure you have written the name of each of the seven plates on your map as well.

4 Thinking about what you observed

a. Find the Himalaya Mountains on your map. These mountains are continuing to grow taller. What does this tell you about the type of plate boundary between the Eurasian plate and the Indo-Australian plate?

b. The boundary around the Pacific plate is commonly referred to as the "Ring of Fire." Examine the geologic events along this boundary. Why do you think this boundary has been given this name?

c. Find the Mid-Atlantic Ridge on your map. The Atlantic Ocean was once much smaller millions of years ago and has been growing in size to present day. What does this tell you about the type of plate boundary that exists at the Mid-Atlantic Ridge?

d. The boundary between the African Plate and the Eurasian Plate in the Mediterranean Sea is a convergent boundary. Africa is slowly pushing northward towards Europe. What will eventually become of the Mediterranean Sea in the future? Why?

e. The boundaries between the Eurasian Plate and the Indo-Australian Plate as well as the Nazca Plate and the South American plate are convergent boundaries. What is a major difference in the geologic features and events at these two plate boundaries?

f. Lithospheric plates move about as fast as your fingernails grow—2.5 cm/year! Your teacher will help you figure out the main directions to move the plates on your map. Using this information and the rate of movement, figure out what Earth will look like in 50 million years!

20A Earthquakes

What conditions affect the timing, duration, and intensity of an earthquake?

Most earthquakes are associated with plate boundaries. This is because the slow movement of the plates against each other causes stress on the rocks at the boundaries to increase. Stress results in the buildup of pressure and stored energy in objects. *Earthquakes* occur when the rocks that are under stress experience a release of pressure and stored energy. In this investigation, you will use simple materials to simulate an earthquake.

Materials

- Safety goggles
- Sandpaper (2 sheets)
- Masking tape
- 2 paper clips
- Rubber band
- Scissors
- Kite string (30 cm)
- Index card (2 cm × 12.5 cm long)
- Metric ruler
- Hardcover book
- 4 sugar cubes

1 Setting up

1. Cut two pieces of sandpaper in half. Each half will be 9 × 5.5 inches. Turn the four halves smooth side up and place them short-end-to-short-end. Tape the halves together so that they form a 36-inch strip.

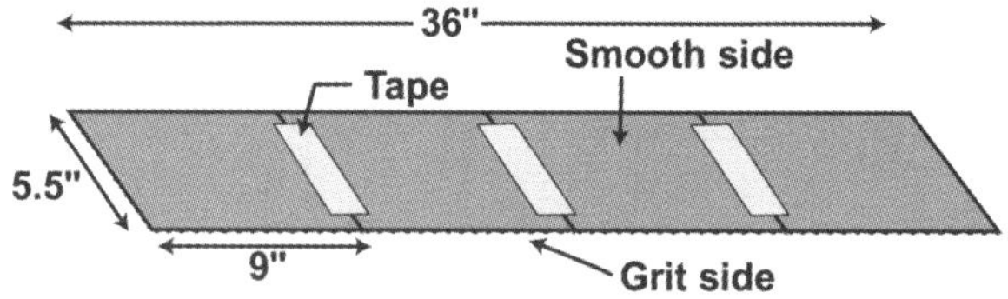

Sandpaper strip

2. Tape the sandpaper strip to your desk so the smooth side is down and the grit is up.
3. Put on safety goggles. Use caution with rubber bands. Do not shoot or overstretch them.
4. Make a strain gauge by hooking two paper clips into a rubber band. Tie a long piece of kite string to one of the paper clips.
5. Now, you will make a paper ruler with a piece of index card that is 2 cm wide by 12.5 cm long. First, place the blank card under the paper clip that is not attached to the string. Tape the card to this paper clip as shown in the diagram. The strain gauge and the paper ruler will move together in the model.

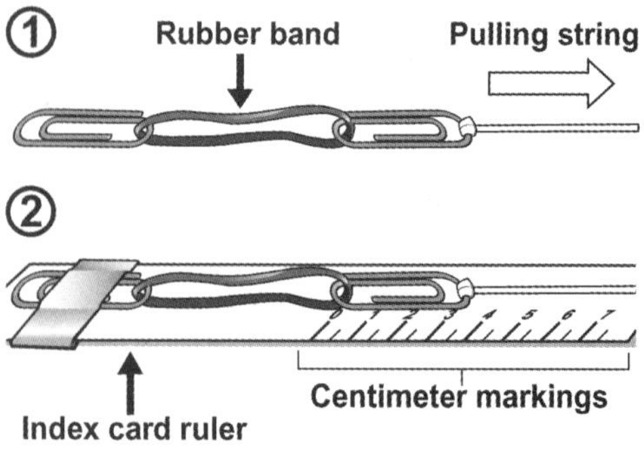

Strain gauge setup

6. Next, starting with zero at the far end of the rubber band, mark off centimeters using a metric ruler until you get to the end of the card. Make your strain gauge look like the diagram. Make sure you have at least 7 centimeters marked on the paper ruler.

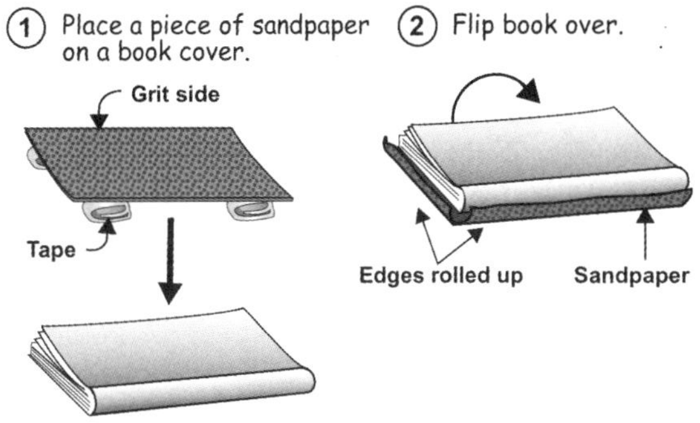

7. Using masking tape, attach a piece of sandpaper on the outside of one cover of a book. The sandpaper should be a little larger than the cover. Once the sandpaper is on the book, fold up the overhanging sandpaper.
8. Tie a piece of kite string to the paper clip that is taped to the paper ruler. Then, place this string down the center of a page in the middle of the book. Use a very small piece of tape to hold the string in place on the page. Close the book. Tie the overhanging kite string to another paper clip so that it serves as an anchor to keep the string from pulling through the book. You may want to tape this paper clip to the top of the book with masking tape.
9. Place the book on the sandpaper strip at one end of the sandpaper strip. Make sure your setup matches with the setup of the earthquake model shown in the diagram below.

Earthquake Model

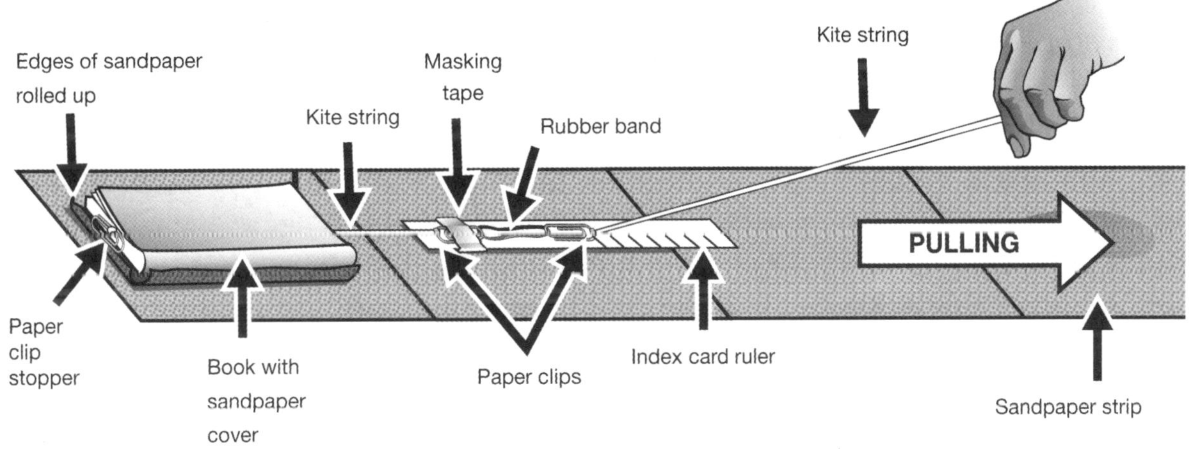

2 Stop and think

The graphic at right shows an earthquake occurring near a fault. Use this graphic to identify how the earthquake model represents an earthquake. Then, fill in Table 1.

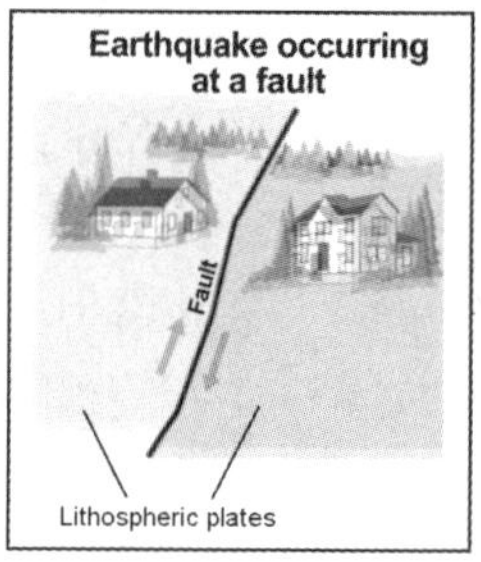

Table 1: Earthquake model

Material in setup	What does it represent?
The book	
The sandpaper strip	
The boundary between the book and the sandpaper	

3 Doing the experiment—working with the model

Safety tip: Wear goggles!

1. Place the book at one end of the sandpaper strip.
2. Gently pull on the string until the book moves a little. If the book does not move straight, move the string inside the book a little off center, and re-tape the string to the page.
3. Make adjustments to the setup as needed until it works well. However, do not pull the book along the sandpaper too many times. Doing this will wear down the sandpaper and change your results.
4. Now, place the book back at the end of the strip, and pull the string until the edge of the rubber band moves from zero on the paper ruler to the 1-centimeter mark. This distance is related to how much stress is in the rubber band.
5. Answer questions **a** and **b** in Part 8: Thinking about what you observed.

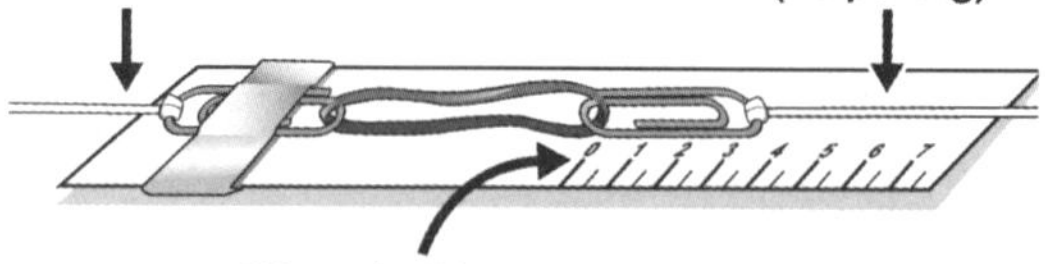

4 Doing the experiment—simulating the timing of an earthquake

To simulate an earthquake, you will pull the string until the book suddenly moves along the sandpaper strip. This sudden movement, called a "slip" or a "failure," is the release of stress between the book and the sandpaper. It is also a release of energy.

1. Place the book at the far end of the sandpaper strip.
2. Pull the string slowly and smoothly until the book slips (moves suddenly). Record the stretch just as the book slips. This is the *slip stretch*.
3. Without moving the hand holding the string, record the remaining stretch after the book stops. This is called the *stop stretch*.

4. Repeat this process twice. Start with the book at the far end of the sandpaper strip for each trial. Record all data in Table 2. Find the average slip stretch and stop stretch lengths. Then, answer questions **c** and **d** in Part 8.

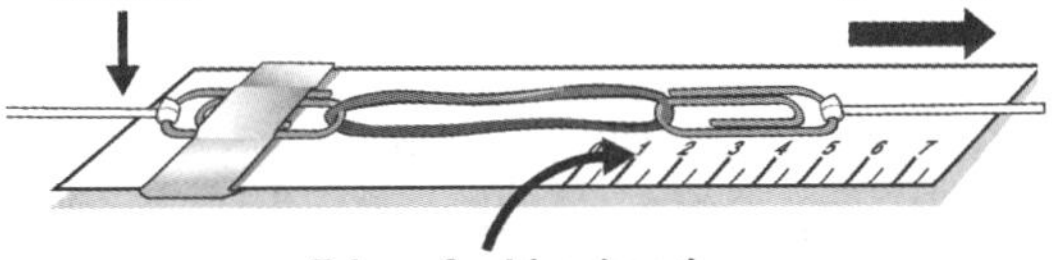

Table 2: Earthquake timing simulation

Trial Number	Length of slip stretch (cm)	Length of stop stretch (cm)
1		
2		
3		
Average		

5 Doing the experiment—simulating the duration of an earthquake

1. In this part of the investigation, you will simulate another earthquake. As you did in the previous simulation, pull the thread to move the book. However, when the book slips, continue to pull the thread just hard enough to keep the book moving. Practice your technique.
2. When you have perfected your technique, begin collecting data in Table 3. Read the stretch measurement after the book first slips but while the book is still moving. Run three trials, record your data below, and find the average.
3. Answer questions **e** and **f** in Part 8.

Table 3: Earthquake duration simulation

Trial number	Length of stretch while the book is moving (cm)
1	
2	
3	
Average	

6 Doing the experiment—simulating the intensity of an earthquake

1. You and your group will simulate another earthquake by having team members drum on the tabletop with their fists. Each person should drum on the table with gentle to medium force.
2. While drumming is taking place, simulate an earthquake with your model.
3. Depending on what you observe in this experiment, record slip and stop stretches or stretches while the book is moving. Record your data in Table 4 and then average your data.
4. Answer questions **g–i** in Part 8.

Table 4: Earthquake intensity simulation

Trial number	Length of slip stretch (cm)	Length of stop stretch (cm)	Length of stretch while book is moving (cm)
1			
2			
3			
Average			

7 Doing the experiment—simulating the damage caused by an earthquake

1. Perform each of the experiments with sugar cubes stacked on the book. Place one sugar cube alone, then have stacks of two, three, and four cubes. These stacked sugar cubes represent one-, two-, three-, and four-story building,
2. In the spaces of Table 5, record whether the cubes moved or the stacks fell during each simulation.
3. Answer questions **j–l** in Part 8.

Table 5: Earthquake damage simulation

Earthquake experiment	One cube	Two cubes	Three cubes	Four cubes
Timing simulation				
Duration simulation				
Intensity simulation				

8 Thinking about what you observed

a. Did the book move when the stretch was 1 centimeter?

b. What does the movement of the book on the sandpaper strip represent in this investigation?

c. The movement of lithospheric plates occurs all the time, but earthquakes do not. Why doesn't plate movement cause continual small earthquakes? Why do earthquakes occur every once in a while? Explain your answer.

d. Did all of the energy stored in the rubber band release when the book slipped? Do you think an earthquake releases all of the stored energy when it occurs?

e. How does the data from "timing of an earthquake" compare with the data from "duration of an earthquake"? Why do you think this is?

f. Earthquakes last longer than a few seconds. They do not simply start and quickly stop. Explain the relatively long duration of earthquakes based on the results of the experiment.

g. How does the data from "timing of an earthquake" or "duration of an earthquake" compare with the data from "intensity of an earthquake"? Why do you think this is so?

h. How did the drumming affect the intensity of the "earthquake" in the model?

i. Do you think one earthquake can cause another earthquake? Explain your answer.

j. Which sugar-cube building experienced the most damage? Why do you think this happened?

k. Which experiment resulted in the most damage? Why do you think this happened?

l. Given the results, suggest one safety tip that would reduce damage to buildings during an earthquake.

9 Exploring on your own

a. Find out what areas in the United States have the highest seismic risk.

b. In terms of plate tectonics, what is happening at the San Andreas Fault?

c. How fast is the plate movement at the San Andreas Fault?

20B Volcanoes

How are volcanoes and plate boundaries related?

Mount St. Helens in Washington State erupted violently in 1980, sending ash and dust high into Earth's atmosphere. The winds in the atmosphere blew this ash and dust around the world. All active volcanoes erupt and release material like lava, ash, and dust that is very hot and therefore dangerous. Some volcanoes, such as Mount St. Helens, are especially dangerous because of the sudden, violently explosive nature of their eruptions. Other volcanoes, such as Mauna Loa in Hawaii, are less explosive. Less explosive volcanoes spew lava fountains and streams of melted rock, but in a gentler manner. In this investigation, you will discover key differences between gentle and explosive volcanoes and will discover a pattern in their geographic distribution.

Materials

- Bathymetric map
- Red and blue markers
- Ruler (to help plot latitude and longitude points)

1 The Volcanic Explosivity Index

The *Richter scale* has been used by geologists for more than 50 years to measure the strength of an earthquake. The Richter scale is a number scale that describes the magnitude of an earthquake. Geologists have also used a number scale that describes volcanic eruptions. This number scale is called the Volcanic Explosivity Index, or VEI. The higher the VEI, the more explosive or violent the eruption of a volcano. Explosive eruptions are associated with high plumes of lava and ash escaping from the top of the volcano, such as Mount St. Helens. Volcanoes with low VEI numbers have gentle eruptions. The plumes of these eruptions are not very high and not as much lava is released when the volcano erupts, such as Kilauea in Hawaii. Table 1 provides a list of volcanoes and their VEI ratings.

Table 1: Examples of volcanoes and VEI ratings

VEI	Plume height	Volume (m^3)	Average time interval between eruptions	Example
0	<100 m	≥ 1000	one day	Kilauea
1	100–1000 m	≥ 10,000	one day	Stromboli
2	1–5 km	≥ 1,000,000	one week	Galeras, 1992
3	3–15 km	≥ 10,000,000	one year	Ruiz, 1985
4	10–25 km	≥ 100,000,000	≥ 10 years	Galunggung, 1982
5	>25 km	≥ 1,000,000,000	≥ 100 years	Mount St. Helens, 1981
6	>25 km	≥ 10,000,000,000	≥ 100 years	Krakatoa, 1883
7	>25 km	≥ 100,000,000,000	≥ 1,000 years	Tambora, 1815
8	>25 km	≥ 1,000,000,000,000	≥ 10,000 years	Toba, 71,000 years ago

a. What is the relationship between the VEI and the average time interval between eruptions?

b. The plume height is the height the erupted material rises from a volcano. What is the relationship between the plume height and the VEI?

c. What rating does Kilauea in Hawaii have on the VEI? What rating does Mount St. Helens in Washington state have on the VEI?

2 Connecting volcanoes to plate boundaries

1. Table 2 provides the position by latitude (Lat) and longitude (Long) for eight volcanoes represented by two-letter symbols. The volcanoes in the left-hand group are known to erupt violently. The right-hand group volcanoes are known as gentle or less violent.
2. Plot the locations of all volcanoes on your bathymetric map.
3. Represent each volcano with its two-letter symbol. Represent violent volcanoes with a red marker and gentle volcanoes with a blue marker.

Table 2: Locations of volcanoes

Violent volcanoes		Gentle (less violent) volcanoes	
PN	***Lat*** 15.1° N, ***Long*** 120.4° E	MA	***Lat*** 19.5° N, ***Long*** 155.5° W
KR	***Lat*** 16.7° S, ***Long*** 105.4° E	FE	***Lat*** 0.4 S, ***Long*** 91.6 W
KA	***Lat*** 58.3° N, ***Long*** 155.0° W	ER	***Lat*** 13.6° N, ***Long*** 40. 7° E
BE	***Lat*** 56.1° N, ***Long*** 160.7° E	PT	***Lat*** 21.2° S, ***Long*** 55.7° E

a. How do the locations of the two kinds of volcanoes relate to the locations of plate boundaries?

b. Volcanoes that form near convergent plate boundaries release magma that is viscous. Viscous magma is very thick and gases become trapped inside, building up pressure over time. What types of volcanoes are located near convergent plate boundaries?

c. Near divergent plate boundaries the magma in volcanoes is not viscous. Magma that is not viscous is not as thick and does not contain as much gas. This causes the magma to flow more easily, causing less pressure to build in the magma before it flows out as lava. What types of volcanoes are located near divergent plate boundaries?

3 Connecting the location of volcanoes to volcanic rock

Table 3 provides information about the magma composition and the type of volcano that you plotted in Part 2. Study the table, and using your map, think about how the magma composition relates to a volcano's location.

Table 3: Type of volcano and magma composition

Volcano	Type	VEI	Volcanic Rock (Magma composition)
Pinatubo (PN)	composite	6	rhyolite
Krakatoa (KR)	composite	7	rhyolite/andesite
Katmai (KA)	complex composite	3	andesite
Bezymianny (BE)	complex composite	2	andesite
Mauna Loa (MA)	shield	0	basalt
Fernandina (FE)	shield	2	basalt
Erta Ale (ER)	shield	2	basalt
Piton de la Fournais (PT)	shield	1	basalt

a. What type of volcanic rock is associated with more explosive eruptions (higher VEI)? What type of volcanic rock is associated with less explosive eruptions (lower VEI)?

b. What is the relationship between the volcanic rock and the type of plate boundary (convergent or divergent)?

c. Using your bathymetric map, where do you typically find volcanoes that contain basalt? Where do you typically find volcanoes that contain rhyolite or andesite?

d. Imagine you are asked to investigate a newly discovered volcano to find out whether it will produce a gentle or violent eruption. Develop a research plan for studying the volcano. What evidence will you need to be able to identify the nature of the eruption? Assume that you can use any resources you need.

21A Solubility

What does it mean to dissolve?

The water you drink is not pure water. There are many things dissolved in drinking water, such as minerals, oxygen, and sometimes flavorings. A solution is a homogeneous combination of different compounds, such as water and sugar. Carbonated soda is a solution containing water, sugar, dissolved carbon dioxide gas, and flavorings. Solutions are essential for life because your body's chemical processes happen in solutions. This investigation will look at some of the properties of solutions.

Materials

- 250 mL Graduated cylinder
- Electronic scale (or triple-beam balance)
- Granulated sugar
- Hot water
- Cold water
- 2 sugar cubes
- Data Collector
- Stirring rod
- Water soluble markers: red and blue
- Two 100 mL beakers
- 2 foam cups

1 Mass and volume in solutions

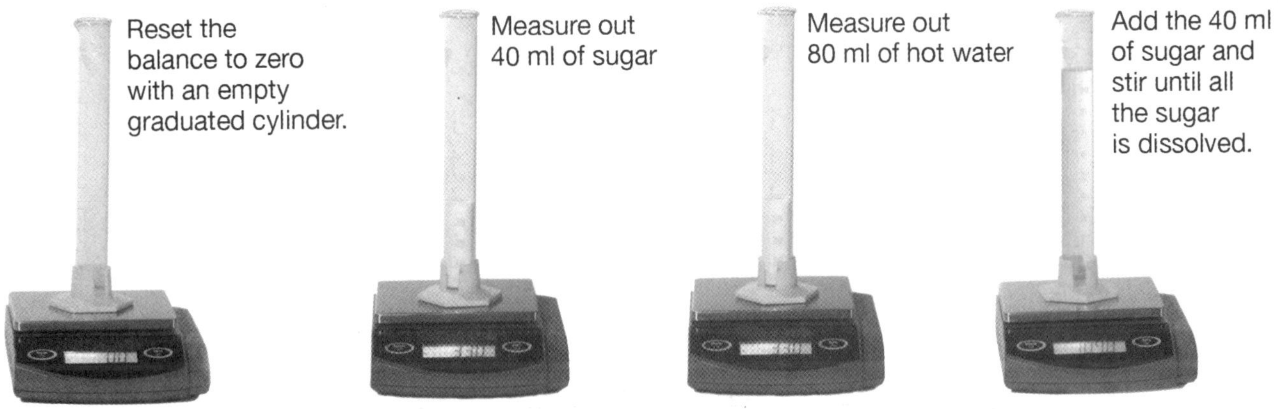

1. Put a clean, dry graduated cylinder on the balance and reset it to zero.
2. Put 40 ml of sugar into the graduated cylinder. Record the mass and volume of the sugar in Table 1 (you zeroed out the cylinder in step 1 by hitting reset on the balance while the graduated cylinder was on there, so the mass displayed by the balance will just be the sugar). Pour the sugar into a foam cup.
3. Pour 80 ml of hot water into the graduated cylinder and measure the mass.
4. Remove the graduated cylinder from the balance. Pour the 40 mL of sugar from the foam cup into the graduated cylinder containing the 80 mL of hot water and stir to dissolve all the sugar. Put the graduated cylinder back on the balance and record the mass and volume of the solution.
5. Use your mass and volume data to calculate the density of the sugar, the water, and the sugar and water solution.

Table 1: Mass and volume data for solutions

Substance	Mass in grams	Volume in mL	Density in g/mL
sugar		40	
water		80	

Substance	Mass in grams	Volume in mL	Density in g/mL
solution of sugar and water			

2 Stop and think

a. As the sugar dissolved, you couldn't see it any more. How do you know it is all still there?

b. Compare your calculated density of sugar with your calculated density of water. Do your calculations match up with your observations? Hint: Think about density and buoyancy.

c. What happened to the volume of the water and sugar solution compared to the volume of water and sugar you started with? Can you explain this?

d. Based on the volume of water displaced by the sugar, calculate the actual density of sugar? Does this new density match up with your observations?

3 Temperature and dissolving rate

For this part of the investigation, you are going to see how sugar cubes dissolve in water of different temperatures.

1. Color one face of a sugar cube with red water-soluble marker. Put the sugar cube in an empty 100 ml beaker with the colored side facing up.
2. Color one face of a second sugar cube with blue water-soluble marker. Put the sugar cube in an empty 100 ml beaker with the colored side facing up.
3. Start the Data Collector's stopwatch and carefully pour the hot or cold water down the edge of each beaker until it covers the sugar cube by about one centimeter. Do not pour the water directly on the sugar cubes!
4. Watch the colored face of each sugar cube. Which one dissolves fastest?

4 Stop and think

a. Temperature is a measure of how fast the molecules of a substance are moving. Are water molecules moving faster in hot water or cold water?

b. When a substance dissolves, molecules of the liquid crash into the solid molecules and knock some of the solid molecules loose. Explain how this concept fits with the data you observed on the dissolving sugar cubes.

21B Solubility Curve of KNO_3

What is a solubility curve?

Solubility refers to the amount of solute that can be dissolved in a certain volume of solvent under certain conditions. In this experiment, you will be examining the relationship between temperature and the solubility of potassium nitrate (KNO_3). The solvent will be water. Using class data, you will construct a solubility curve for KNO_3.

WARNING — This lab contains chemicals that may be harmful if misused. Read cautions on individual containers carefully. Not to be used by children except under adult supervision.

Materials

- 5 test tubes
- Test tube holder (a spare 250 mL beaker works well)
- Potassium nitrate (KNO3)
- 10 mL graduated cylinder
- Water
- Two 400 to 600 mL beakers
- Ice
- Data Collector
- Temperature probe
- Electronic scale (or triple-beam balance)
- Hot plate

1 Stop and think

Based on what you know about dissolving substances in water, do you think the solubility of KNO_3 will increase or decrease with temperature?

2 Doing the experiment

1. Fill a 400 mL beaker ¾ full with water. Place it on a hot plate. Allow it to heat and go on to the next step.
2. Your teacher will assign you a specific amount of KNO_3. Use a balance to obtain the exact mass of potassium nitrate you will be using and put it in a test tube. Record the mass in Table 1.
3. Add 5 mL of water to your test tube containing your KNO_3. Record this volume in Table 1.
4. Place the test tube with the KNO_3 and water into your hot water bath. Allow it to heat until all the KNO_3 dissolves, while stirring. The temperature of the water may need to reach approximately 85°C. You can check its temperature with the temperature probe once the probe is connected to the Data Collector.
5. Once all solid has dissolved, remove the test tube from the hot water bath and allow it to cool.
6. Place a temperature probe in the test tube and watch for the first signs of crystallization. (Hint: Stirring occasionally may help you see the small crystals that form). When you start to see your first crystal gather at the bottom of the test tube, record your temperature in Table 1.
7. If crystals still do not form when the temperature cools down to 30 degrees, place the test tube in a beaker of room temperature water.
8. If the crystals still do not form when the temperature cools down to 25 degrees, take the test tube out of the room temperature water and place it in a beaker of ice water (provided by your teacher).

9. Show your teacher your results. If your teacher gives you permission, place your test tube back in the hot water bath and re-dissolve the solid. Flush the solution down the drain with plenty of hot water.
10. Rinse and clean all remaining apparatus and put it away.

Table 1: KNO_3 crystal formation sample data

Mass of KNO_3 (g)	Volume of water (mL)	Mass of water (g)	Temp. when crystals first appeared (°C)

3 Calculations

a. In the lab you measured out 5.0 mL of water. What is the mass of 5.0 mL of water? Explain.

b. Knowing how many grams of your solid can dissolve in 5.0 g of water, how many grams could dissolve in 100.0 g of water? Report this value to the class (Hint: Set up a ratio). Record your results and those of the other groups in Table 2.

Table 2: Group data

	Group #1 1 g	Group #2 2 g	Group #3 4 g	Group #4 6 g	Group #5 8 g
temperature (°C)					
solubility (g/100g H_2O)					

4 Thinking about what you observed

a. Using the class data, construct a solubility curve. Plot the temperature (0°C–100°C) on the x-axis and the solubility of potassium nitrate (g KNO_3/ 100g H_2O) on the y-axis. Make sure your graph has a title and the axes are labeled including units. Connect all points with a smooth curve when complete.

b. In your own words, explain how solubility of KNO_3 varies with temperature.

c. From your solubility curve, predict the solubility of KNO_3 at

1. 65°C
2. 50°C
3. 25°C

d. Is this solubility curve useful for temperature values above 100°C?

e. How many grams of KNO_3 can be dissolved in 200.0 mL of water at 35°C?

22A Modeling a River

Which variables affect the formation of a river?

When an experiment is done, usually one *variable* is tested to see how it affects another variable. In this investigation, you will learn about variables. You will manipulate variables in an experiment that uses a stream table.

A stream table is a model of a river system. You cannot easily change variables that affect a real river. You can for a model river in a stream table! What variables might you be able to change? You will find out in this investigation.

Materials

- Stream table stage
- Stream table
- Bucket with spigot
- Bucket without spigot
- Displacement tank
- Data Collector (optional)
- Water

1 Setting up

1. Fill the top 2/3 of the stream table with sand. Pour some water onto the sand and mix it all up so it is completely wet. The sand should fill the tray up so it reaches just about 2 centimeters below the edge of the tray.
2. Smooth the surface of the sand out as much as you can so it is nice and flat.
3. Place the stream table on one of the rungs of the stand.
4. Move the spigot on the bucket to the closed position. Then, fill this bucket with water. Place it on top of the stage with the spigot pointing over the distribution trough that channels the water into the stream table.
5. Place the empty bucket (the one without the spigot) below the spigot of the stream table to catch water that will run out of the tray once water starts to run into the table.
6. Make sure the spigot on the stream table is in the closed position.

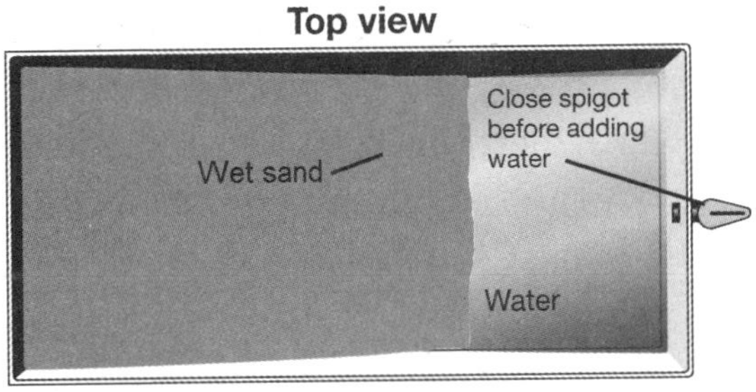

2 Looking at a system

What is a system? A system is a group of objects and the variables that affect those objects. We'll be looking at the stream table as a system. Examine the stream table setup with your group and answer the following questions;

a. What are the objects that make up the stream table?

b. What are the variables that affect the objects in the system?

c. What are some variables that probably do not affect the system?

3 Doing the activity

1. Look at the lists of variables that affect the objects in the stream table system you made in Part 2.
2. With your group, observe how these variables affect the process of making a stream, and the final effects on the sand in the tray.
3. Make sure your spigot at the bottom of the sand tray is closed. Place the bucket without the spigot under the spigot on the sand tray.
4. Open the spigot on the bucket that is holding the water and let the water flow into the trough, and down onto the stream table.
5. Carefully observe what happens as the water flows onto the sand and down the tray. Make notes of your observations with your group.
6. Close the spigot when your observations are over.
7. If water gets to the point where it may spill over, open the spigot on the tray and drain some water in the bucket below the spigot.

4 More about variables

Variables are factors that affect a system. In Part 2 you listed variables that affected the stream table system. In Part 3, you made observations. You watched what happened, and made notes about how you and your group thought the variables affected the system. In an experiment, the investigator usually wants to see how one variable affects the system. Variables in an experiment can be categorized into independent variables, control variables, and dependent variables. These different types of variables are explained below.

Variables in an experiment

Independent variable –	Control variable –	Dependent variable –
A variable that is changed in an experiment.	A variable that is held constant in an experiment.	A variable that is affected by the change to the independent variable.

How does a change in the independent variable affect the dependent variable?

5 Posing a scientific question

Every experiment begins with a question:

- How large is Earth?
- How fast do mountains grow?
- What is the best fertilizer for roses?
- How old is this glacier?

Asking questions helps to focus on what we want to know. Once we find out what it really is that we want to know, we consider the possibilities based on what we have observed. Then, we make a prediction about what we think is true. In science, this is called a hypothesis. The purpose of an experiment is to test if a hypothesis is true.

1. Examine your list of variables you made in part 2, and your observations you made in Part 3.
2. With your group, look at your list of variables and decide on a question that involves the formation of a stream in your tray and how changing one of those variables will change the stream.
3. There should be a cause and effect relationship between the variable and its effect on the system. For example; Does making an object heavier make it fall to the ground faster when it is dropped?
4. What is your group's question?

6 Hypothesis and variables

Once you have your question formed, it is time to make a hypothesis about what you and your group think will happen to your dependent variable when you change your independent variable. Your hypothesis is a possible answer to this question.

Table 1: Your hypothesis, variables, and experiment

What is your question?	
What is your hypothesis?	
What is your independent variable?	
What are your control variables?	
What is your dependent variable?	

7 Designing your experiment

It is time to plan what you will do to test your hypothesis. This is what designing an experiment is all about. Answer the questions below and present your proposal to your teacher for final approval.

a. What are you going to change about your independent variable?

b. How will you measure and record this change? (Hint: make a data table)

c. What are you going to measure concerning your dependent variable?

d. How will you measure and record any change?

e. What will you do to ensure your control variables remain the same?

f. What is the procedure for your experiment?

g. How will you know if your hypothesis is correct or not?

8 Setting up and performing your experiment

Once you have approval from your teacher, carefully follow your group's plan and perform your experiment. Be sure to take notes about observations, things you may or should have done differently, or things you did well as your experiment is underway. Good luck!

9 Evaluating your data

a. Was your experiment able to answer your question?

b. Did this experiment lead you to think of any other scientific questions? What are they?

c. What would you have done differently next time you design and perform an experiment?

d. What was successful that you would try again in your next experiment?

e. Can you present your data graphically, or in some other way that would help you to communicate your findings?

10 Conclusions

a. Was your hypothesis correct?

b. If not, can you tell why it was not correct?

c. Did you make any interesting observations that were not part of the original experiment?

11 Presenting your findings

As time permits, gather your data together, make any graphs, charts, or diagrams that would be helpful, and present your findings to your class as a group. All members should have some part in the presentation. Write up a report that summarizes your experience.

22B The Water Cycle

How is a model of the water cycle in your classroom like Earth's water cycle?

Earth has large quantities of water. The Sun's energy evaporates water from Earth's surface into the atmosphere. This water is then recycled as it cools and condenses into clouds and then precipitates back to the ground as rain, sleet, or snow. The processes of evaporation and condensation are very important in the *water cycle*.

In this investigation, you will learn how to build and observe a model of the water cycle in your classroom.

Materials

- Water (preferably bottled water)
- Beaker (for pouring water
- Table salt
- Electronic scale (or triple beam-balance)
- 2 GeoBoxes
- 2 collection containers (shorter than height of GeoBox)
- 2 weights
- Plastic wrap (50 cm long piece)
- Masking tape
- Metric ruler

1 Setting up

1. Pour fresh water to the 3-centimeter mark in one GeoBox and pour salt water up to the 3-centimeter mark in the other GeoBox. The salt water will be approximately as salty as ocean water.
2. Put a collection container that is shorter than the height of the GeoBoxes in the middle of each of the GeoBoxes. A drinking glass works well for this purpose.
3. Cover the GeoBoxes with plastic wrap. Seal the edges using masking tape all the way around box rim. Do not stretch the wrap tightly.
4. Place a weight in the middle of the plastic wrap on each GeoBox so the lowest part of the plastic wrap is centered over the drinking glass.

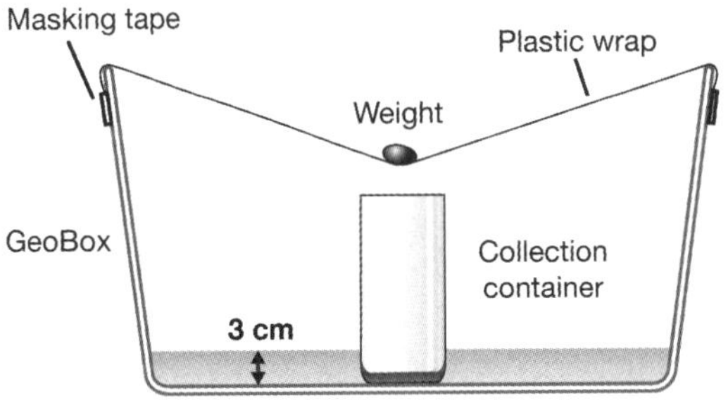

Fresh water

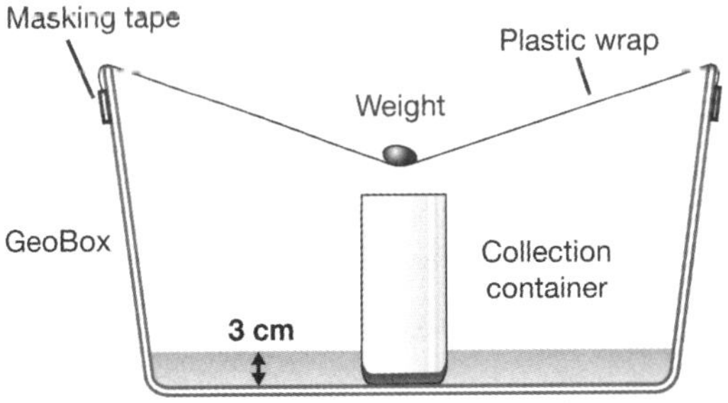

Salt water

2 Doing the experiment

1. Place the GeoBoxes in a warm, sunny window. You have now made a *solar still*. A solar still allows you to collect pure water using the Sun's energy. You will see how this is done in the investigation.
2. Check the GeoBoxes periodically.
3. Record your observations and what you think is happening inside the GeoBoxes in Table 1.

Table 1: Observations

Day	Freshwater GeoBox		Saltwater GeoBox	
	Observation	What you think is happening	Observation	What you think is happening
1				
2				
3				
4				

3 Thinking about what you observed

a. What happened to the fresh water level in the first GeoBox?

b. What happened to the salt water level in the second GeoBox?

c. What happened to the water level inside the collection containers in each box?

d. Where is evaporation happening inside the GeoBoxes?

e. Where is condensation happening inside the GeoBoxes?

f. How did water get into the collection containers?

g. How do you think what you observed happening in the GeoBoxes relates to the water cycle?

h. You will find a water cycle diagram in Chapter 4 of your textbook. Compare this diagram to what took place in the GeoBoxes. Draw this diagram at the right and label where each stage of the water cycle is taking place inside the GeoBox.

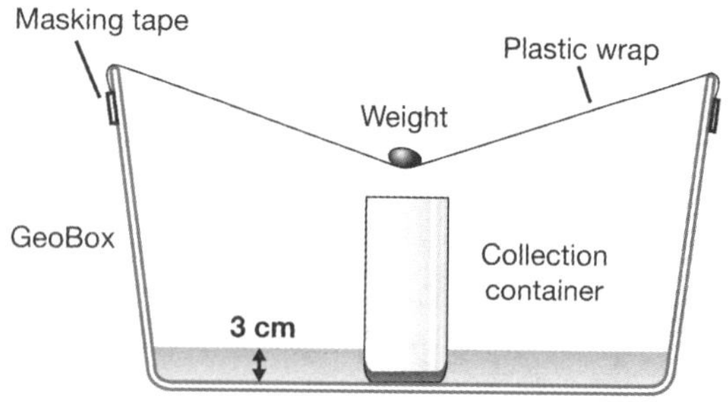

i. Describe a situation where a solar still would be useful for obtaining fresh water.

j. With teacher supervision only, test the water in the collection containers. If you have permission to do so and the water has been collected in fresh plastic cups, you may taste the water. If not, allow the collected water in each collection container to evaporate. If the water is fresh, you will not see any salt crystals when all the water has evaporated.

23A Water Systems

How does running water shape rivers and the landscape?

Running water on Earth's surface is a powerful force that shapes the landscape. The power of running water in a river is related to the slope of the river, the shape of the river, and the volume of water flowing in the river. In a river, water erodes sediment from the bottom and sides of the river and moves the sediment farther downstream. In this investigation, you will model how running water moves different sized sediment particles and shapes rivers.

Materials

- Stream table setup (includes stage, two buckets, table, removable tray, and grit)
- Plastic Petri dish (for collecting grit at each level of the stream table)
- Permanent marker
- Beaker
- Tweezers
- Small hand shovel (optional)

1 Setting up

1. Set up your stream table as shown in the picture at right. Make sure the table is set on the stage's second rung. The bucket on the stage should have a spigot.

2. Using your hands (or a shovel), smooth the sediment (grit) in your stream table so that the bottom third of the table is empty and clear of particles. The bottom third of the table represents the ocean. The sediment should slope from the "ocean" to just under the base of the tray.
3. Use your hands to make a meandering stream channel in your sediment. Dig down only halfway to the bottom of the sediment. Make sure your stream goes all the way to the "empty ocean" in your stream table.
4. Close the drain spigot at the end of the stream table. Place the bucket without the spigot under the drain spigot.
5. Fill the upper bucket with water.
6. Begin SLOWLY running water into the top of the river channel. Watch the river gradually fill with water and the sediment particles move downstream. Also note the formation of a delta at the mouth of the stream near your "ocean."

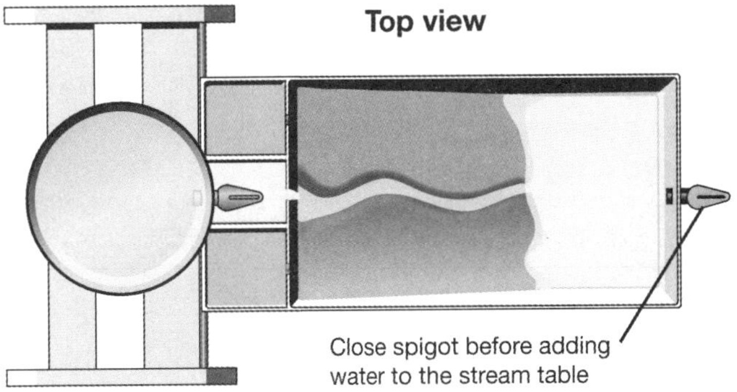

7. Stop the flow of water when your delta has grown into about half of your ocean, or when the water level in the ocean rises too close to the top of the stream table wall.
8. Make sure the lower bucket is lined up with the drainage hole. Open the drain spigot and allow the water to drain from the table into the lower bucket. Observe the general sizes of the particles remaining near the source of the river, the middle, and at the delta.
9. Use a permanent marker to divide a petri dish into 6 equal sections, labeling them from 1 through 6.

10. Using your tweezers, take small samples of the sediment from the following locations and place them in the numbered sections of your petri dish. Use the last page of this investigation to find out what a river delta looks like!
    - **Section 1:** The river source
    - **Section 2:** The end of the stream before the delta
    - **Section 3:** The middle of the delta
    - **Section 4:** The end of the delta
    - **Section 5:** The middle of the ocean bottom
    - **Section 6:** The far end of the ocean bottom

2 Stop and think

Diagram 1: River system

a. Make a drawing of your river channel, delta, and ocean. Label the following river features: meanders, river source, floodplain, delta, ocean.

b. Number your diagram showing the location of each of your six sections. For example, label your river source "1." Draw to scale the relative size of the sediment particles found at each location.

c. How does the size of the particles change as you go from the river source to the far end of the ocean? Why do you think the sediment is sorted this way?

d. Where was the speed of the water fastest? Where was it the slowest? How can you explain this?

3 Doing the experiment

1. Repeat steps 1 through 10 in the procedure. But this time, change the slope of the stream table so that it is steeper than your first trial.
2. Repeat steps 1 through 10 in the procedure. This time, change the slope of the stream table so that it is not as steep as the first trial.

4 Thinking about what you observed

a. How does increasing the slope of a river affect the rate of erosion by running water?

b. How does decreasing the slope of a river affect the rate of erosion by running water?

c. How do you think the width of the river channel affects the erosion of sediment in a river?

d. Based on your knowledge of sedimentary rocks, where would you likely find samples of conglomerate, sandstone, and shale in a river system?

5 Exploring on your own

1. Use your stream table to model the formation of the following river features. The images on the next page will help you with your work.
2. Your teacher will check off each feature as your team correctly models its formation. Your teacher can ask anyone in your lab group to model and/or describe how the feature forms. Therefore, all members need to be prepared to answer. Your team will earn the most points toward your lab grade if you can demonstrate the knowledge on the first try.

Table 1:

Try 1	Try 2	Try 3	
			v shaped valley upstream
			meanders
			delta
			floodplain
			alluvial fan
			oxbow lake
			waterfall

River delta: The mouth of a river that flows into an ocean or lake.

V-shape valley: When a river cuts a mountain it forms a V-shaped valley (a U-shaped valley is formed when a glacier moves through mountains).

Meander: S-shaped curves in a river.

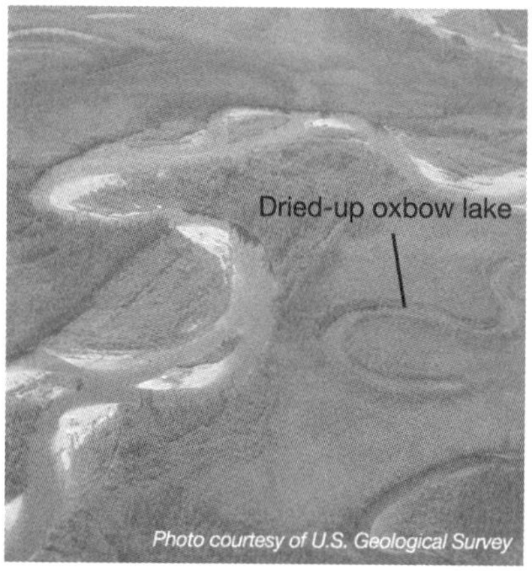

Oxbow lake: A meander that breaks off from the main river channel.

Alluvial fan: A fan-shaped area of sediment caused by a fast-flowing stream slowing down as it flows onto flatter land.

Flood plain: Flat land nearest a river that usually occurs at a distance from the source of the river. A flood plain is very good land for growing plants because the flooding of the river deposits nutrients in the land. However, flood plains are not good places for building because of the flooding.

Waterfall: Falling water that results when a river flows from a high to a low place in its path (such as over a cliff).

23B Coastal Erosion

How do objects and structures affect erosion?

Weathering and erosion by wind, water, and waves constantly changes coastal environments as sediment is transported from one location to another. Coastal erosion increases as wind, water, and waves increase such as during hurricanes. Human-influenced alterations can also cause the rate of erosion of coastal areas to increase at a faster rate than normal. In this investigation, you will explore how the construction of hard structures such as breakwaters and sea walls can impact coastal areas. In addition to how these alterations impact the shape of our coastlines, you will also explore how building homes on beach areas affects erosion.

Materials

- Stream table with buckets and stage
- Water
- Sediment
- Rocks of various sizes
- Sheet of heavy-duty aluminum foil
- Clay
- Toothpicks

1 Setting up

1. Set up your stream table so that it is on the lowest rung of the stand.
2. Fill the top third of the stream table with a layer of sand that is nearly to the top of the stream table's walls.
3. Adjust the spigot at the end of the stream table so that water will not drain from it.
4. Use the buckets and fill the bottom portion of the stream table with water so that it just reaches the sand. Note: The water level should not be more than half the thickness of the sand.
5. Shape out a unique coastline in the stream table using your hands. If necessary wet the sand a bit to help shape your coastline.

Safety note: Do not stand on the stage! This piece of equipment can only support up to a mass of 10 kilograms.

2 Making predictions

a. Predict what will happen to your shoreline if you were to create waves in your "ocean."

b. How do you think the rate of erosion will change as the amount and size of waves increases?

3 Getting started

1. Draw the shape of your original coastline in Table 1.
2. Use the plastic trough and generate waves in the stream table. Observe and draw the resulting eroded coastline in Table 1.

Table 1: Coastal erosion observations

Sketch and description of coastline before waves	Sketch and description of coastline after waves

4 Breakwaters

Offshore breakwaters are structures made out of hard materials like rock or concrete that are built a distance off the shore in the coastal waters. Humans build breakwaters, which are generally, but not always, parallel to shore, to reduce erosion caused by waves. Investigate what happens to a coastline when a breakwater is present.

1. Using the rocks, create a wall that extends just above the level of the water and runs parallel to half of your coastline. (Make your breakwater about 3–4 cm away from the shoreline.
2. What do you think will happen to the coastline when you start generating waves? Make a prediction about the coastline behind the breakwater, and the coastline not behind the breakwater in Table 2.
3. Use the plastic trough and begin generating waves. Watch what happens to the shoreline behind the breakwater and also along the coastline that is not behind the breakwater. Record your observations in Table 2.

4. Sketch the new eroded shape of the coastline with the model breakwater in your stream table in Table 2.

Table 2: Coastal erosion observations with the breakwater

Prediction of what will happen to coastline	Observation of what happens to coastline
Sketch of coastline with breakwater before waves	**Sketch of coastline with breakwater after waves**

5 Seawalls

1. Remove your breakwater. Refer to your sketch in Table 1 to reshape the coastline as close as possible to the original coastline.
2. The piece of aluminum will serve as your model seawall. Fold it over upon itself several times until your piece is about 10 centimeters tall. Insert the seawall down into the sand so that at least 5 centimeters of aluminum sticks up above the sand. Place the seawall 15 cm landward from the edge of the water.

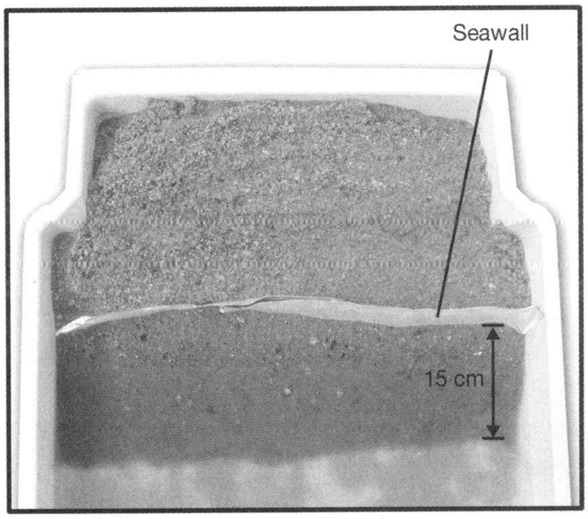

3. Make a prediction and sketch a side view (not a birds-eye view as used with the breakwater example) of what will happen to the part of the coast in front of the seawall and the beach behind the seawall. Record your prediction in Table 3.
4. Use the plastic trough and begin generating waves.
5. Observe and record what happens to the coast in front of the seawall and the beach behind the seawall in Table 3.
6. Sketch a side view of the new eroded shape of the coastline with the seawall in your stream table in Table 3.

Table 3: Coastal erosion observations with the seawall

Prediction of what will happen to coastline	Observation of what happens to coastline
Side view sketch of coastline with seawall before waves	**Side view sketch of coastline with seawall after waves**

6 Houses

1. Remove your seawall. Reshape the coastline as close as possible to the original coastline you shaped.
2. Make 2 to 3 miniature houses with the clay. Stick toothpicks in each corner of the bottom of your houses. Insert these "stilt houses" near the edge of your coastline.
3. Make a prediction and sketch a side view of what will happen to the sand on the coastline that is supporting the three stilt houses, and the sand where there are no houses. Record your prediction in Table 4.

4. Use the plastic trough and begin to generate waves in your model. Observe the erosion in areas where houses have been built in the coastline and areas where there are no houses. Record your observations in Table 4.
5. Sketch a side view of the new eroded shape of the coastline with houses in your stream table in Table 4.

Table 4: Coastal erosion observation with houses

Prediction of what will happen to coastline	Observation of what happens to coastline
Side view sketch of coastline with houses before waves	**Side view sketch of coastline with houses after waves**

7 Thinking about what you observed

a. How does the rate of erosion of coastal areas change when the size and frequency of waves increases?

b. What natural causes can create large and more frequent waves?

c. How did the erosion of the coastline differ behind the breakwater compared to the part of the coastline without the breakwater in front of it? Explain the reasoning behind any differences.

d. Explain the effects on the coastal areas between the seawall and the water. How would this process impact the sea life that lives in this area? How would it effect the way this land may be used by people?

e. What do you think happens to the density of the sand and other sediments near the coast where houses are built? How would this affect the houses?

f. Summarize how the building of hard structures such as seawalls, breakwaters, and houses can impact the natural erosion of coastal areas due to waves.

24A Harmonic Motion

How do we describe the back-and-forth motion of a pendulum?

Harmonic motion is motion that repeats in cycles. Many important systems in nature and many useful inventions rely on harmonic motion. For example, the phases of the Moon and the seasons are caused by Earth's harmonic motion. This Investigation will explore harmonic motion using a pendulum. The concepts you learn with the pendulum will also apply to other examples of harmonic motion.

Materials

- Data Collector and 2 photogates
- Physics stand
- Pendulum
- Graph paper

1 Setting up the pendulum

Attach the pendulum to one of the top holes in the stand.

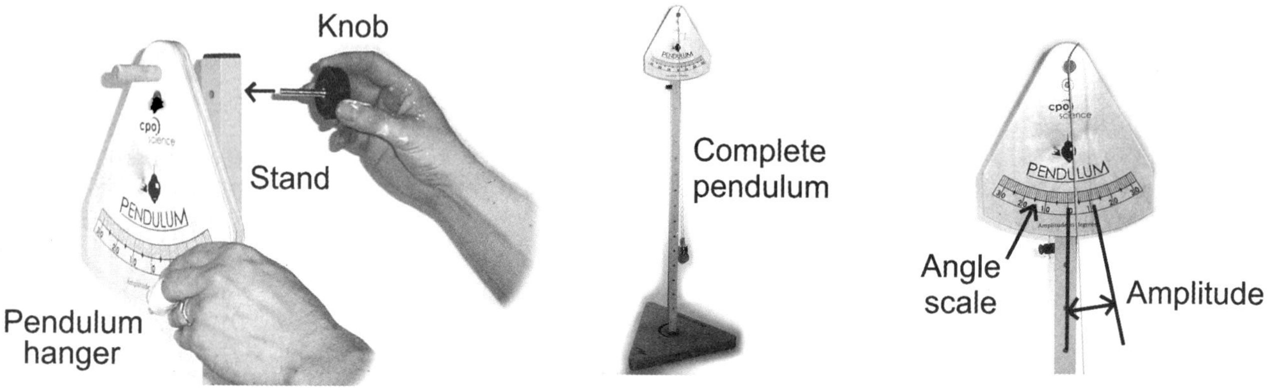

Start the pendulum swinging and watch it for a minute. Think about how to describe the motion.

a. Write one sentence about the motion using the word "cycle."

b. The *amplitude* is the maximum amount the pendulum swings away from its resting position. The resting position is straight down. One way to measure amplitude is the angle the pendulum moves away from center. Write one sentence describing the motion of your pendulum using the word "amplitude."

c. Draw a sequence of sketches that describe one complete cycle using arrows to indicate the direction the pendulum is going at that point in the cycle.

2 Oscillators and period

a. Use the stopwatch function of the Data Collector to measure the period of your pendulum. Time ten cycles. Do three trials and use Table 1 to record your data.

b. Divide the average time for ten cycles by 10 to get the period.

c. Write a one sentence description of how you measured the period.

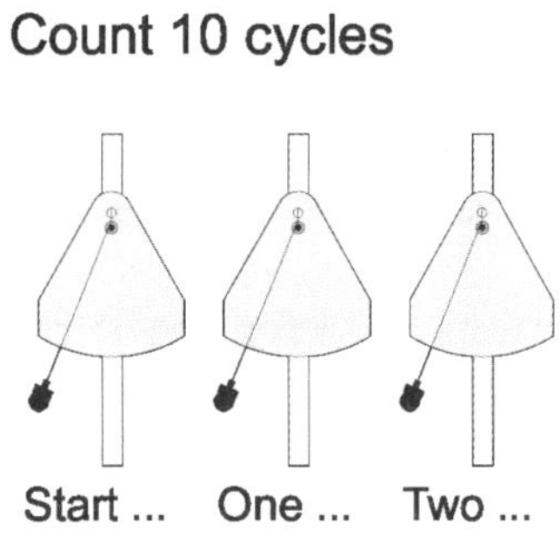

Divide time by 10

example
period = 15.20 sec ÷ 10
= 1.52 seconds

Table 1: Pendulum period data: Time for 10 cycles (sec)

Trial 1	Trial 2	Trial 3	Average
		Period of pendulum (average divided by 10)	

3 Measuring period with a photogate

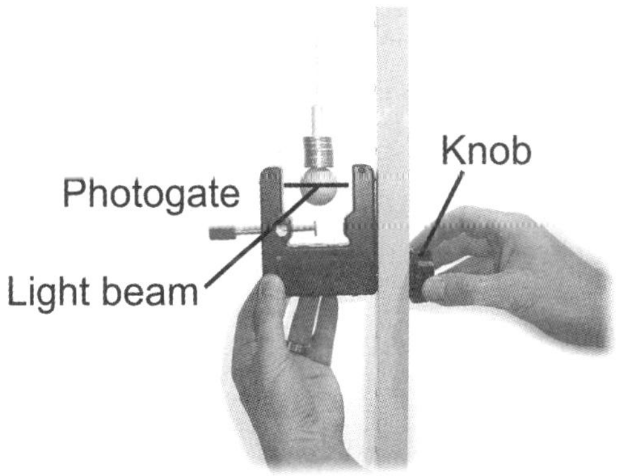

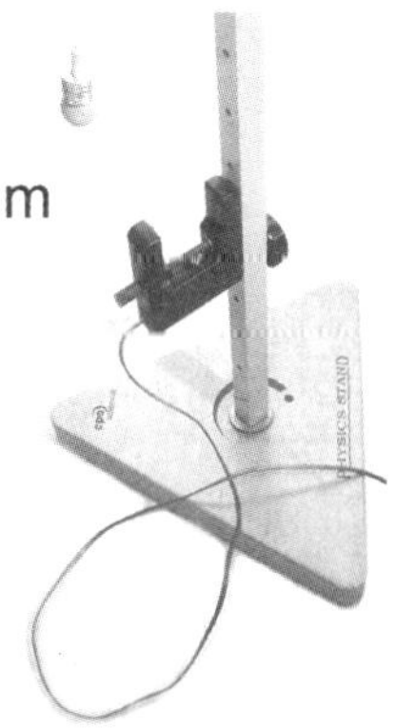

1. Attach the photogate as shown in the diagram. The pendulum breaks the light beam when it swings through the photogate. Try to keep the string length close to the length you used in part 2.
2. Put the Data Collector in period mode and let the pendulum swing through the light beam.
3. If you press the GO button once the display freezes allowing you to write down a number before it changes. Pressing GO a second time starts another measurement.

4 Thinking about what you observed

a. Write down the time measurement you get from the Data Collector.

b. Is the time you get from the Data Collector the period of the pendulum? Explain why the time is or is not the period of the pendulum (hint: compare to your results from part 2).

c. Explain how the time measured by the Data Collector is related to the period of the pendulum.

5 What variables affect the period of a pendulum?

In this experiment, the period of the pendulum is the only dependent variable. There are three independent variables: the mass of the bob, the amplitude of the swing, and the length of the string.

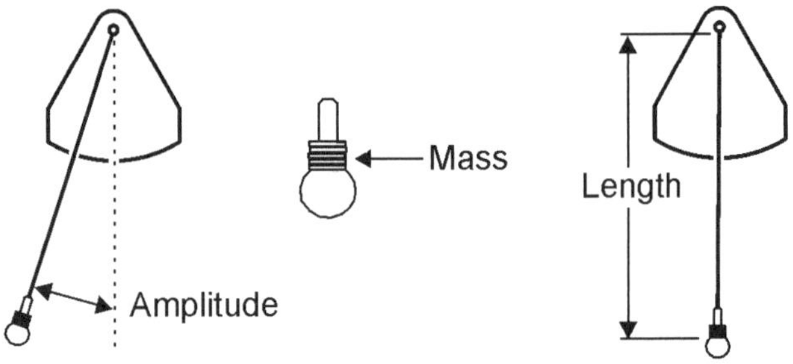

1. The amplitude can be changed by varying the angle that the pendulum swings.
2. There are washers that you can use to change the mass of the bob.
3. The length of the string can be changed by sliding it through the slot in the peg. Measure the length of the string from the bottom of the string peg to the bottom of the washers.

Design an experiment to determine which of the three variables has the largest effect on the period of the pendulum. Your experiment should provide enough data to show that one of the three variables has much more of an effect that the other two. Be sure to use a consistent technique that gives you consistent results.

a. Think of three experiments you can do to see what variables affect the period of the pendulum. Write down one sentence describing each experiment.

b. Do the three experiments and record the measurements you make to assess the effect of changing each variable.

6 Analyzing the data

a. Of the three things you can change (amplitude, mass, and string length), which one has the biggest effect on the pendulum, and why? In your answer you should consider how gravity accelerates objects of different mass.

b. Split up your data so that you can look at the effect of each of the three variables by making a separate graph showing how each one affects the period. To make comparison easier, make sure all the graphs have the same scale on the y-axis (period). The graphs should be labeled like the example below.

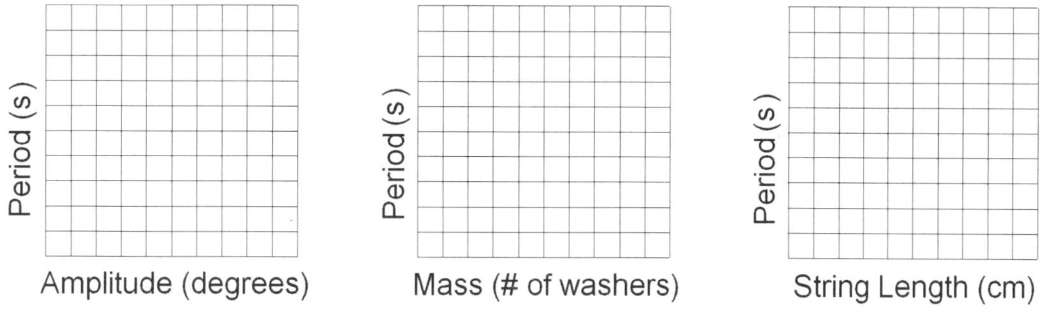

7 Applying what you know

Pendulum clocks were once among the most common ways to keep time. It is still possible to find beautifully made pendulum clocks for sale today. To make a pendulum clock accurate, the period must be set so a certain number of periods equals a convenient measure of time. For example, you could design a clock with a pendulum that has a period of 1 second. The gears in the clock mechanism would then have to turn the second hand 1/60th of a turn per swing of the pendulum.

a. Using your data, design and construct a pendulum that you can use to accurately measure a time interval of 30 seconds. Test your pendulum clock against the electronic stopwatch.

b. Mark on your graph the period you chose for your pendulum.

c. How many cycles did your pendulum complete in 30 seconds?

d. If mass does not affect the period, why is it important that the pendulum in a clock is heavy?

e. Calculate the percent error in your prediction of time from your pendulum clock. The percent error is 100 times the difference between your prediction and 30 seconds, divided by 30 seconds.

f. You notice in a magazine that a watch manufacturer advertises that its quartz watch loses no more than 5 seconds per month. Assume that the watch loses the maximum amount (5 seconds) in 31 days. Calculate the percent error of the quartz watch by comparing 5 seconds to the number of seconds in a month.

24B Waves in Motion

How do waves move?

Waves are oscillations that move from one place to another. Like oscillations, waves also have the properties of frequency and amplitude. In this investigation, you will explore waves on strings and in water. What you learn applies to all other types of waves as well.

Materials

- Metal Slinky® toy spring
- Meter stick
- Wave tray
- Food coloring
- Water

1 Making a transverse wave pulse

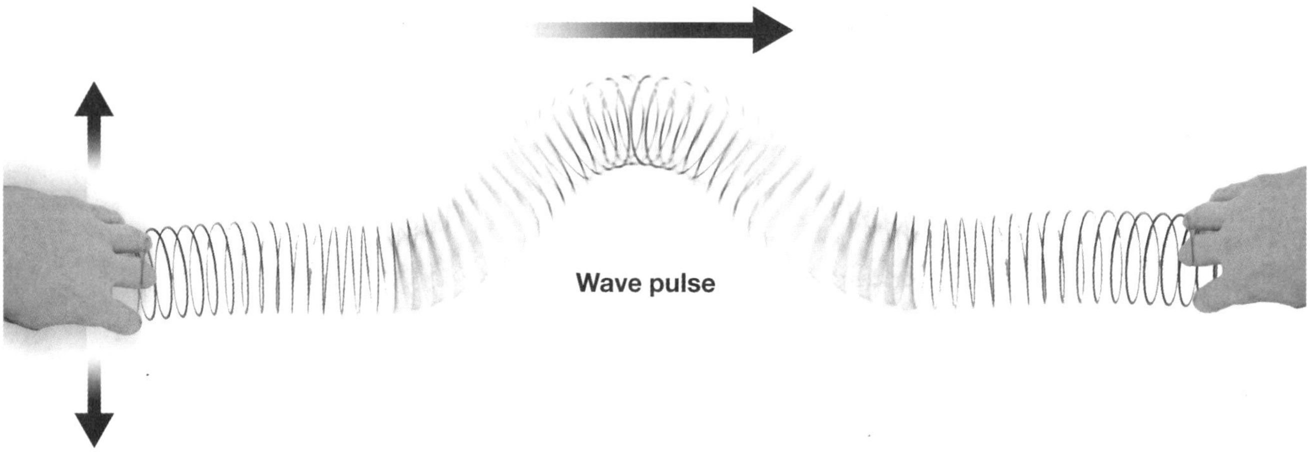

2. It takes two students to do this experiment. Each student takes one end of the spring.
3. Bring the spring down to the floor. Stretch it to a length of about 3 meters while keeping the spring on the floor.
4. One student should jerk one end of the spring rapidly to the side and back, just once. Make sure both ends of the spring are held tight and do not move once the wave is in motion. A wave pulse should travel up the spring.
5. Watch the wave pulse as it moves up and back. Try it a few times.

2 Thinking about what you observed

a. How is the motion of a wave pulse different from the motion of a moving object such as a car? (HINT: What is it that moves in the case of a wave?)

b. What happens to the wave pulse when it hits the far end of the spring? Watch carefully. Does the pulse stay on the same side of the spring or flip to the other side? Use the word "reflect" in your answer.

c. Imagine you broke the spring in the middle. Do you think the wave could cross the break? Discuss the reasoning behind your answer in a few sentences.

d. Why does the wave pulse move along the spring instead of just staying in the place you made it?

3 Making a longitudinal wave pulse

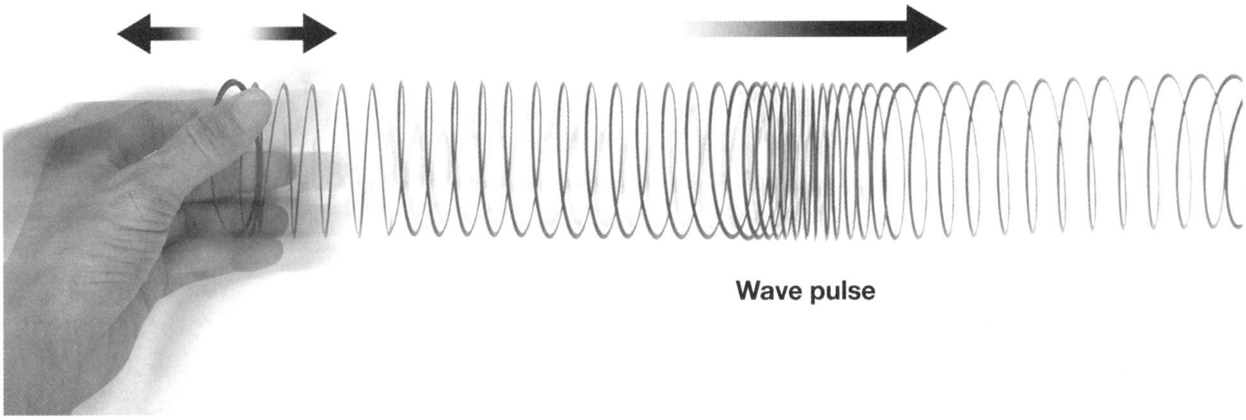

1. Just like the wave you made in Part 1, each student takes one end of the spring.
2. Bring the spring down to the floor. Stretch it to a length of about 3 meters while keeping the spring on the floor.
3. One student should jerk one end of the spring rapidly forward and back, just once. Make sure both ends of the spring are held tight and do not move once the wave is in motion. A wave pulse should travel up the spring.
4. Watch the wave pulse as it moves up and back. Try it a few times.

4 Thinking about what you observed

a. How is the motion of the longitudinal wave pulse different from the transverse wave pulse you made in Part 1? (HINT: How is the motion of the spring itself different?)

b. What happens to the wave pulse when it hits the far end of the spring? Does it behave like the transverse wave, or much differently? Use the word "reflect" in your answer.

c. Why do you think longitudinal waves are also sometimes called "compressional waves"?

d. Do you think a wave can be made by only stretching the Slinky® instead of compressing it? Make a prediction then try it and see if you were right.

5 Waves in water

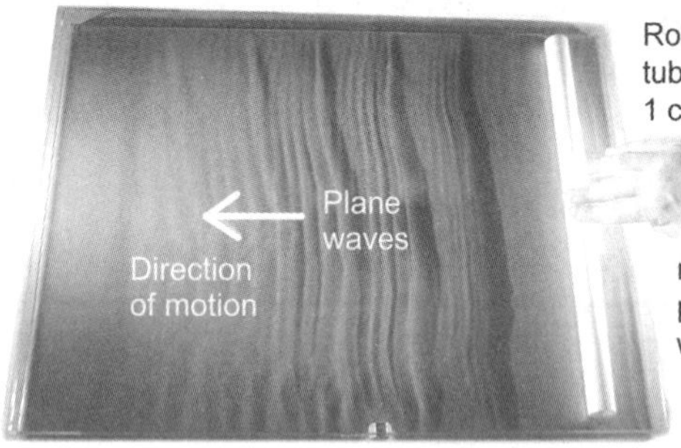

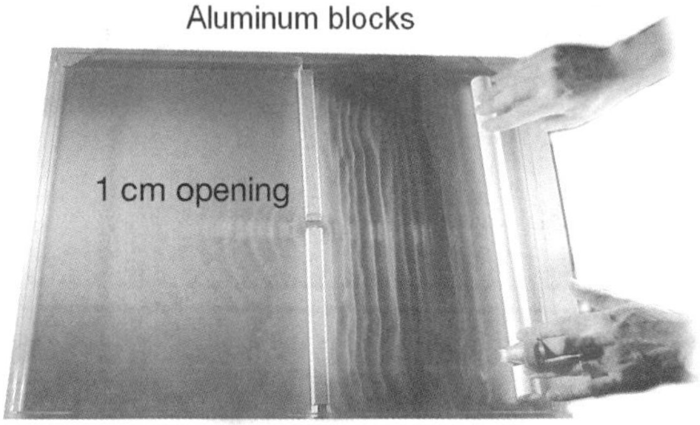

1. Fill a flat tray with about one-half centimeter of colored water. The color helps you see the waves.
2. Roll the wave tube forward about 1 cm in a smooth motion. This launches a nearly straight wave called a *plane wave* across the tray.
3. Next, poke the surface of the water with your fingertip. Disturbing a single point on the surface of the water makes a *circular wave* that moves outward from where you touched the water.
4. Arrange two aluminum blocks so they cross the tray leaving a 1 cm opening between them.
5. Make a plane wave that moves toward the blocks. Observe what happens to the wave that goes through the opening.

6 Thinking about what you observed

a. Draw a sketch that shows your plane wave from the top. Also on your sketch, draw an arrow that shows the direction the wave moves.

b. Is the wave parallel or perpendicular to the direction the wave moves?

c. Draw another sketch that shows the circular wave. Add at least four arrows that show the direction in which each part of the wave moves.

d. At every point along the wave, are the waves more parallel or perpendicular to the direction in which the circular wave moves?

e. Sketch the shape of the wave before and after passing through the 1 cm opening.

f. Does the wave change shape when it passes through the opening? If you see any change, your answer should state into what kind of shape the wave changes.

g. Are the waves you made in the water transverse or longitudinal waves, and why?

25A Color

What happens when you mix different colors of light?

All the colors of visible light can be created artificially using a combination of three primary colors: red, blue and green. In this Investigation, you will use a white light source and color filters to discover what happens when you mix different colors of light. You will also learn how the filters work.

Materials

- Optics with Light & Color
- Colored pencils

1 Sources of light

a. Compare the light from a light bulb with the light from the same bulb when seen in a mirror. In both cases, describe the path of the light from the source to your eyes.

b. Look at your clothes. Does the light reaching your eye from your clothes originate in your clothes? Or does the light originate somewhere else?

c. Turn off all the lights, and shade the windows so it is completely dark. Can you see your clothes in the dark? What does this experiment tell you about whether your clothes give off their own light or reflect light from somewhere else?

d. Turn on a television or computer screen in a dark room. Can you see the TV or computer screen in the dark? What does this experiment tell you about whether the TV or computer screen give off their own light or reflect light from somewhere else?

2 Making colors

1. Slide all three flashlights into their own light stand.
2. Turn the flashlights on.
3. Connect the red and green flashlights by sliding their stands together using the rail and slot connectors on the side.

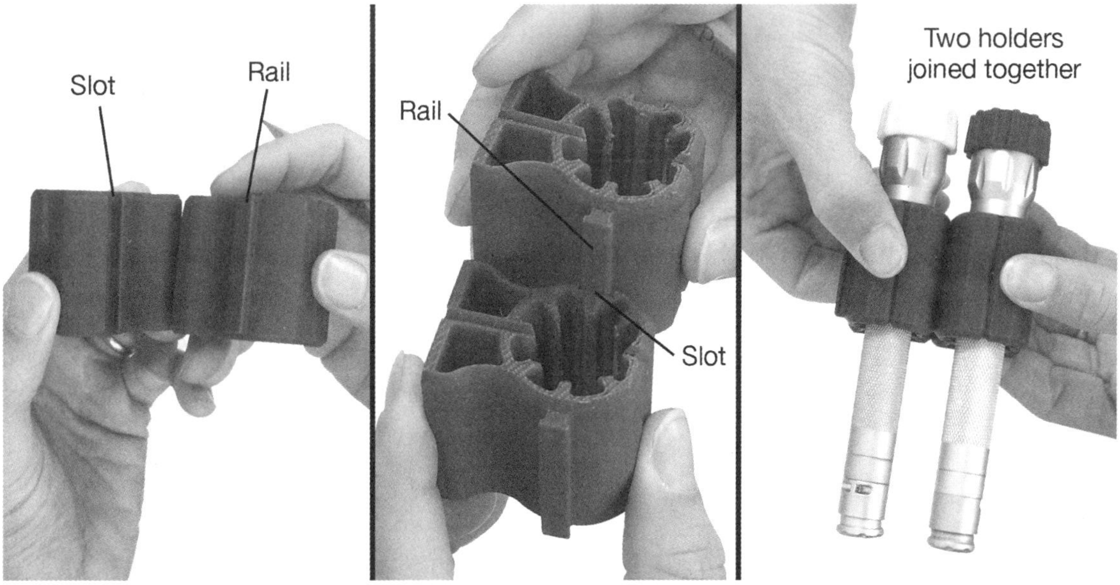

4. Place the blue flashlight on top of the red and green lights, making a small pyramid stack. Set the blue light on top of the other two with the stand on its side, so that the rail on the stand fits in the small groove created between the stands of the red and green lights.

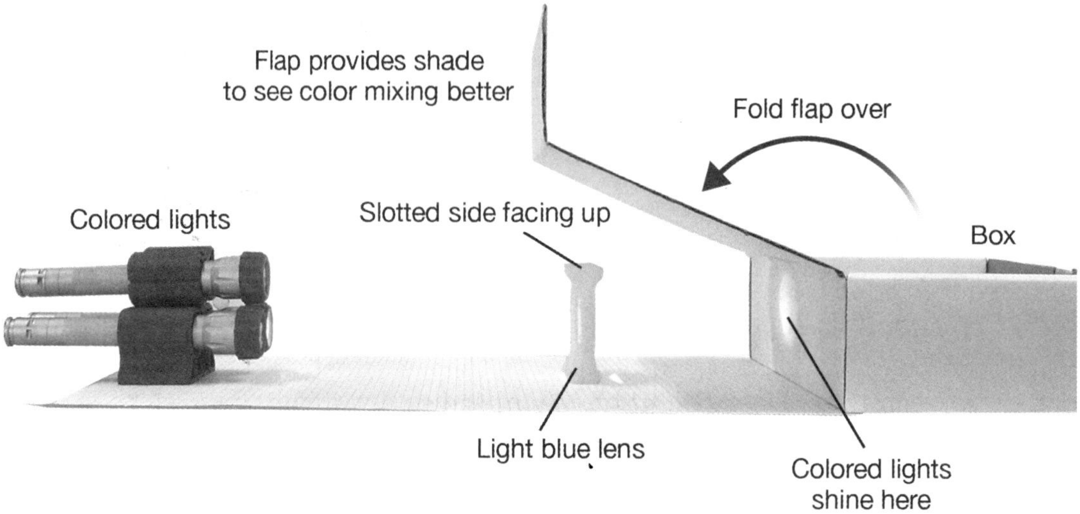

5. Set the light blue lens just in front of the lights so they shine through it. Place the lens so the slotted side is facing up. This will increase the visibility of the color mixing we are looking for.
6. Set the white box that the Optics with Light & Color kit comes in on opposite side of the paper from the lights. Fold the top of the box over to shade the area where the three colored lights are shining on the box.
7. Slowly move the lens away from the lights and toward the box until you see the three spots of color (red-green-blue) overlap on the box.

3 Thinking about what you observed

a. What color do you see when red and green light mix?

b. What color do you see when red and blue light mix?

c. What color do you see when blue and green light mix?

d. What color is produced when all three colors of light equally mix?

Table 1: Mixing primary colors of light

Color combination	Color you see
Red + Green	
Green + Blue	
Blue + Red	
Red + Green + Blue	

e. Research and explain the following terms from the diagram below: cone cells, rod cells, retina.

f. Research and explain how the eye sees white light in terms of the photoreceptors in the eye.

Photoreceptors in the eye

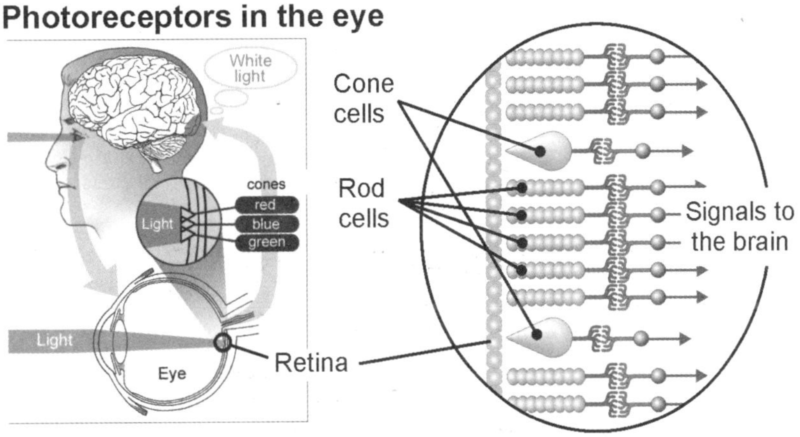

4 How does a color filter work?

A red laser can produce a single pure color of red which can be useful in some applications. For example, making three-dimensional images, called holograms, requires a very pure source of light which a laser can provide.

In the Optics with Light and Color kit, you have three different sources of colored light; but just how "pure" are these colors? In this part of the Investigation you will examine the light produced by each colored light, and learn how a **color filter** works. You will use the diffraction glasses to make your observations.

1. Examine the red light with the diffraction glasses.
2. Using colored pencils, sketch what you see in the appropriate column of Table 2.
3. Repeat steps 1 and 2 for the green and blue lights.
4. Remove one of the color filter from one of the color lights. Examine the light produced by the white LED and record your observations in the table.

Table 2: Examining colors

Red light	Green light	Blue light
White light		

5 Drawing conclusions

a. Compare the colors in the red light to those in the green light. What are the similarities and differences in the range of colors?

b. Compare the colors in the green light to those in the blue light. What are the similarities and differences in the range of colors?

c. How do the colors in the white light compare to the red, green, and blue lights combined?

d. The red light consists of a white LED covered with a red filter. What does the red filter do?

e. What do color filters do?

f. If you wanted to get yellow light, what part of the spectrum would the color filter have to absorb?

25B Reflection and Refraction

How does light behave when its path is changed?

Looking in a mirror we see a twin of ourselves reversed left-to-right. A fish underwater appears in a different place from where the fish really is. Both of these illusions are caused by the bending of light rays. This Investigation explores reflection and refraction, two processes that bend light rays. You will also use refraction to calculate the focal point of a lens.

Materials

- Optics with Light & Color
- Protractor & metric ruler
- Index card
- Thin tipped white board marker
- Graph paper (2 pieces)
- Sheet of blank white paper

1 Observing the law of reflection

How to trace the beam of the laser

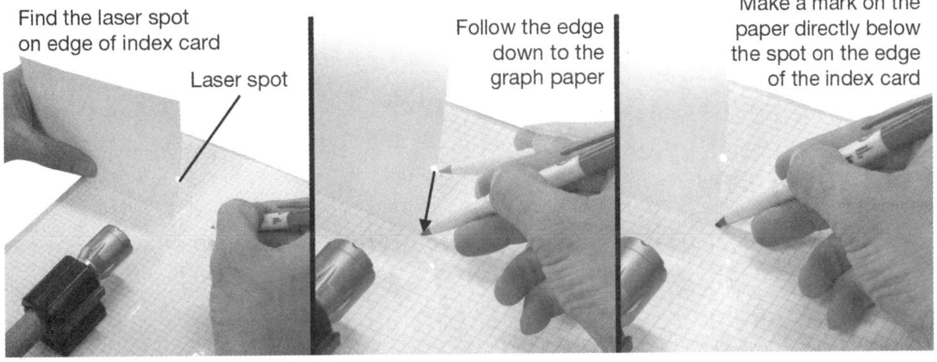

2. Set a sheet of graph paper on your lab table.
3. Slide the laser into a light stand, turn the laser on and put it on the graph paper.
4. Align the laser so the beam follows one horizontal line across the paper.
5. To line up the beam with the horizontal line on the graph paper, place an index card on its side onto the paper. Keep its long side flat on the graph paper and slowly slide the card over until you can just start to see the beam on the very edge of the card. Follow the vertical edge of the index card down and make a mark on the paper. The mark tells you where the beam is, and you can use this method to trace the path of the beam in all the Optics with Light & Color activities.

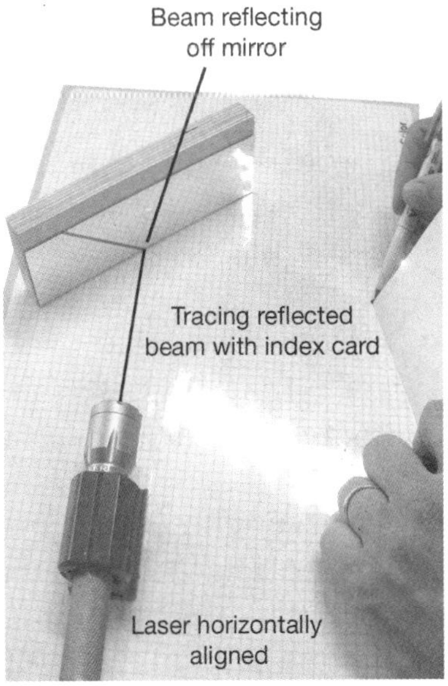

6. Follow the beam as it moves across the graph paper and make sure it lines up with one of the horizontal lines on the paper.
7. Set the mirror on the graph paper so the light beam from the laser hits its shiny front at an angle. The mirror should be placed so its long side is down on the paper.
8. Draw a line on the graph paper marking the front face of the mirror.

9. Use a pencil and the index card to trace the light rays going toward and away from the mirror. Label it ray #1.
10. Draw small arrows every couple of inches indicating the direction the beam is traveling.

2 Thinking about what you observed

a. A diagram showing how light rays travel is called a *ray diagram*. Lines and arrows on a ray diagram represent rays of light.

b. Look at your ray diagram showing the surface of the mirror and the light rays before and after the mirror.

c. Which is the incident ray? Label it on your ray diagram.

d. Which is the reflected ray? Label it on your ray diagram.

Drawing the ray diagram

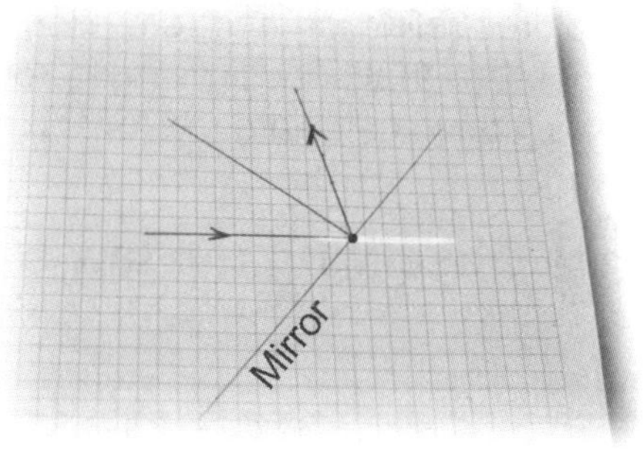

3 The law of reflection

a. Move the laser and the mirror to a new location on the graph paper. Make sure the beam hits the mirror at a slightly different angle and repeat steps 7–10. Be sure to label each ray. Continue shining and tracing the beam from a total of 4 different locations. Label the rays 1–4.

b. For each ray diagram, draw a line perpendicular to the mirror surface at the point where the rays hit. This line is called the *normal line*.

c. Use a protractor to measure the angle between the normal and the incident and reflected rays. Record your measurements in Table 1.

d. Write down your own statement of the law of reflection, describing the relationship between the angles you measured.

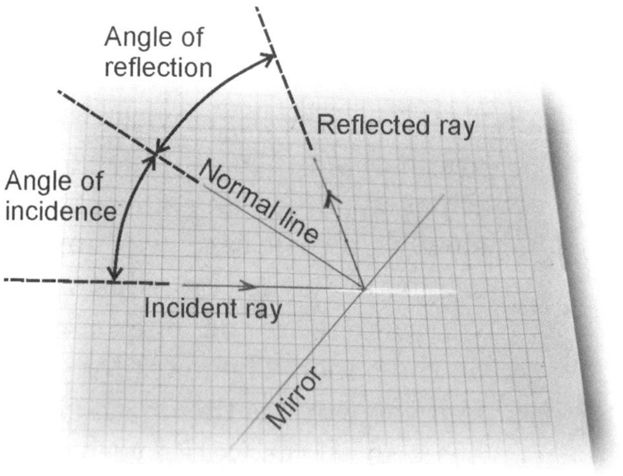

e. A laser shines at a mirror at an angle of incidence of 75 degrees. Predict its angle of reflection. After predicting, test your prediction. Were you right?

Table 1: Angles of incidence and reflection

	Diagram #1	Diagram #2	Diagram #3	Diagram #4
Angle of incidence				
Angle of reflection				

4 Light rays going through a prism

A prism is a solid piece of glass with polished surfaces. Prisms are useful for investigating how light bends when it crosses from one material into another, such as from air into glass or glass into air.

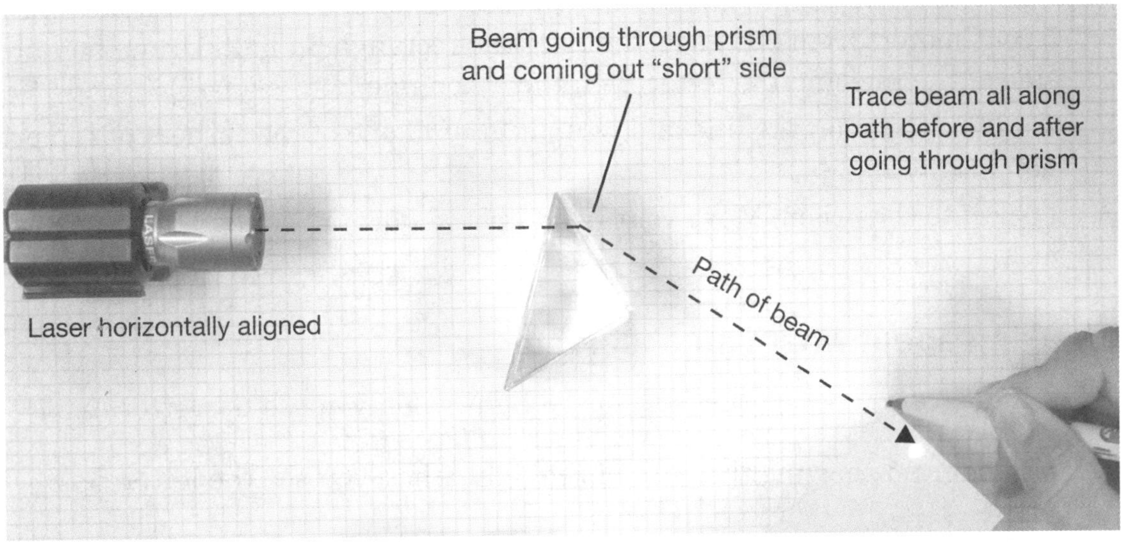

1. Set the laser on the left side of the paper. Turn the laser on. Face the laser so it is shining horizontally across the paper.
2. Place the prism in the middle of the paper (as shown in the picture above) into the laser beam so the beam comes out the opposite short side.
3. Rotate the prism in the beam and observe where the beam comes out.
4. Keep rotating the prism until you can see the beam refracted and reflected at the same time.

5 Sketching what you observed

a. Draw at least one ray diagram showing a laser beam that is refracted after passing through the prism. The *refracted ray* is the ray that comes out of the prism at a different angle than it entered.

b. Draw a ray diagram showing a laser beam that is reflected.

c. Draw a ray diagram showing a laser beam that is both refracted and reflected.

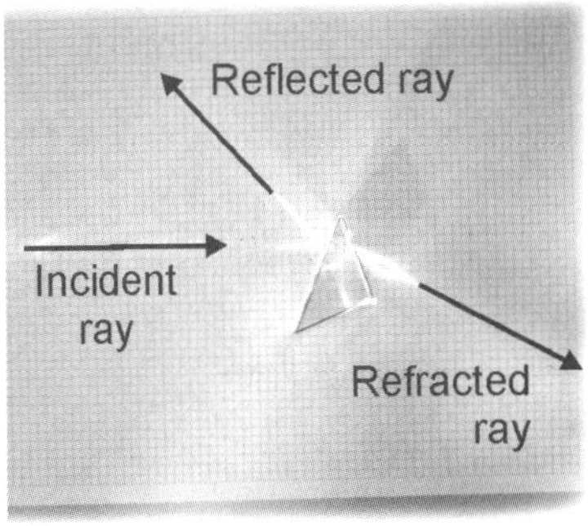

6 Seeing reflection and refraction at the same time

Both refraction and reflection often occur when light hits a boundary between materials such as the boundary between glass and air. The amount of light reflected or refracted depends on the angle at which you are looking relative to the surface.

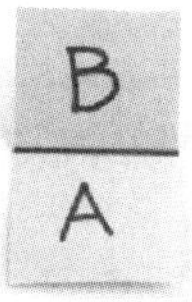

Fold a paper card marked with A and B

The image in the prism changes as you move you head!

1. Take a piece of graph paper about the size of a business card and draw a line about 5 centimeters from one edge, dividing the rectangle in half. Draw the letter A on one side of the line and the letter B on the other side.
2. Fold the paper on the line and wrap it around one of the corners of the prism that is not a right angle.
3. Look into the prism. Move your head up and down to change the angle at which you look.

7 Thinking about what you observed

a. Draw a diagram showing the path of the light when you see the letter A.

b. Draw a diagram showing the path of the light when you see the letter B.

c. Is the image in the prism always reflected or refracted or can there be both reflection and refraction at the same time?

8 Refracting light through a lens

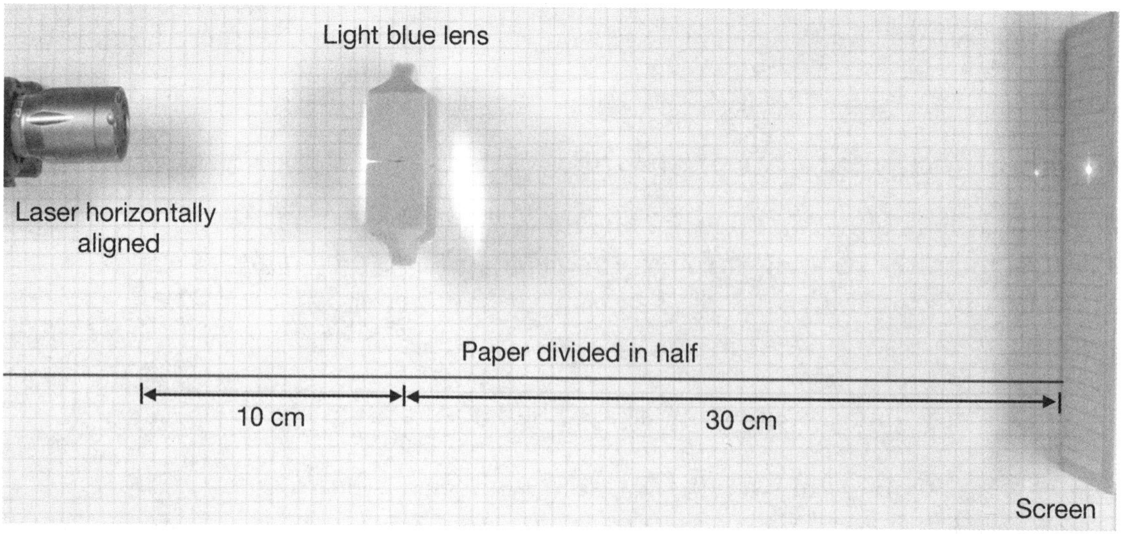

1. Divide a new sheet of graph paper in half horizontally. For the first part of this activity use the top half of the new sheet of graph paper. Place the laser on the edge of the paper and shine the laser so it follows a horizontal grid line across the paper.
2. Place the light blue lens 10 cm to the right of the laser with the slot facing up. Line the lens up vertically using the grid lines on the graph paper. It is important that the beam is perpendicular to the lens. Make sure the beam of the laser is lined up with the middle of the lens. There are lines on the side of each lens indicating the middle of the lens.
3. Trace around the base of the lens so it can be removed and put back in place in case you need to move it to complete ray tracing.
4. Shine the laser through the lens so the beam passes off-center, almost at the very outer edge of the lens. Set the block with the mirror about 30 cm on the right side of the lens so the beam hits it after passing through the lens. Use the side of the block with the graph paper on it so you can clearly see where the beam hits the block.
5. Trace the laser beam before and after it passes through the lens. Be sure to always carefully mark the points that the beam exits the laser, enters the lens, exits the lens, and then hits the block. Connect all these points to see the path of the beam.
6. Realign the laser with a different horizontal grid line parallel to the original beam and closer to the center of the lens. Again, trace the path of the beam before and after it passes through the lens.
7. Realign the beam so it passes directly through the center and trace the beam again.
8. Trace two more beams passing through the lens on the other side of the center of the lens for a total of five beams. Label the beams 1-5 on both sides of the lens.
9. Repeat steps 1–8 with the dark blue lens using the bottom half of your graph paper.

9 Thinking about what you observed

a. Feel the glass surface with your fingers and note the shape of the lenses. How are they different?

b. Draw a quick sketch of the shape of each lens itself with no stand from a side view. Label each lens.

c. Describe the paths of the rays before and after they traveled through the light blue lens. Include the words refract, converge and diverge in your description.

d. What is the focal point of a lens? Mark the focal point on the first ray diagram.

e. What is the focal length of the lens? Measure the focal length of the light blue lens.

f. Describe the paths of the rays before and after they traveled through the dark blue lens. Include the words refract, converge and diverge in your description.

g. How are the two lenses different?

h. One lens is referred to as a diverging lens, and the other a converging lens. They are also sometimes referred to as convex or concave. Research these terms and explain which is which.

10 Making an image with a lens

Certain types of lenses can make an image of a distant light source. The image forms about one focal length away from a lens when the object is far away.

1. Find a wall at least 5 meters away from a lamp or sunlit window. Tape a piece of white paper to the wall to create a screen for seeing the image.
2. Get the light blue lens. Hold the lens at different distances parallel to your screen and window or light source. Try distances between 15 and 25 centimeters.
3. Move the lens until you see a sharp image of the lamp or window on the screen. An image is produced when your lens is about one focal length away from the screen.
4. Measure the distance from the lens to the wall and record in Table 2. Use this technique to determine the focal lengths for both lenses.

Table 2: Focal lengths of lenses

	Focal length (cm)
Light blue Lens	
Dark blue lens	

Images can be smaller or larger than the object that created them. Images can also be right side up or inverted.

a. Was the image created by the light blue lens smaller or larger than the object?

b. Was the image right side up or was it inverted?

11 Projecting an image with a lens

You can think about a lens as collecting a cone of light from each point on an object.
For a perfect lens all the light in the cone is bent so it comes together at a point again to make the image. This is how movie projectors take an image on film and project it onto a screen.

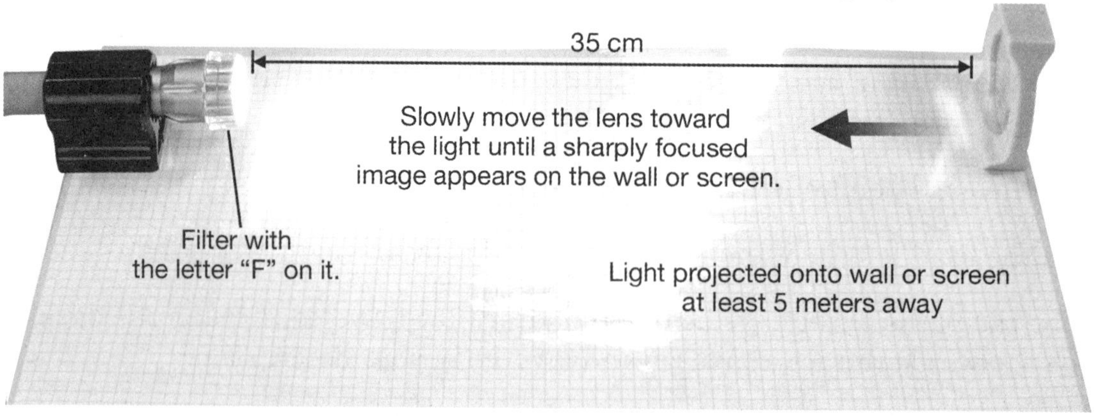

1. Place the clear filter with the letter "F" onto the flashlight. Turn on the flashlight and shine it horizontally.
2. Take the light blue lens and set it on the graph paper 35 cm away from the light.

3. Shine the light at a distant light colored wall at least 5 meters away. If one is not available, affix a piece of paper to the wall as your projection screen. Slowly move the lens toward the light until you see a sharp image of the "F" on the wall or screen. Have one group member check the projected image closely while the lens is slowly moved to find the exact place the lens needs to be to make it come into focus.

   At what distance from the light does the lens produce a sharply focused image?

4. Take a paper card and use it to block some of the light from the lens. The card should be 2–3 cm from the lens and on the same side as the light.

12 Thinking about what you observed

a. Describe the characteristics of the image formed by the lens. Characteristics include whether the image right-side-up, inverted, larger, or smaller.

b. Discuss with your class why blocking part of the lens makes the image dimmer, even though you still see the entire image.

c. CHALLENGE! The thin lens formula is used to calculate the exact focal length of a lens. Calculate the focal length of your light blue lens using the thin lens formula. You will need to measure the distance from the lens to your projection screen or wall where the image became sharply in focus. How does it compare with the other 2 focal lengths of the light blue lens you observed? Was the method from section 8 or 10 more accurate compared to the focal length you just calculated using the thin lens formula?

f = focal length of lens

d_i = distance of image from lens

d_o = distance of object from lens

The thin lens formula

$$\frac{1}{f} = \frac{1}{d_i} + \frac{1}{d_o}$$

26A Phases of the Moon

What causes the lunar cycle?

Why does the Moon appear to change shape over the course of a month? In this investigation you will model the lunar cycle and figure out if there really is a "dark side of the Moon."

Materials

- Flashlight
- 2- or 3-inch foam ball
- Pencil or craft stick
- 1 yellow sticker
- 1 black sticker

1 Think about it

a. When was the last time you noticed the Moon? Describe or draw its shape.

b. You have probably noticed that the Moon doesn't look the same every time you see it. But have you ever thought about how it changes over the course of a month? In the chart below, draw the shapes of the Moon in the order that you think you would see them. Two views have been drawn to help you get started. How would you represent a "full Moon?"

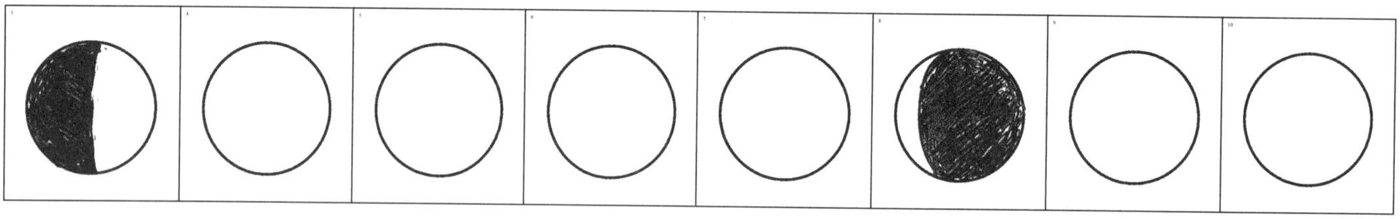

2 Make a hypothesis

You've spent some time thinking about **what** the lunar cycle looks like. Now, can you explain **why** the Moon appears to change shape?

a. Create a hypothesis to explain why the Moon appears to change shape. Why doesn't it look the same every time you see it?

b. What variable or variables do you think affect the way the Moon appears?

3 Model the lunar cycle

Figure 1

1. Place a foam ball on a pencil or stick. This ball represents the Moon.
2. Have another student hold a flashlight. The flashlight represents the Sun. Your head represents Earth.
3. Hold the ball slightly above your head, at arm's length from your face. Stand about one meter from the flashlight, which is held at the same level as the ball.
4. Observe the Moon in each of the positions shown in Figure 2. Face the ball at each position.
5. For each position, indicate how much of the ball is dark and how much is illuminated, in the graphic below. Use a pencil to show the shaded region. Leave the illuminated part white.

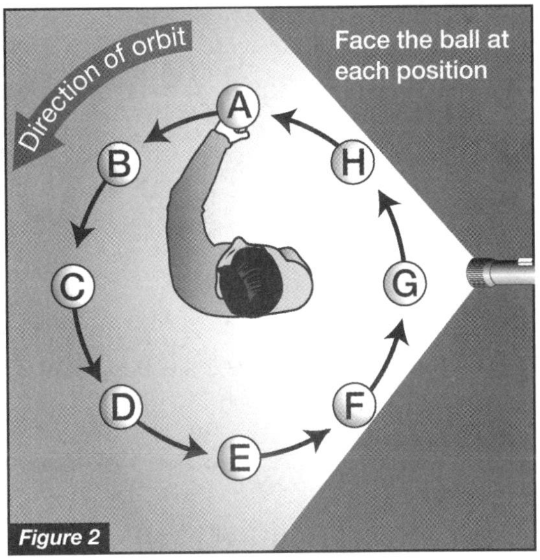

Figure 2

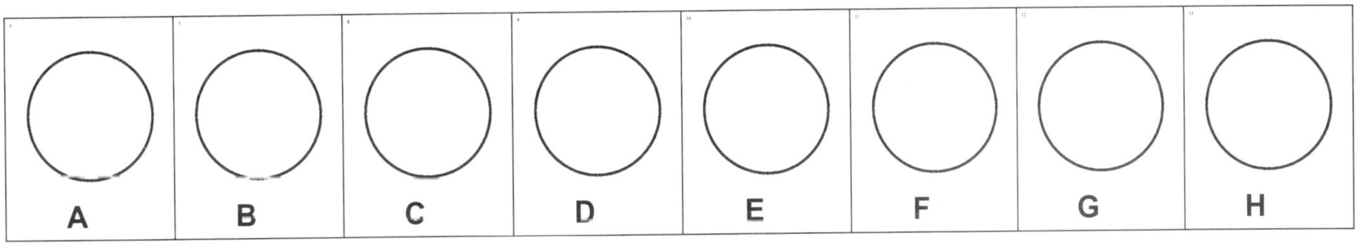

4 Thinking about what you observed

If you could see the Sun-Moon-Earth system from above, you would see that half of the Moon is almost always illuminated by sunlight.

Sun-Moon-Earth System as Seen from Above

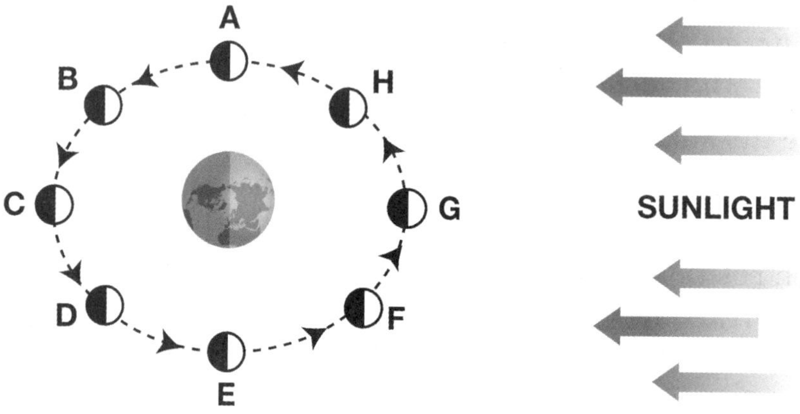

Here on Earth, the fraction of the Moon that is illuminated appears to change as the Moon revolves around Earth.

a. When you see a full Moon in the sky, at which position (A–H) is the Moon in its orbit?

b. At which position (A–H) would the Moon not be visible from Earth?

c. Reread the hypothesis you wrote in step 2. If the activity changed your thinking about why the Moon appears to change shape, revise your statement. Otherwise, write "no change."

d. If you hold the ball at eye-level, do the Moon phases appear as they do in real life? What does that tell you about the Moon's actual orbit around Earth?

5 Is there a "dark side" of the Moon?

You may have noticed that the Moon has some darker areas (known as maria) on its surface. Because the maria are always in the same location when we view the Moon, we know that we always see the same side of the Moon from Earth. Think back to your lunar cycle model as you answer the questions below.

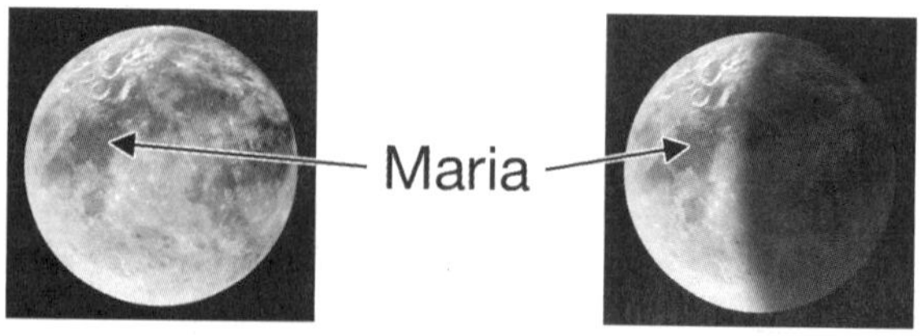

a. If we always see one side of the Moon, does that mean that there is another side of the Moon that is always dark?

b. Why is it that we always see the same side of the Moon? Do you think the Moon rotates as it orbits Earth? Why or why not?

6 Test your ideas

In order to determine whether or not the Moon rotates, you will mark one side of your Moon model with a yellow sticker. You need to stand outside the Sun-Earth-Moon model, at point 1, 2, 3, or 4 in the diagram below, in order to detect rotation. If the sticker always faces the same classroom wall, you will know the moon doesn't rotate. If the sticker faces different walls at different times, you will know that the Moon does, in fact, rotate.

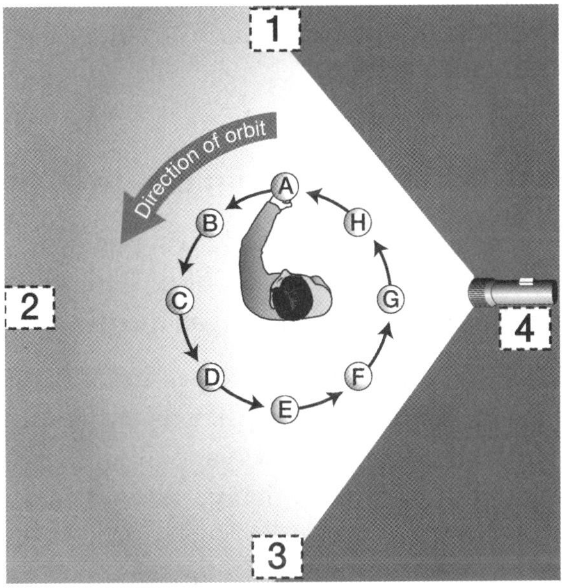

1. Place a yellow sticker on one side of the styrofoam ball. This sticker represents the side of the ball that we can see from Earth. Place a black sticker on the opposite side.
2. You will play the role of an observer in space, outside of the Sun/Moon/Earth system. Choose position 1, 2, or 3 as shown in the diagram at right.
3. Have another member of your group play the role of the Sun as before.
4. A third group member should hold the Moon model with the yellow sticker facing him/her. This student will model the lunar cycle just as you did in part 3.
5. Observe the yellow sticker as the Moon moves in its orbit. Notice whether or not the yellow sticker faces the same classroom wall throughout the activity.
6. Watch the lunar cycle model one more time. This time, watch the black sticker. Notice whether or not the black sticker remains in the dark throughout the cycle.
7. Switch roles within your group until everyone has observed the lunar cycle again.

7 Thinking about what you observed

a. Does the Moon rotate? How do you know?

b. If so, how many complete rotations do you think the Moon makes during one orbit of Earth?

c. Is one side of the Moon forever in darkness? Why or why not?

d. Which is a better name for the side of the Moon we cannot see from Earth:
The *dark side* of the Moon, or
The *far side* of the Moon?
Why?

26B Earth's Seasons

What causes the seasons?

Why do the seasons occur? In the summertime, it is hotter, the days are longer, and the sunlight is more intense. In wintertime, it is colder, the days are shorter, and the sunlight is less intense. What causes these variations in the heating and cooling of Earth? In other words, why do we have seasons?

Materials

- Globe
- Solar (PV) cell
- Multimeter
- Metric tape measure
- Masking tape

1 What is the main cause of the annual cycle we call seasons?

The graphic at right shows you what Earth's orbit around the Sun looks like. The distance between Earth and the Sun changes during the year.

Earth's North–South axis is its axis of rotation. This axis is always tilted in the same direction, at the same angle, and always points to the North Star, which is very far away. As a result, if you were standing on Earth's North Pole, the North Star would always be directly overhead.

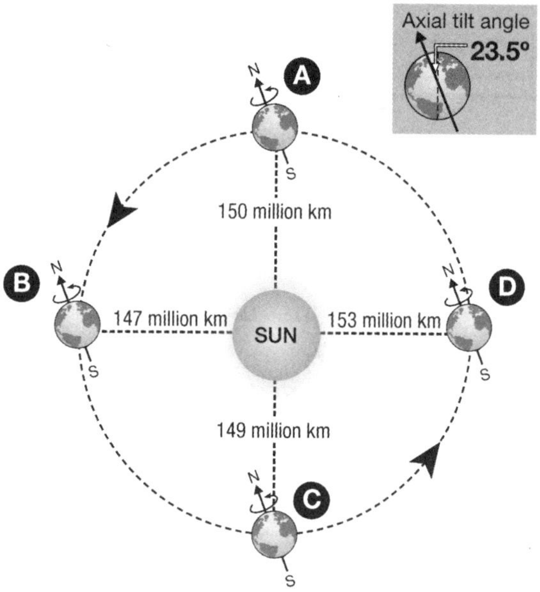

a. To start the investigation, come up with a hypothesis stating why you think the seasons occur. Do you think they are caused by Earth's changing distance from the Sun? Do you think Earth's tilt causes the seasons? Do you think both factors play a role? Or do you think other factors cause the seasons?

b. Which quarter of the diagram (A to B, B to C, C to D, or D to A) do you think represents summer in the Northern Hemisphere? Why do you think this is so?

2 Setting up a model of Earth's orbit

In part 1, you read that Earth's distance from the Sun varies slightly as it orbits the Sun. Now you will create a model that represents the changes in Earth-Sun distance.

1. It is impossible to measure millions of kilometers in your classroom, but you can use a scaled distance in which 1 centimeter represents 1 million kilometers. Therefore, a distance of 150 million kilometers is represented by 150 centimeters.
2. Using the scale distance of 1 centimeter = 1 million kilometers, determine the scale distance for positions B, C, and D. Write the scale distance in the third column of Table 1.

Table 1: Scale distances

Position	Distance from the Sun (km)	Scale distance from the Sun (cm)
A	150,000,000	150
B	147,000,000	
C	149,000,000	
D	153,000,000	

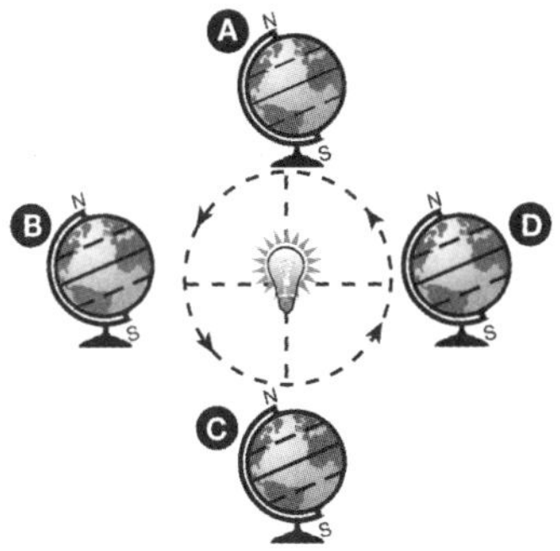

3. Your teacher will set up a light source in the middle of the room and choose a point high on one wall to represent the north star. Once the Sun is in place DO NOT MOVE IT.
4. Gather these materials: Earth globe, solar cell, tape measure, masking tape.
5. Your globe will represent one position in Earth's orbit—A, B, C, or D (diagram, right). Your group will carefully place your globe at one of the 4 positions; A, B, C, or D. To do this, part of your team will have to move the globe and part will have to operate the tape measure.
6. Extend your tape measure out so you have about 5 cm more tape than the length you need to measure. Lock the tape measure into place and while holding the locking end in your hand allow the tape to stick out directly at the globe.

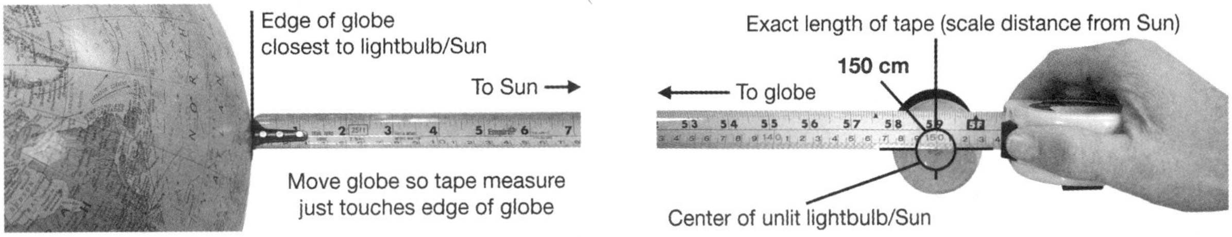

7. Place the exact length of tape (scale distance from the Sun) you are measuring directly over the center of the unlit light bulb representing the Sun. Move your globe until the center of the globe touches the extended end of the tape measure. The center of your globe should be aligned with the center of the light bulb.
8. All four globes should have their axes pointing in the same direction. Use the diagram and the North Star sign your teacher puts up as your guide.
9. Once your globe is exactly the correct distance from the Sun and has its axis pointing in the right direction tape the globe's base to the table. The globe should be able to withstand a bump or two without moving.

a. What do the four globes in the model represent? HINT: There are not four separate Earths revolving around the Sun!

b. Why do the north-south axes of each globe have to point in the exact same direction?

3 Examining the intensity of the light that falls on the globe

Now you are ready to look at variations in the light intensity that falls on each globe. The greater the light intensity, the more electricity your solar cell produces. Measuring the solar cell output allows us to find differences in light intensity at different places on the globes.

Use the same solar cell and digital meter throughout the investigation. Your group will move around the room from globe to globe, following Earth's path around the Sun.

1. Attach the leads of your digital meter to the solar cell. Set the digital meter to measure direct current in milliamps. You will take 2 measurements at each position.
2. The first measurement will measure how the distance between the Sun and Earth affects light intensity. Measure the milliamps of current produced by the solar cell at the middle of your globe. This way, no tilt is involved, and only distance will impact the current produced by the solar cell. Hold your solar cell flat against the globe and directly facing the Sun. Record your data in Table 2.

Measuring the Effect of Distance

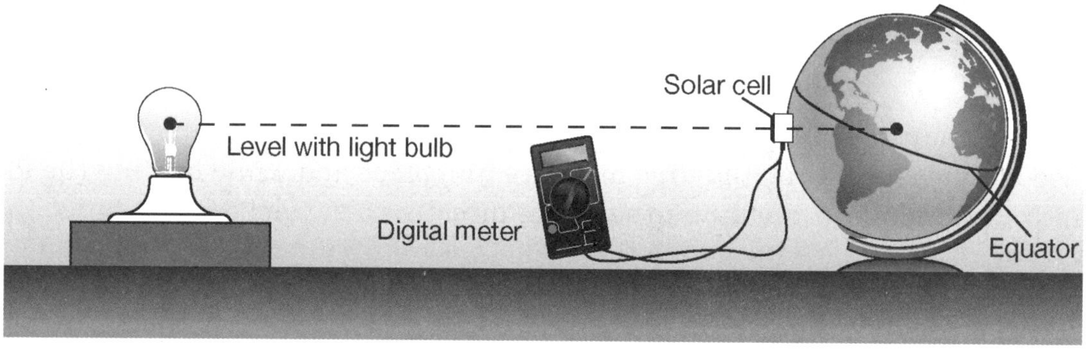

3. The second measurement will measure how Earth's tilt affects light intensity. Move the solar cell up to the Tropic of Cancer and measure the milliamps produced there. Firmly hold the solar cell flat against the surface of the globe, and make sure that while tilted, it is still facing in the direction of the Sun. Record your data in Table 2.

Measuring the Effect of Tilt

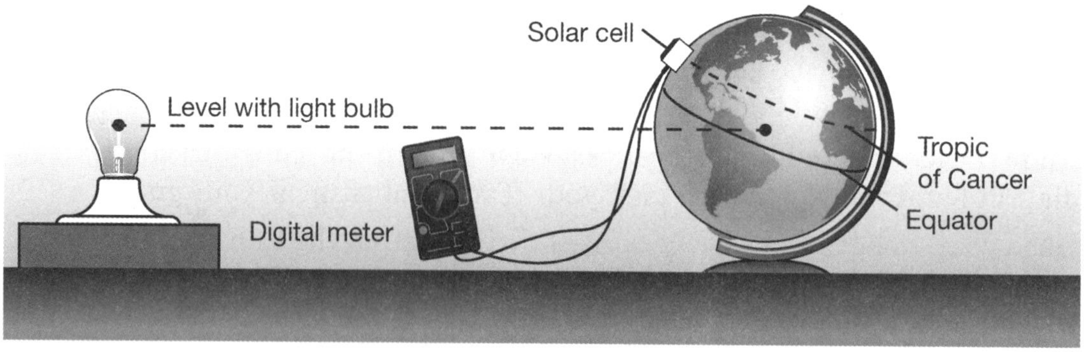

4. Once you have taken both measurements, move to the next globe. *Take your solar cell and digital meter with you.* Measure and record the current at the center of the globe and at the Tropic of Cancer again in Table 2.
5. Repeat both measurements at the other two globes and record the data in Table 2.

Table 2: Globe position data

Position	Distance from the Sun (km)	Scale distance from the Sun (cm)	Current at middle of globe (mA)	Current at Tropic of Cancer (mA)
A	150,000,000	150		
B	147,000,000			
C	149,000,000			
D	153,000,000			

4 Analyzing your data

a. Why was it important to measure the current produced at the middle of the globe and also at the Tropic of Cancer?

b. The largest change we see in distance from Earth to the Sun is from 147,000,000 kilometers to 153,000,000 kilometers. While 6,000,000 kilometers is not a short distance, find the percent change in distance using the following formula:

$$\text{Percent Change} = \frac{\text{final value} - \text{initial value}}{\text{initial value}} \times 100$$

Does this percent change seem large or small to you?

c. Based on your data, how much does light intensity change as these distances change?

d. How much did the angle of the solar cell change at the four different positions you measured on the Tropic of Cancer? Did the difference in angle seem large or small to you?

e. According to your results, how much does light intensity change as the angle changes?

5 Conclusions

a. Of the two factors—distance from the light source and axial tilt—which has a greater effect on light intensity?

b. Based on your light intensity results, which of the two factors plays the most significant role in causing the seasons? Was your hypothesis supported by your results?

c. Based on your results, which position (A–D) represents the first day of summer in the Northern Hemisphere? Which position represents the first day of winter in the Northern Hemisphere? Explain your answer.

d. At which position (A, B, C, or D) does the Northern Hemisphere receive the most intense light? The least intense light?

e. Which quarter of Earth's orbit represents summer in the Northern Hemisphere (from A to B, B to C, C to D, or D to A)? Explain your answer based on the data you collected.

f. Which quarter represents summer in the Southern Hemisphere? Explain your reasoning.

27A Stars and Spectroscopy

How can we use a spectrometer to tell what stars are made of?

With the exception of the Sun, stars appear as small specks of light in the night sky. Astronomers use a technique called <u>spectroscopy</u> to analyze the light emitted by stars. Using spectroscopy, they can determine a star's temperature and the elements from which it is made. In this investigation, you will learn how to analyze light using spectroscopy. You will determine which elements are present in different light sources. You will then analyze the light emitted by the Sun.

Materials

- Spectrometer
- Colored pencils
- Incandescent light source
- Fluorescent light source

1 Using the spectrometer

A spectrometer splits light into a spectrum of colors and displays the different colors of light along a scale. The scale measures the wavelengths of different colors of light in nanometers (nm).

Safety Tip: Never look directly into any light source—especially the Sun!

Hold the spectrometer so that the printed side is facing upward. In a well-lighted room, hold the spectrometer so that one eye is looking through the diffraction grating and the other eye is closed. You should see a scale, as illustrated below. The bottom scale measures wavelengths in nanometers. You should also notice colors at various places inside the spectrometer. This is caused by light entering the spectrometer from different sources.

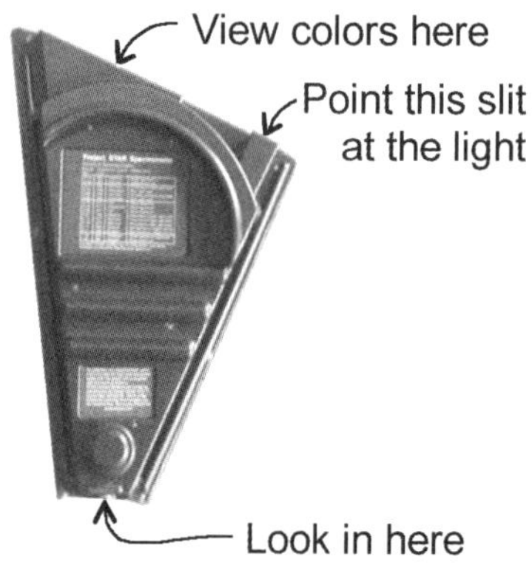

Notice that the plastic disk that is attached to the diffraction grating can be turned. Looking into the spectrometer, rotate the disk until you see colors in a horizontal stripe to your left. The colors should appear between the two lines of numbers on the scales.

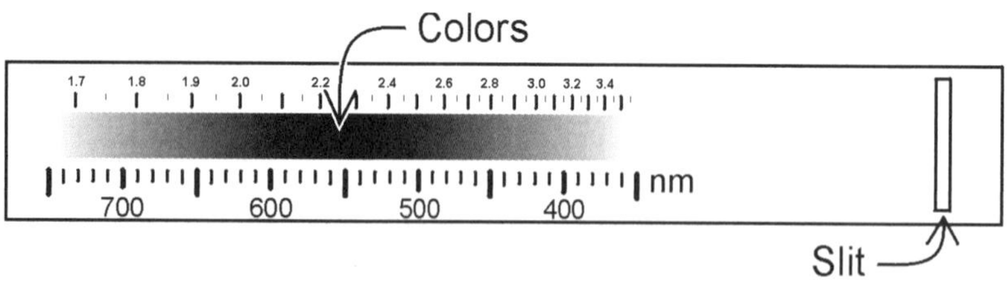

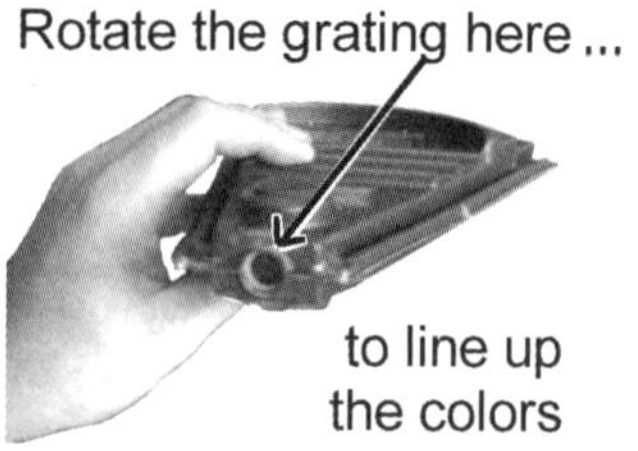

a. While looking through the eyepiece, point the slit of the spectrometer directly at an incandescent bulb. Use colored pencils to show where the different colors of light appear on the spectrometer scale.

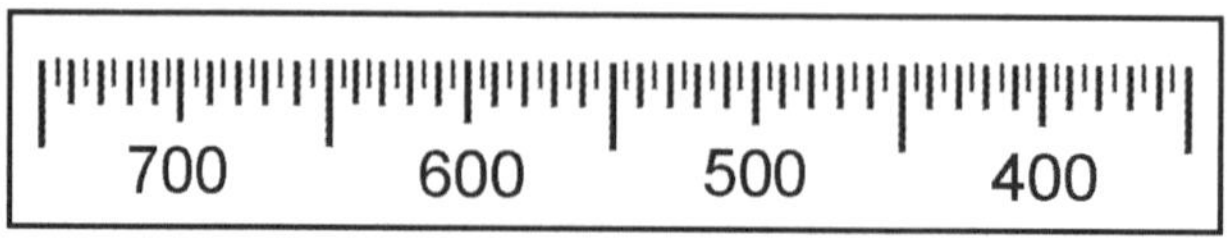

b. Blue light has the highest energy and red light the lowest. Based on your observations with the spectrometer, what is the relationship between wavelength and amount of energy?

2 Using a spectrometer to identify elements in a fluorescent light

1. Use the spectrometer to examine a fluorescent light source (most likely the ones that illuminate your classroom). This time, you will see vertical lines (called spectral lines) of different colors instead of a smooth spectrum like you observed with the incandescent light.
2. You should see a green line at 546 nanometers on the scale. If the green line is not at 546 nanometers, ask your teacher to calibrate the spectrometer for you.
3. Use colored pencils to sketch the lines you observe. Be very precise in your sketch by placing the lines you see in the exact positions on the scale below.

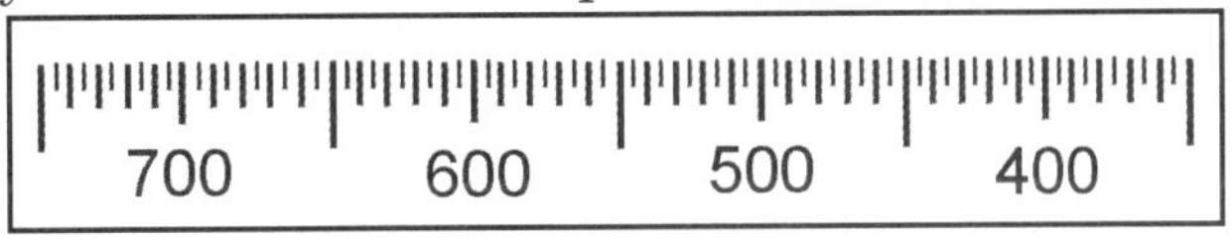

4. Identify the wavelength of each spectral line, from left to right, then fill in Table 1.

Table 1: Spectral lines produced by a fluorescent light

Line number	Spectral line color	Spectral line wavelength (nm)
1		
2		
3		
4		

What do the lines mean?

When elements are heated until they are hot enough to emit light (like those elements that make up stars), they produce characteristic spectral lines. Each element produces a pattern of spectral lines that is like a fingerprint. Shown to the right are some examples of the spectral lines produced by four different elements. Each line has a specific wavelength (these values are not shown in the diagram).

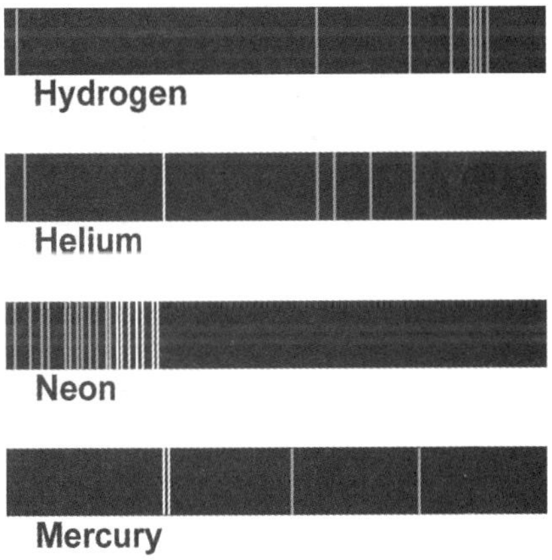

The light produced by a fluorescent source is created when electric current is passed through a gas inside of the tube. This gas, which is made of only one element, absorbs energy, and emits light.

a. The light produced by the fluorescent tube you observed contains only one element. Compare the spectral lines you observed with the ones shown in the diagram above. Which element does it contain?

b. Fluorescent tubes have special instructions for disposal and must not end up in a landfill. Based on your spectral analysis of the gas inside the tube, why is this so?

c. When astronomers use a spectrometer to analyze the light produced by stars, they observe the combined spectral lines of all of the elements present in the star. What specific information would an astronomer need to know in order to determine which elements are present in a star?

3 Analyzing light from different sources

Identify five different light sources to observe with the spectrometer. Examples include gym lights, street lights, security lights, monitor screens, plant growing lights, and glow sticks. Write the types of light sources in the first column of Table 2. Follow the steps below for each light source.

1. Use the spectrometer to analyze the light emitted by the light source.
2. Use colored pencils to draw the position of each spectral line in column 2 of Table 2.
3. Record the wavelength, in nanometers, of each spectral line in the third column of Table 2.
4. Table 4 on the next page lists the values, in nanometers, of the spectral lines produced by various elements. Use these values to identify the elements found in each light source you observed and write them in the last column of Table 2.

Table 2: Spectral lines produced by different light sources

Light source	Spectrometer scale (nm)	Position of each vertical line (nm)	Elements present
	700 600 500 400		
	700 600 500 400		
	700 600 500 400		
	700 600 500 400		
	700 600 500 400		

a. If one light source displays more spectral lines than another, does that mean that it contains more elements? Explain your answer.

b. Which light source contains the greatest variety of elements?

c. Which light sources contain only one element?

4 Analyzing the light from a star

So far, the light sources you observed contain only a few elements. However, stars' atmospheres contain many elements and are much more complex.

1. Use the spectrometer to analyze the light of our closest star—the Sun. Do not point the spectrometer directly at the Sun. Instead, point it at reflected sunlight, off a cloud, for example, or a patch of blue sky.
2. Record in Table 3 the color and value, in nanometers, of each spectral line you observe.
3. Use Table 4, shown below, to identify the elements present in the Sun's atmosphere and record them in the third column of Table 3.

Safety Tip: Do not point the spectrometer directly at the Sun. Point it only at reflected sunlight. NEVER look directly at the Sun!

Table 3: Analyzing light from the Sun

Spectral line color	Spectral line wavelength (nm)	Element present

Table 4: Spectral lines and elements present

Spectral line (nm)	Element present	Spectral line (nm)	Element present
393	calcium	527	iron
397	calcium	546	mercury
405	mercury	577	mercury
434	hydrogen	579	mercury
436	mercury	589	sodium
486	hydrogen	590	sodium
517	magnesium	656	hydrogen
517	iron	687	oxygen

a. Explain why the Sun's light produces more spectral lines than the light sources you observed in Parts 2 and 3.

b. Where do elements in the Sun's atmosphere come from? Explain your answer in detail.

c. The Sun is a middle-aged star. If you could analyze the light from a much older star, what would you expect to see? Justify your answer using your knowledge of the star life cycle.

27B Star Life Cycle and the H-R Diagram

How can we use an H-R diagram to know where a star is in its life cycle?

Just like humans, stars go through a life cycle. Over the course of their lives, stars change in ways that make each stage different and recognizable. A useful way to organize this information is in a Hertzsprung-Russell diagram, also called an H-R diagram. In this investigation, you will learn how to identify stars of different types and ages using the information in an H-R diagram.

Materials

- None required

1 The H-R diagram

For astronomers, a graph that displays a star's luminosity on the y-axis and its surface temperature on the *x*-axis sets up an extremely useful diagram called a Hertzsprung-Russell, or H-R diagram. In 1910 Ejnar Hertzsprung and Henry Norris Russell discovered that when all of the known stars were put on their graph, several obvious groups became apparent. By examining the differences in these groups, later astronomers were able to realize that the groups were best described as stars in different periods in their life cycle, rather than completely different types of stars. Just as infants, adolescents and adults are in different stages of the human life cycle, main sequence, red giants, supergiants, and white dwarfs are stars in different stages of the life cycle of stars.

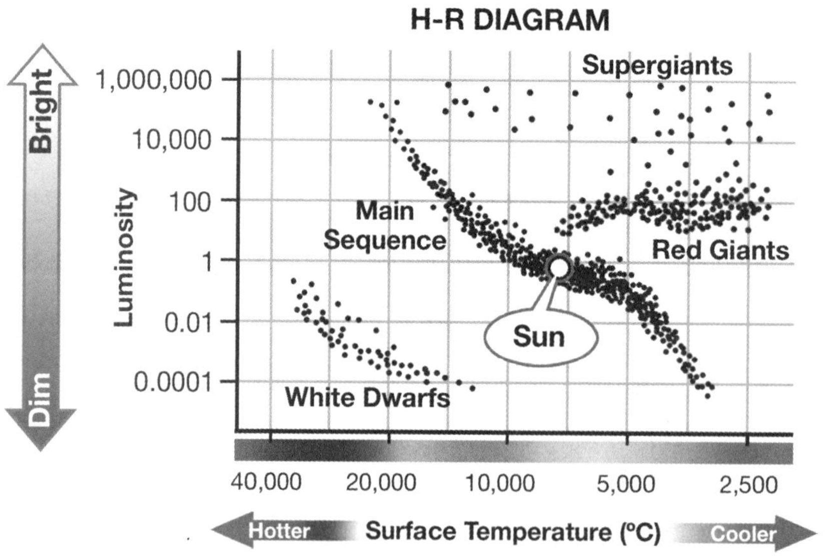

2 Making observations

a. Describe the qualities and the location of white dwarf stars based on the H-R diagram.

b. Describe the qualities and the location of supergiant stars based on the H-R diagram.

c. Describe the qualities and the location of main sequence stars based on the H-R diagram.

d. Describe the qualities and the location of red giant stars based on the H-R diagram.

3 Graphing star data

Table 1 contains data concerning the luminosity and surface temperature of 12 stars. Plot each star's data on the grid and label each star (A–L) on Table 2.

Table 1: Luminosity and surface temperature data

	Name	Luminosity	Surface temperature (°C)
A	Proxima Centauri	0.0017	3,100
B	Rigel	66,000	11,000
C	Sirius A	25.4	9,500
D	Sirius B	0.026	25,000
E	Aldebaran	350	3,600
F	Antares	65,000	3,200
G	Arcturus	250	4,700
H	Betelgeuse	135,000	3,300
I	Procyon A	8	6,200
J	Procyon B	0.0005	7,500
K	Pi Andromedae A	800	16,000
L	40 Eridani B	0.013	12,000

Table 2: Star data graph

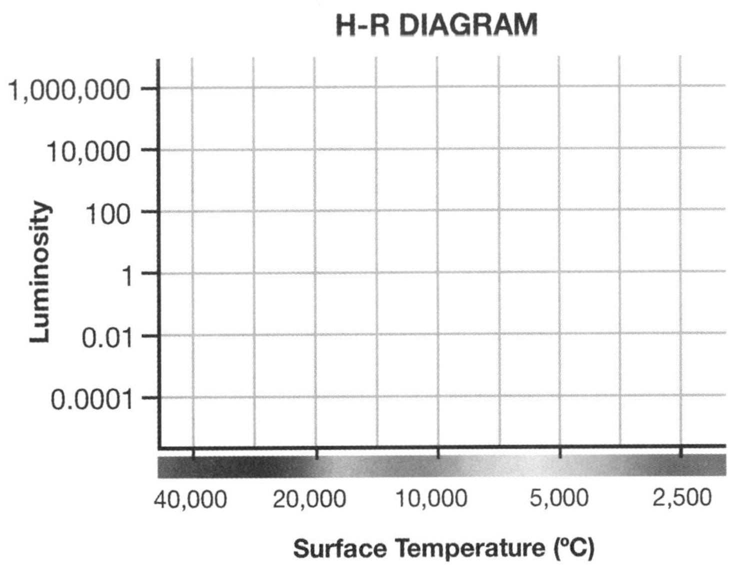

4 Examining your graph

Compare your graph of star data to the H-R diagram from part 1. Look for groupings on your graph that are similar to the H-R diagram and answer the following questions.

a. Which stars on your graph are located in the area associated with main sequence stars?

b. Which stars on your graph are located in the area associated with supergiant stars?

c. Which stars on your graph are located in the area associated with red giant stars?

d. Which stars on your graph are located in the area associated with white dwarf stars?

e. Can you identify which group a star will fit into just by looking at its temperature alone? Its luminosity alone? Explain your answer.

5 Groupings on the H-R diagram and the life cycle of stars

Even though a star is not alive, it goes through a "life cycle" similar to living things. The groups that appear in different sections of the H-R diagram are in different phases of their life cycle.

a. In what part of their life cycle are main sequence stars; beginning, middle, or end, or some combination of those stages?

b. What characteristics are common to main sequence stars?

c. How are red giant and supergiant stars similar? How are they different? Include how each one fits into the life cycle in some part your answer.

d. In what part of their life cycle are white dwarf stars; beginning, middle, or end, or some combination of those stages?

e. What characteristics make white dwarf stars different from main sequence stars?

f. What is the difference between a white dwarf star and a red giant star?

g. Where is the Sun located on the H-R diagram and what group is it in?

h. Where would a Black Hole be located on an H-R diagram?

28A How Does A Telescope Work?

What does a telescope do that makes things far away able to be seen?

A *telescope* is a device that makes distance objects appear closer. Telescopes work by collecting the light from a distant object with a lens or mirror and bringing that light into a concentrated point, called a *focal point*. The bright light from the focal point is then magnified by another lens (the eyepiece) so that it takes up more space on your retina (the light-sensing membrane of the eye). This makes the object appear to be larger and closer. In this investigation, you will identify the parts of a simple telescope and learn how it works.

Materials

- Light blue lens from Optics Kit
- Hand lens
- CPO Telescope
- 3 × 5 inch index card
- Metric tape measure (at least 10 meters)
- Ruler with millimeter divisions
- Graph paper
- Physics Stand

1 Making a simple refracting telescope

Magnification by a telescope is accomplished by changing the direction of light rays leaving a distant object by refraction (with lenses), reflection (with mirrors), or some combination of both. There are many different designs for telescopes of all shapes and sizes. In this investigation you will make a simple refracting telescope with two convex lenses.

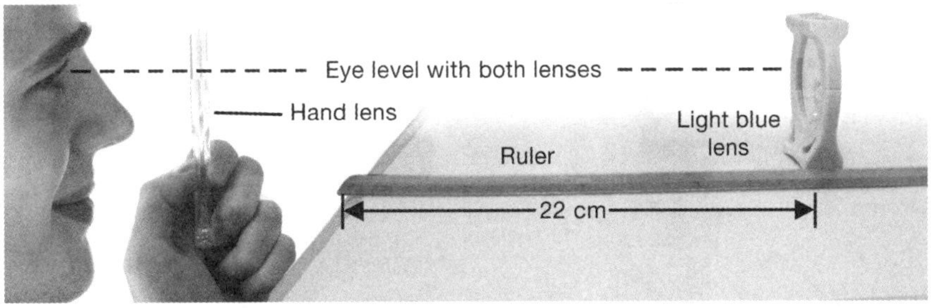

1. Place a ruler on the table so the zero mark is at the edge and the light blue lens is at the 22 cm mark. This will be your objective lens. Its job is to gather the light from distant objects. Choose an object to view about 3 m away from the light blue lens.
2. Orient the light blue lens so you can look directly through it by moving your head to about 2–3 cm away from the edge of the table.
3. With your eye at the same level as the light blue lens, raise the hand lens up to your eye and look through the large magnifying part of the lens directly at the light blue lens on the table. The hand lens will be your eyepiece. It magnifies the image of distant objects passing through the objective lens. Be careful not to bump the table with your face or press the hand lens into your eye. Concentrate on the image in the light blue lens as you look through the hand lens.
4. With the lens up to your eye, move your head slowly away from the table until the object 3 m away comes into focus.
5. You have just made a simple refracting telescope! Try looking at something a different distance away and adjust the distance between the hand lens and blue lens until it comes into focus.

6. CHALLENGE: See if you can look through the smaller magnifying lens on the handle of the hand lens and get your telescope to work. HINT: Try placing the blue lens at 20 cm to start. What is the difference between the image you saw using the larger magnifying lens and the smaller magnifying lens?

2 Thinking about what you observed

a. Which lens was the objective lens in the simple refracting telescope you made? Which lens was the eyepiece?

b. Was the image you observed through the telescope right-side up, or upside down?

3 Learning the parts of a refracting telescope

For this part of the investigation you will use the CPO refracting telescope. This type of telescope uses a convex lens to gather light into a focal point, and another convex lens to magnify the image, just like the one you made in part 1. Use the diagram below to identify the parts of your telescope. The function of each part is listed next to the diagram. Place the letter of each function next to the correct part.

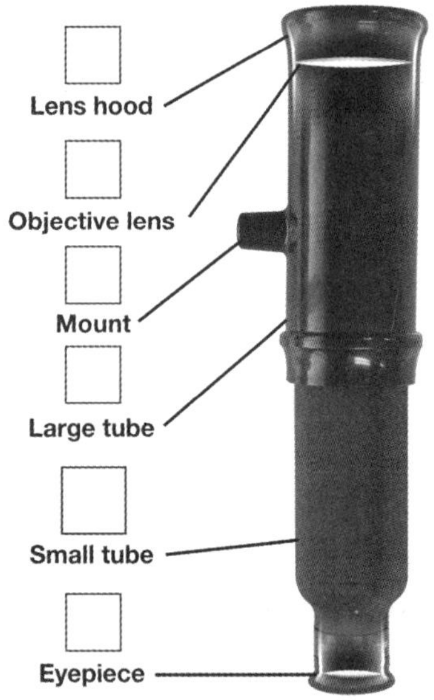

A. Moves in or out to focus image

B. Magnifies image

C. Reduces glare

D. Holds the light-gathering lens

E. Gathers light

F. Holds threaded knob in place for mounting to stand

Once you have matched up all the parts and their functions, use the CPO telescope to look at the 3 m away object you viewed in part 1, then answer questions a and b below.

a. What do you notice about the orientation of the object 3m away and the image you see through the telescope? Is it similar to or different than the simple refracting telescope you made in part 1?

b. How does the image change as you move the telescope when you try to aim it at an object?

4 Focal Length

The **focal length** of a lens is the distance between the center of the lens and the focal point. Here is how the focal length of a lens is measured:

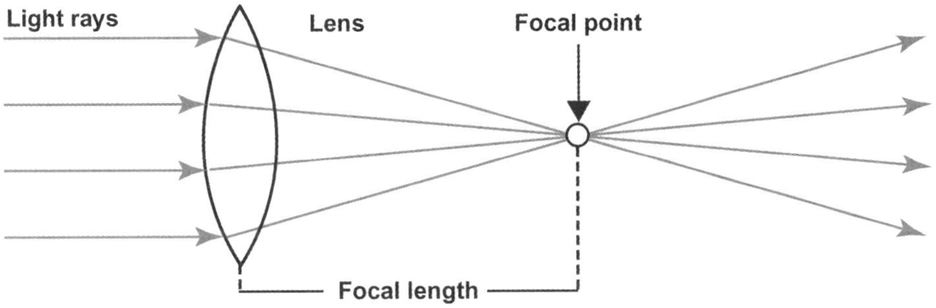

a. On the hand lens, how do the focal lengths of the large magnifying lens and the small magnifying lens compare? Which is longer? Which is shorter? HINT: Try forming an image of an overhead or far away light source on the table and compare these distances between the two lenses.

b. What happens to the image seen through a telescope when you use an eyepiece with a shorter focal length?

c. Look through the CPO Telescope. Its objective lens is identical to the lens in the light blue lens. Use the telescope to look at the object 3 m away that you viewed in part 1. How does that image compare to the other two images you saw, first with the large, then with the small magnifying lenses of the hand lens?

d. Would you say the eyepiece of the CPO Telescope has a shorter or longer focal length compared to both magnifying lenses of the hand lens?

5 Focusing a telescope

Safety Tip: Never look at the Sun or a bright object through your telescope! You will cause damage to your eye.

1. Mount your telescope onto the Physics Stand by sliding the blue threaded knob through one of the holes in the stand, out the opposite side, and finally into the mount of the telescope. Screw it in firmly, but not too tight; you may need to adjust the telescope up and down. Be sure to choose a height and placement on the stand that allows you to easily look through the telescope. Mounting the telescope on the stand will make it easier to aim and keep it steady when you look through it. Sitting in a chair or kneeling when using the telescope on the stand makes it easier to look through than stooping over.

2. Pull the small tube all of the way out of your telescope and place a piece of masking tape lengthwise along the tube's flat part ending at the small tab you'll use to measure distance.

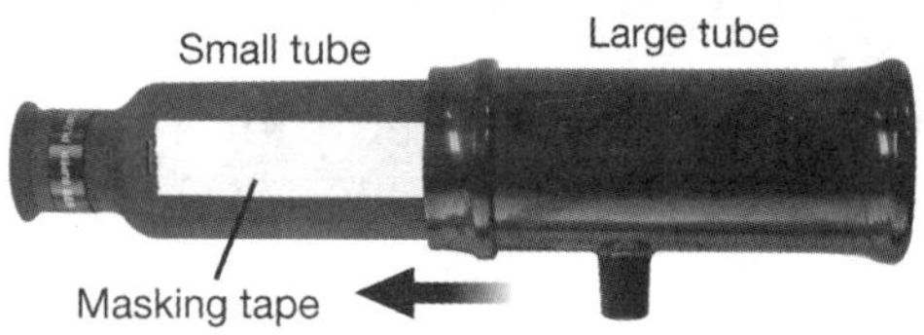

3. Write the letter "e" to measure 1 centimeter tall on a 3-by-5-inch card. Place the card at eye level when looking through the telescope on a wall in a large room (your classroom, the library, etc.).
4. Starting at the wall, lay a tape measure on the floor. Pull 10 meters of tape out from the wall. This will be your distance scale, 0–10, with zero at the wall.

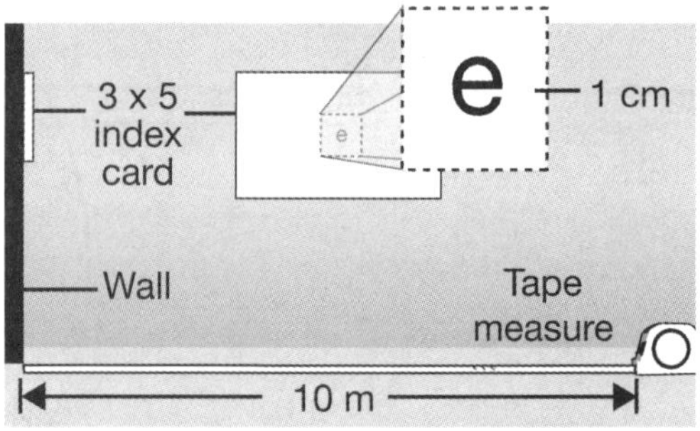

5. Stand at the 10-meter mark and use your telescope to observe the letter on the card on the wall. Pull the small tube in or out until the letter comes sharply into focus. Record your observations in the table below. For your observations, describe the appearance of the image in terms of orientation and apparent size.

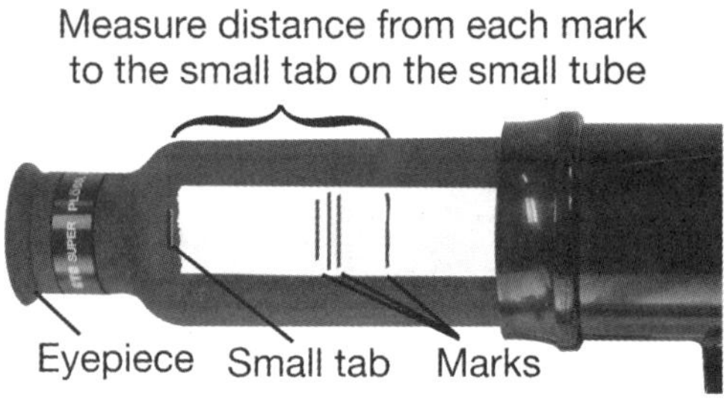

6. Draw a pencil line on the masking tape where the short tube enters the long tube. Write "10 m" (your distance from the card) next to the mark.
7. Repeat steps 4 and 5, moving closer to the wall by one meter for each trial. Continue to move 1 meter closer to the card, while repeating steps 4 and 5, until you cannot bring the letter into focus.
8. Use a ruler with millimeter markings to measure the length of the small tube, from the edge of the eyepiece to your marks on the tape, as shown. Record your measurements in Table 1.

Table 1: Focusing a telescope

Distance from card (m)	Observations	Length of small tube (mm)
10		
9		
8		
7		
6		
5		
4		
3		
2		
1		

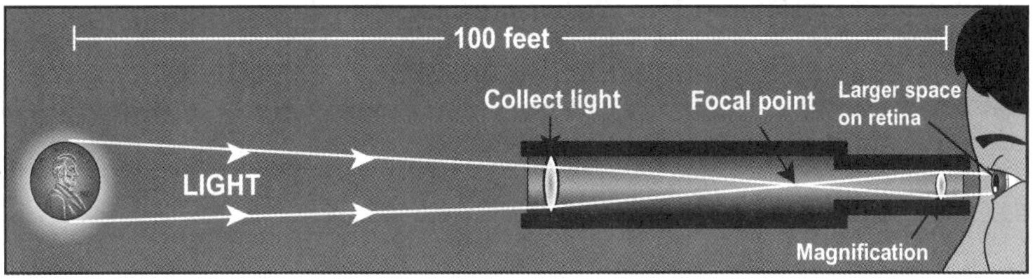

6 Stop and think

a. In general, as you move closer to an object, how should you adjust the small tube?

b. When you view the letter "e" through your telescope, why does the image appear upside down? (HINT: Trace the light rays coming from the penny in the picture above.)

c. Make a graph of distance versus length of the small tube. Would you describe the relationship as direct or inverse? Explain your answer.

7 Determining the magnification of a telescope

A telescope's magnification is a measure of the number of times the image being viewed is enlarged. A telescope magnification of 10x means the image of an object is magnified 10 times. The magnification of a telescope is the relationship between the objective lens and the eyepiece you are using. Since eyepieces in most telescopes can be interchanged, the magnification can vary. The magnification of a telescope is determined using the following equation:

$$\text{magnification} = \frac{\text{focal length of objective lens}}{\text{focal length of eyepiece}}$$

Popular focal lengths for eyepieces are 10 millimeters and 25 millimeters. Suppose your telescope has an objective lens with a focal length of 200 mm and an eyepiece with a focal length of 25 mm. What is the magnification of your telescope?

$$\text{magnification} = \frac{200\text{ mm}}{25\text{ mm}} = 8\text{x}$$

8 Thinking about what you observed

a. If you wanted to increase the telescope's magnification in the problem above, which eyepiece would you choose—10 mm or 40 mm? Explain your answer and show your solution to the problem.

b. Calculate the magnification of the following telescopes:

Objective lens	Eyepiece	Magnification
1200 mm	10 mm	4
1200 mm	25 mm	
800 mm	20 mm	

c. A telescope with an 800 mm objective lens can gather more light than a telescope with a 200 mm objective lens. Does this mean that the 800 mm telescope has greater magnification than the 200 mm telescope? Explain your answer.

d. The focal lengths of two telescopes are shown below. Which telescope would have the greater magnification? Through which telescope would you expect to see a brighter image? Explain your answers.
Telescope A: 800 mm objective lens; 20 mm eyepiece
Telescope B: 200 mm objective lens; 5 mm eyepiece

28B Astronomy and Light

How do we measure the distance to stars and galaxies?

The *brightness* of a star depends on both the star's distance and on how much light the star actually produces (its *luminosity*). When astronomers know a star's brightness and luminosity they can calculate the distance to the star. In this investigation, you will discover the mathematical relationship between how bright a light source appears and it's distance.

Materials

- Solar cell
- Small box
- Light fixture or lamp
- 100 W light bulb
- Multimeter

1 Doing the experiment

For the experiment, you will use a 100-watt light bulb as a light source. You will measure the brightness of the light source at various distances from the light bulb using a solar cell and electric meter.

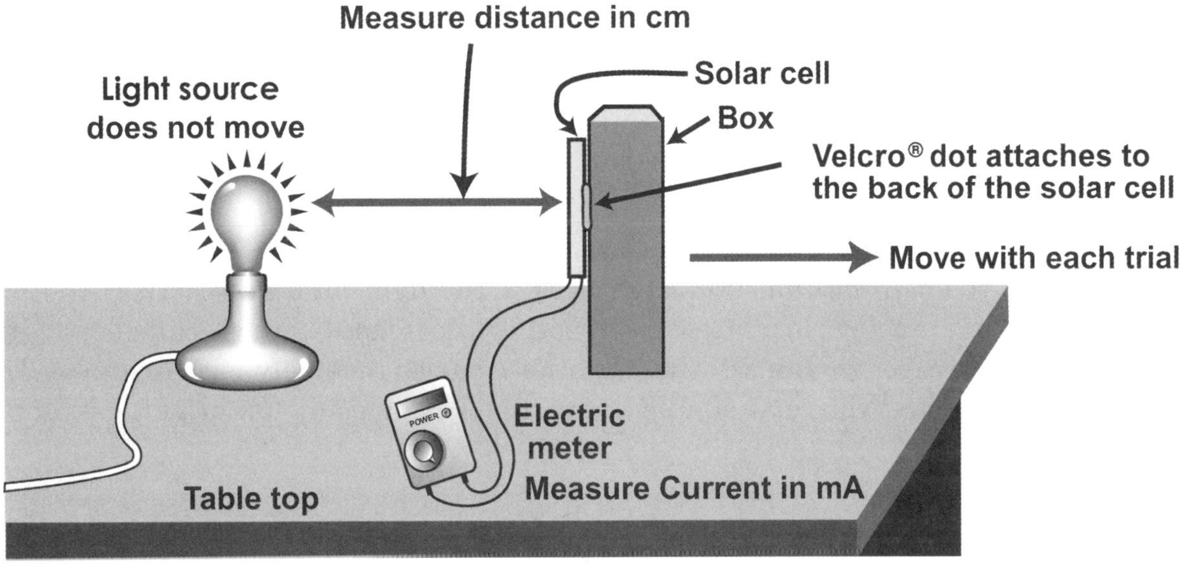

1. Turn on your light source and turn off the overhead lights in the classroom.
2. Place the solar cell exactly 10 centimeters from the light bulb. Make sure the solar cell is directly facing the light bulb and is not at an angle. Make sure the electric meter is set to measure current in milliamps (mA).
3. Measure the light's brightness in mA and record your results in Table 1.
4. Measure the brightness at 10-centimeter intervals and record your data in Table 1.

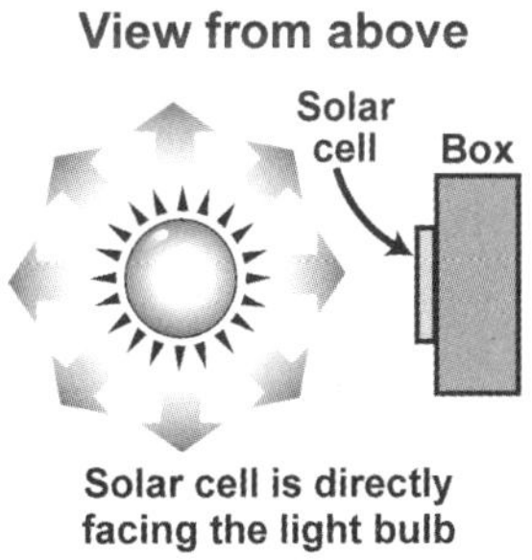

Table 1: Brightness and distance data

Distance (cm)	10	20	30	40	50	60	70	80	90	100
Brightness (mA)										

2 Stop and think

a. Does this experiment measure the *luminosity* or the *brightness* of the light source? Explain your answer.

b. Is the power rating of the light source (100 watts) a measure of the light bulb's *luminosity* or *brightness*? Explain your answer.

c. What effect do you think increasing the distance will have on your measurements of brightness?

3 Analyzing your data

Make a graph of brightness versus distance. Plot brightness on the y-axis and distance on the x-axis.

a. Is your graph increasing or decreasing from left to right?

b. Describe the shape of the curve on your graph. Have you seen a curve like this before?

c. Is there a mathematical relationship between brightness and distance from your graph? Explain your answer.

4 Inverse relationships

From your graph, you can tell that the brightness of the light bulb *decreases* as the distance *increases*. When one factor increases as another decreases, it is called an **inverse relationship**. In this case, we can say the brightness varies inversely with distance. This can be shown as a mathematical relationship using the variables B for brightness and D for distance, where:

Brightness → $B = \frac{1}{D}$ } Inverse of distance

Your graph shows an inverse relationship between brightness and distance, but can you tell from the curve exactly by how much brightness will decrease as distance increases? Here are three possible mathematical relationships:

$$B = \frac{1}{D} \quad \text{or} \quad B = \frac{1}{D^2} \quad \text{or} \quad B = \frac{1}{D^3}$$

The first equation says the brightness is inversely related to the distance. The second equation says the brightness varies with the inverse of the distance *squared*.

a. What does the third equation state?

b. Assuming that the correct equation is one of the three above, how could you figure out which is the correct one?

c. Which equation do you think is the correct one? Explain your reasoning.

5 Identifying the correct inverse relationship

To figure out which equation is correct, you will need to further analyze your data from Part 2.

1. Enter your distance data from Table 1 into the first row of Table 2 below. You will need to convert centimeters to meters. The first two values are done for you.
2. Calculate 1/D and enter the results in the second row of Table 2. The first two are done for you.
3. Calculate $1/D^2$ and $1/D^3$ and enter the values in rows three and four of Table 2.
4. Enter your brightness data from Table 1 into the fifth row of Table 2.
5. Make the following three graphs:
6. Graph 1: *Brightness vs. 1/D*
7. Graph 2: *Brightness vs.* $1/D^2$
8. Graph 3: *Brightness vs.* $1/D^3$

Table 2: Analyzing your distance and brightness data

Distance (cm)	10	20	30	40	50	60	70	80	90	100
Distance (m)	0.10	0.20								
1/D	10	5								
$1/D^2$	100	25								
$1/D^3$										
Brightness (mA)										

6 Reaching a conclusion

a. Which graph identifies the correct inverse relationship between brightness and distance? Explain your choice.

b. Write down the correct formula for the relationship between brightness and distance.

c. Test your formula by following these steps:

(1) Choose a distance for which you did not measure brightness (for example, 45 centimeters).

(2) Calculate brightness using your formula. This is your *predicted brightness.*

(3) Move the solar cell to the distance you chose and measure the brightness of the light. This is your *actual brightness.*

(4) Calculate your percent error using the following formula:

$$\left(\frac{\text{predicted brightness - actual brightness}}{\text{predicted brightness}}\right)\times 100 = \text{percent error}$$

d. Your actual brightness should be fairly close to your predicted brightness. What are some possible reasons for differences between predictions and measurements in this experiment?

7 Applying your knowledge: How bright would the Sun appear on Neptune?

You have learned that the average distance of Earth from the sun is 150 million kilometers. This is also defined as 1.0 astronomical unit (AU). Using the relationship between brightness and distance that you discovered in this investigation, we can determine that the brightness of the Sun from Earth is equal to 1.0 solar brightness units (SBU):

$$\text{Brightness of the sun from Earth} = \frac{1}{(1.0\ \text{AU})^2} = 1.0\ \text{SBU}$$

Use what you have learned in this investigation to complete Table 3 below. Then answer the questions.

Table 3: Apparent brightness of the sun from the planets

Planet	Average distance from the sun (AU)	Brightness (SBU)
Mercury	0.37	
Venus	0.72	
Earth	1.0	1.0
Mars	1.5	
Jupiter	5.2	
Saturn	9.5	
Uranus	19.2	
Neptune	30.0	

a. How much brighter is the sun viewed from Mercury compared with its brightness viewed from Earth?

b. How much fainter is the sun viewed from Neptune compared with its brightness viewed from Earth?

c. CHALLENGE! Alpha Centauri is 4.1×10^{13} km from Earth. How bright would the Sun appear in SBU from Alpha Centauri? (Hint: You must first convert kilometers to astronomical units.)

8 Using light to measure distances to stars and galaxies

In the investigation, you determined that the brightness of a light source decreases as the distance increases. However, the brightness of the light source also depends on how much light it actually emits (its luminosity). The inverse square law of brightness shows the relationship between brightness (how bright an object appears from a certain distance), luminosity (how much light an object actually emits), and distance:

INVERSE SQUARE LAW

$$B = \frac{L}{4\pi D^2}$$

Brightness (B), Luminosity (L), Distance (D), Constant (4×3.14)

a. Which variables would an astronomer need to know in order to determine the distance to a faraway galaxy or star? Describe how an astronomer could determine each variable.

b. In the diagram to the right, which galaxy is the farthest from Earth? Explain your reasoning.

B = 3.0
L = 18.0
Galaxy A

B = 6.0
L = 9.0
Galaxy B

Optional Investigations

The investigations in this section are designated as C Level. They are not required as part of the course but may be used to further explore and extend the concepts in the chapter.

1C Conversion Chains

How can you use unit canceling to solve conversion problems?

$$115 \text{ km} \times \frac{1 \text{ mi}}{1.6 \text{ km}} = 72 \text{ mi}$$

Suppose you are traveling in Canada or Mexico, and you see a sign that says the next city is 115 km away. How many miles is that? You could use the dimensional analysis process to figure it out. Dimensional analysis is a method of using conversion factors and unit canceling to solve unit conversion problems. In this activity, you will use conversion chains to start and solve conversions.

Materials

- Conversion Chain card deck
- Conversion chain table for recording chains and solutions
- Calculator

1 Conversion Chain cards

The deck of cards contains:

12 START cards, 12 SOLVE cards, 36 conversion factor cards

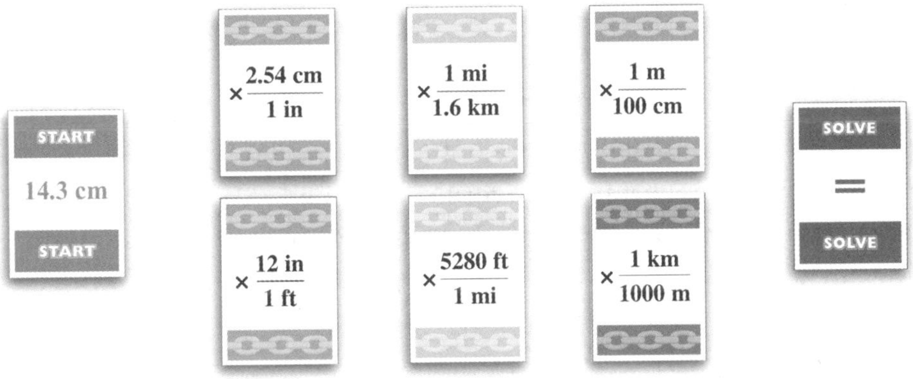

1. Look at the cards to identify what START, SOLVE, and conversion factor cards look like.
2. Shuffle the cards thoroughly!

2 Creating Conversion Chains

The object of the game is to have the most points after solving 6 conversion chains.

1. Deal 4 cards to each player (works best with 3–5 players).
2. Choose a player to go first. This player must play a START card draw one from the deck. If the player draws a START card, it may be played; otherwise play passes to the next player. Play passes until a START card has been played. Anytime a START card is played the player scores 5 points. (Each group must assign a scorekeeper).
3. The next player may play another START card or a conversion card that will continue the conversion chain. That means that the card must have a denominator unit that matches a START card unit, to cancel the starting unit and continue the chain. Only 6 START cards can be played in a game.

4. If the player has more than one card that could be played on multiple START cards, the player must choose only *one* card to play on each turn.
5. After playing a card, the player *must* draw a card.
6. If the player does not have a suitable card, the player must draw one card from the deck. If the card will play, he can play it. If the card won't play, he must pass his turn.
7. The next player plays a conversion factor card, plays a START card, or draws a card, and play continues.
8. A team can have as many as 3 conversion chains going at once.
9. After a minimum of 3 conversion factor cards have been played on any particular START card, any player is allowed to play a SOLVE card on her turn, and the player ***immediately scores 5 points for each conversion chain card in the chain***.

3 Solving a chain and scoring points

1. Anytime a SOLVE card is played on any conversion chain, all players must record the chain on the answer sheet and calculate the answer with the correct number of significant digits and unit.
2. Players compare answers and determine who has the correct answer. Consult your teacher if there are disagreements.
3. *Any* player with the correct answer scores 25 points.
4. Keep playing until your team has created and solved 6 conversion chains.
5. At the end of the game, after 6 conversion chains have been scored, players must subtract 5 points per card for unplayed cards remaining in the hand.
6. Total up the points, subtract the unused cards, and determine the winner.

4 Reminders and strategy

1. Whenever you play a START card, you score 5 points! If you draw a START card after 6 chains have already been started, you must keep that card in your hand, and it will cost you 5 points in the end.
2. Remember: you always draw a card at the end of *every* turn.
3. When a conversion chain has been solved, place the START, chain, and SOLVE cards to the side in a pile. You might want to look at a solved chain later in the game. Also, this will help your team keep track of how many conversion chains have been solved (the goal is to solve 6).
4. You can play a SOLVE card only after there are at least 3 conversion factor cards in any chain. For beginners, a SOLVE card should be played after 3 chain cards are in a chain. For advanced players, a SOLVE card can be held back and played after the chain is much longer. The SOLVE card player benefits, since 5 points are awarded for each chain card in the chain to that player!
5. Record each conversion chain your team creates in Table 1. There are blanks for recording 6 conversion factors per chain. If a chain has less than 6, just leave remaining boxes blank. If you need more than 6, continue on the back of the page.
6. The first row of Table 1 shows an example chain.

Table I: Conversion Chains

START card							Answer with correct sig figs and unit	Points scored
46.3 m	x $\frac{1\ km}{1000\ m}$	x $\frac{1\ mi}{1.6\ km}$	x $\frac{5280\ ft}{1\ mi}$	× ―	× ―	× ―	153 ft	
	× ―	× ―	× ―	× ―	× ―	× ―		
	× ―	× ―	× ―	× ―	× ―	× ―		
	× ―	× ―	× ―	× ―	× ―	× ―		
	× ―	× ―	× ―	× ―	× ―	× ―		
	× ―	× ―	× ―	× ―	× ―	× ―		
	× ―	× ―	× ―	× ―	× ―	× ―		
							TOTAL	

5 Thinking about the game

a. The conversion chains you create for this game could be quite long, and the actual conversions could be solved with fewer conversion factors. What is the least number of conversion factors you would need to make these conversions, given the cards you have in the deck?

- converting ft to cm
- converting mi to m

b. Because of the limitations of the cards in the deck, to convert from km to cm, you would have to go through English units. What is a much easier way to convert from km to cm? Explain.

c. Explain the math reasoning behind why you are able to “cancel” like units that appear in the numerator and denominator of conversion factors in a conversion chain.

2C Wave Speed

What is the relationship between water depth and wave speed?

In the ocean, most <u>*waves*</u> are driven by the wind. As long as the wave is in deep water, the water depth does not influence the wave very much. When a wave approaches the shore, however, the speed of the wave changes.

In this investigation, you will collect data to observe how water depth affects wave speed.

Materials

- GeoBox
- 1,000 mL beaker
- Masking tape
- Data Collector
- Ruler or measuring tape
- Graph paper
- Wooden block

1 Setting up

Gather your materials including the GeoBox. Fill the 1,000-mL beaker with water.

GeoBox

2 Stop and think

Predict the relationship between wave speed and water depth. Does the wave speed increase or decrease as the water depth increases? Why?

3 Doing the experiment

1. Place water in the GeoBox to a depth of one centimeter.
2. Place the wooden block under one end of the tray. One student will use the Data Collector in Stopwatch mode and be the "timer," while the other student will watch the waves.
3. First practice the following, then measure when you are ready. Pull the wooden block out. This will create a wave as the tray falls to the table. The water will "slosh" lengthwise in the tray.
4. As soon as you begin to move the block, start the stopwatch. The person watching the wave will watch it make four round trips. That means back and forth four times. As soon as it completes the four round trips, stop the stopwatch. Record your data (time) in Table 1. Repeat for the same depth for two more trials.
5. Once you have made three time measurements for the one-centimeter depth, add water up to the two-centimeter mark.
6. Follow the same procedure and continue increasing water depth until you get as high as you can go (probably around six centimeters).

7. Find the average time for each depth. Then calculate wave speed using the formula below.

$$\text{wave speed} = \frac{\text{total distance traveled (cm)}}{\text{average time (s)}}$$

Hint: Be careful when you measure for the total distance traveled —think about how many times the wave goes back and forth.

Table 1: Wave speed at different depths

Depth (cm)	Trial 1 (s)	Trial 2 (s)	Trial 3 (s)	Average (s)	Wave speed (cm/s)
1					
2					
3					
4					
5					
6					

8. Make a graph with depth on the x-axis and wave speed on the y-axis.

4 Thinking about what you observed

a. Describe the pattern of the graph.

b. Why should there be any change in wave speed?

5 Exploring on your own

a. What happens when waves reach the shoreline?

b. What is the difference between swells and whitecaps?

c. Would waves approach the shore in the same manner if the shoreline was gradual or very steep? Explain.

3C Bathymetry of the Sea Floor

How can we tell what kinds of features are on the sea floor?

Did you know that about three quarters of the surface of Earth is covered by water? People have used boats and ships to travel all over the oceans for hundreds, if not thousands of years. During most of this time nobody really knew what was beneath the surface of the ocean. Today we have detailed maps of the ocean bottom. *Bathymetry* is the mapping of the ocean bottom. In this investigation, you will learn how scientists have been able to find out what kinds of features are deep down on the sea floor.

Materials

- Blindfold
- Graph paper
- Colored pencils (red, orange, yellow, green, blue, purple, brown, and black)
- Pencil sharpener
- Side-view depth diagrams A–H

Dolphins use sound waves to navigate, hunt, and visualize their surroundings. They make a series of clicking sounds and wait for the echoes of those sounds to return. This is called echolocation. The direction and volume of the echoes gives the dolphin an idea of what is in their general area. Let's find out what it might be like to be a dolphin and use echolocation.

1 Setting up

1. Clear an area in your classroom that is big enough for everyone in your class to stand in a large circle. You could also use a large open space like a gym, or even a field outside as well.
2. Your teacher will select one person to be the dolphin.
3. Your teacher will select three people to be fish.
4. Your teacher will select two people to be squid.
5. Your teacher will select one person to be a shark.
6. The remaining students will stand in a circle, and gently guide the dolphin back to the middle of the circle if he or she gets to the edge.

2 Doing the activity

1. The dolphin in this activity will wear a blindfold.
2. The dolphin will say "click!" every few seconds. When the dolphin says "click", the fish have to respond by saying "fish", the squid reply by saying "squid", and the shark responds by saying "shark." The replies represent the dolphin's clicks echoing back to the dolphin.
3. The shark swims randomly around the circle. The fish must stay together in a school as they swim around in the circle, by walking together with baby steps. The fishes and the shark must always swim forward at all times. They may not move backwards or stop. The squid can swim around in any direction, stop, and move backwards. The squid and the shark should move at a regular walking speed as they swim around.
4. The dolphin tries to capture his prey, the fish and the squid, and avoid his predator, the shark. The dolphin can stop swimming, but not move backwards.

5. The dolphin captures his prey by tagging them lightly when it sounds like they are near, using echolocation.
6. The dolphin must avoid the shark. If the dolphin bumps into the shark, someone else gets to be the dolphin.
7. The game is over when the dolphin catches all the fishes and both squid.

3 Thinking about what you observed

a. How did the dolphin know where the fishes, squid, and shark were?

b. When a dolphin makes clicking noises under water, how do you think it can tell the difference between a fish and a shark?

c. Why do you think scientists call this process echolocation?

4 Human-made echolocation

In the early 1900's, scientists and engineers tried to use sound waves to measure the depth of the ocean. Sound travels about 340 m/sec in air. That's pretty fast! But in salt water it travels even faster, about 1,500 m/sec. Ships carry devices that use sound waves and echoes in a system called SONAR. SONAR devices send out bursts of sound waves, and then listen for the echoes of the sound waves when they bounce back off the bottom of the ocean.

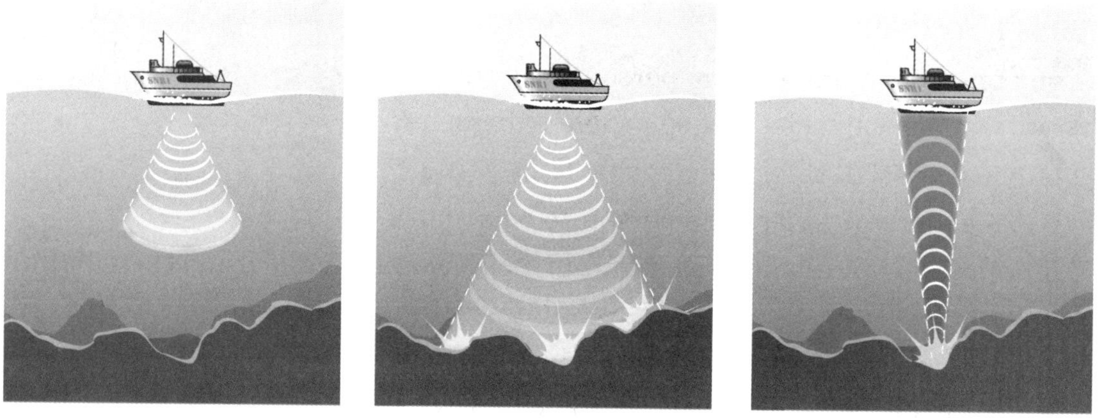

Sound waves sent down from ship | Sound waves bounce off ocean floor | Sound wave echo returns to ship

5 Stop and think

a. How fast does sound travel through salt water?

b. In salt water, how far would a sound wave travel in 1 second? In 2 seconds? In 3 seconds? How about 10 seconds?

c. What is the relationship of the depth of the water to the total distance the sound wave travels?

d. When you find out how far a sound wave has traveled, how can you tell how deep the water is?

6 What's down there?

The use of SONAR has helped identify many different features and regions of the ocean bottom that cannot be seen from the surface. These features have distinctive variations in their shapes and locations.

Table 1: Features and regions of the ocean bottom

Continental shelf

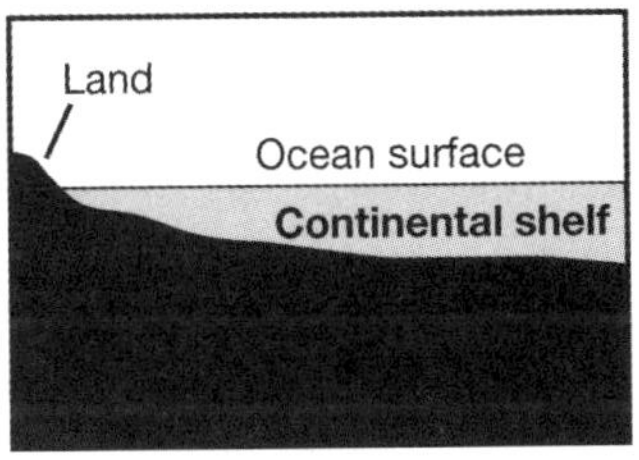

A submerged border of a continent that slopes gradually and extends to a point of steeper descent.

Continental slope

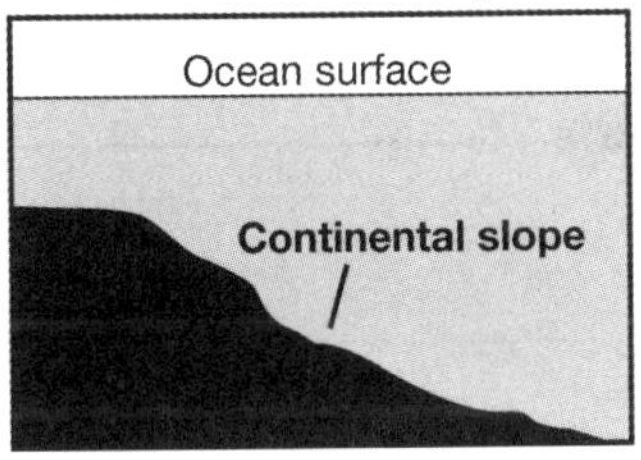

The transitional area between the continental shelf and abyssal plain.

Abyssal plain

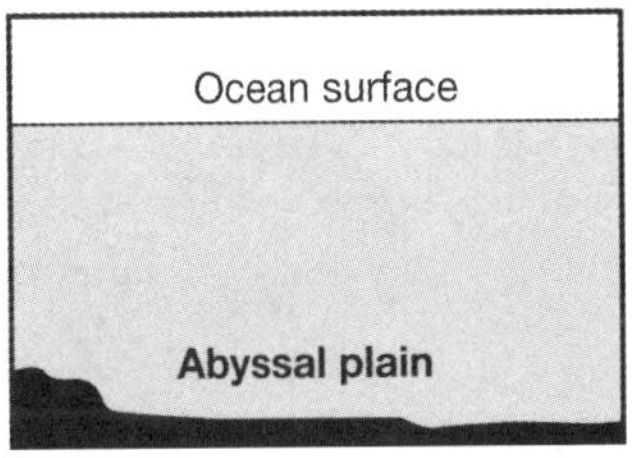

The flat, deep ocean floor. It is almost featureless because a thick layer of sediment covers the hills and valleys.

Seamount

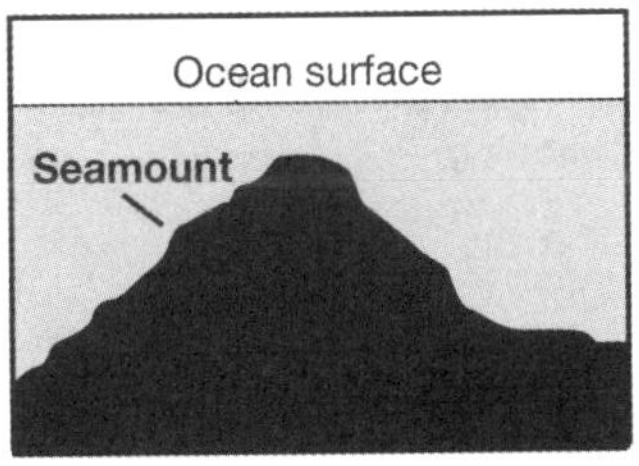

A mountain rising from the ocean seafloor that does not reach to the water's surface.

Guyot

A flat topped seamount that shows signs of being above the water's surface in the past.

Deep-ocean trench

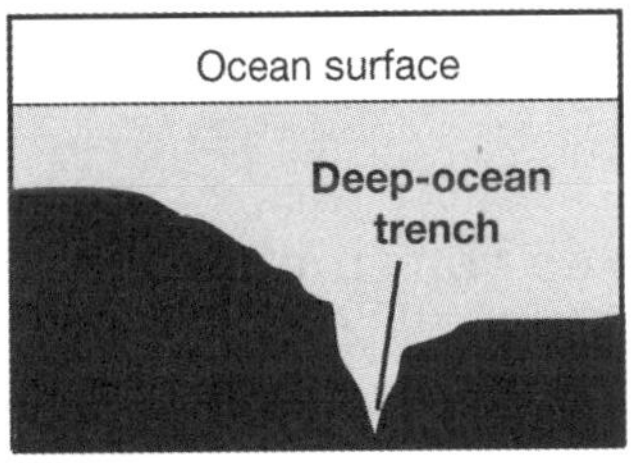

A long, narrow depression of the sea floor having steep sides. The deepest part of the ocean.

7 Stop and think

a. How could you use the information a SONAR system provides to identify the kinds of ocean bottom features and regions in Table 1?

b. What is a seamount that rises above the surface of the ocean called?

c. What do you think causes the flattened top of a guyot?

d. If you were taking SONAR measurements, how would you know that there was a deep-sea trench beneath your ship on the ocean bottom?

e. How would you know you were going over a seamount?

f. What would happen to your SONAR readings as your ship went from the continental shelf into the continental slope?

8 Mapping the ocean bottom

To make a map of the ocean bottom, ships have to travel back and forth in a zig-zag pattern over large areas and take readings from their SONAR equipment all along the way. The sound waves spread out to the side of the boat, so each pass the boat makes can measure depth and indicate features in a wide swath. The beam's width depends on the kind of SONAR equipment the ship has and the depth of the water.

Table 2: Beam width and a zig-zag mapping pattern

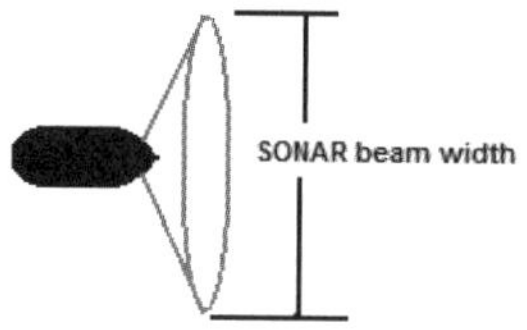

View of ship from above of SONAR beam width

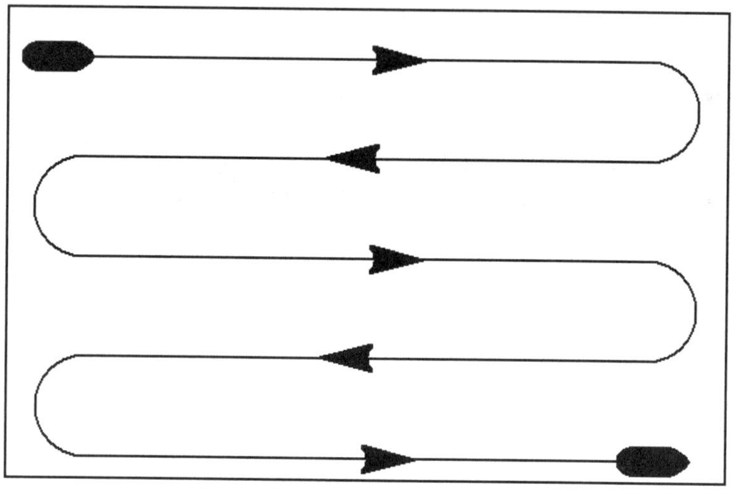

Ship's zig zag pattern for mapping

a. Why do you think a ship follows a zig-zag pattern?

b. How does the width of a ship's SONAR beam affect how many zigs and zags the ship needs to make to get a complete map of an area of the ocean bottom?

9 Readings and depth

1. Your group will be in charge of one set of readings made on a SONAR mapping expedition. You will be given one part (a zig or a zag) of the entire map.
2. The readings indicate how many seconds a SONAR reading took to return to the ship.
3. Your group's task is to create a side view of the terrain under your ship based on the data collected on one pass over part of the whole area.
4. Your teacher will give you a side view depth diagram for your group. Each SONAR reading tells you how long the sound took to return to the ship at that location.
5. Mark the depth of the water based on how many seconds the SONAR sound wave took to return to the ship.
6. The depth will be color coded. Mark each depth according to the color indicated on the depth readout diagram.

Sonar readings

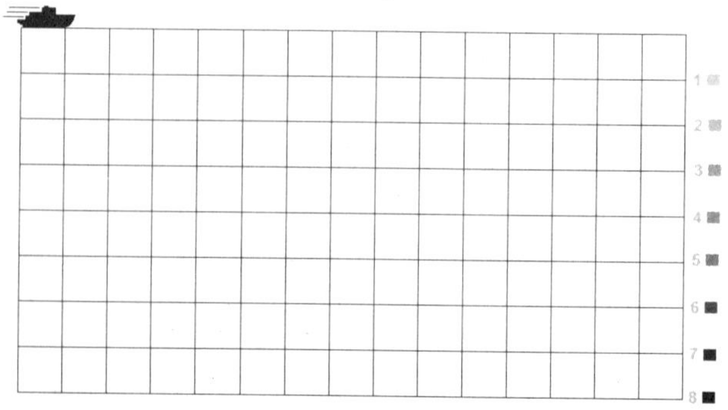

10 Making a bathymetric map from your readings

1. Each group will have a completed color coded side view depth diagram.
2. Your class will make a color coded bathymetric map based on the diagrams produced by all groups.
3. Each group will transfer their color coded depth information, from their side view diagram onto the bathymetric map grid, an overhead view of the area.
4. Once all the groups have transferred their color coded depth information, the bathymetric map is complete.

11 Thinking about your observations

a. What kind of information does your color coded bathymetric map tell you?

b. The readings on your side view depth diagram indicated how many seconds the SONAR signal took to return to the ship. How would you turn this information into actual depth readings?

c. Convert each color into the actual depth it represents on your bathymetric map.

Table 3: Finding the depth of water based on sound travel time

Color	Time sound took to echo off ocean bottom (s)	Speed of sound in salt water (m/s)	Total distance sound traveled (m)	Depth of water (m)
red	1	1500		
orange	2	1500		
yellow	3	1500		
green	4	1500		
blue	5	1500		
purple	6	1500		
brown	7	1500		
black	8	1500		

d. Using information from part 6 of this investigation, what kind of features can you identify on your color coded bathymetric map?

e. Use a piece of graph paper to make your own small color coded bathymetric map that shows a seamount, a guyot, and a deep-ocean trench.

f. **Challenge**: How could you use the information represented by different colors to add contour lines to your bathymetric map?

12 Exploring on your own

a. What does SONAR actually stand for?

b. Near the coast we sometimes see the bottom of the ocean. In deep water, how far does visible light penetrate down into the water? What are the different zones associated with light in the ocean?

c. Research: Who was Marie Tharp and what major project did she help accomplish before anyone else?

d. Research: Besides dolphins, what other animals use echolocation?

4C Studying Two-Part Motion

What happens to the Energy Car as it travels down a hill and across a flat section of track?

In the previous chapter 3 investigations, you explored the motion of the car on a flat track and again on a hill. Now, you can put the track together so it has both a hill and a flat section. What will the speed vs. time graph look like? You will soon find out!

Materials

- Data Collector and 2 photogates
- Energy Car
- Physics stand
- 1 steel marble

1 Predicting

You will set up the track so it has the steeper hill and flat section joined. What will the speed vs. time graph look like?

a. Draw a sketch of what you think the speed vs. time graph will look like. Put *time* on the *x*-axis and *speed* on the *y*-axis.

2 Setting up the experiment

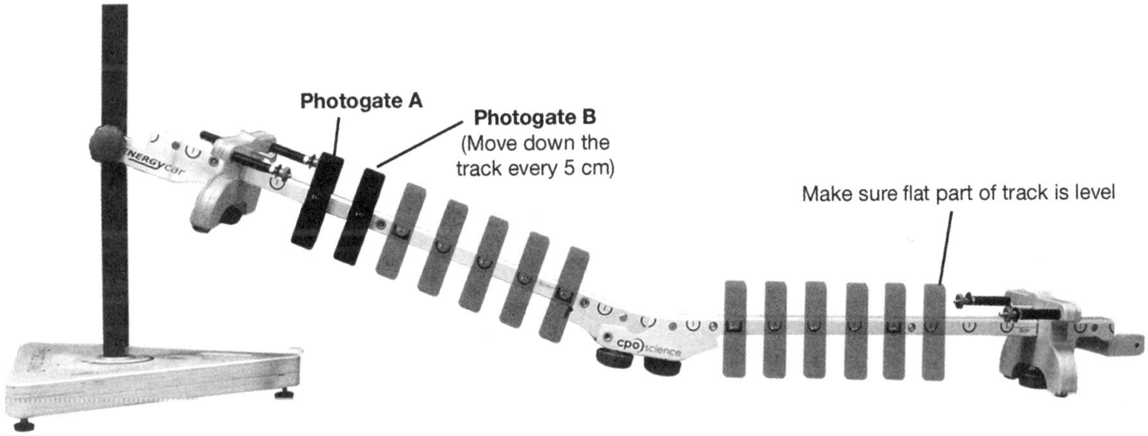

1. Join the steeper hill with the flat track and attach it to the 4th hole from the bottom.
2. Place a stop at the top of the track with the plunger all the way back so the end is flush with the stop.
3. Place the other stop at the bottom of the track; put a lump of clay on the plunger so the car won't bounce back through photogate B when it is near the end.
4. Attach photogate A as shown in the photo, and leave it there.
5. Photogate B will be placed every five centimeters, as shown in the photo.
6. *Make sure the flat part of the track is level.* Use the leveling feet on the physics stand base and on the bottom stop.
7. Place the Energy Car at the top of the track so it rests against the stop's plunger. Put 1 marble in the center of the car. Photogate B should be 5 cm from photogate A.

3 Doing the experiment

1. Release the car and record the time through A, the time through B, the time from A to B, and the distance traveled in Table 1.
2. Move the photogate to the next 5-cm mark and repeat step 1. Do this all along the track until you reach the last mark possible. You will not be able to take time measurements at 35, 40, or 45 cm away from photogate A.

Table 1: Energy Car Data

Distance A to B (cm)	Time A (s)	Time B (s)	Time AB (s)	Speed A (cm/s)	Speed B (cm/s)
5					
10					
15					
20					
25					
30					
50					
55					
60					
65					
70					
75					

4 Finding the speeds

a. Find the speed at A and speed at B for each trial and record in Table 1. Remember: to find the speed of the car, use the width of the flag (1.00 cm) for the distance, and the time through the photogate for the time. ($v = d/t$)

b. Look at the A speeds. What does this data tell you in general about the experiment?

c. Look at the B speeds. What does this data tell you in general about the motion of the car as it moved down and across the track?

5 Graphing the data

a. Create a speed vs. time graph for the data. Put *time AB* (s) on the x-axis and *speed at B* (cm/s) on the y-axis. Draw a best fit line.

b. How does the graph compare to your prediction? Explain.

6 Thinking about what you observed

a. Where is the car accelerating? Justify your answer with evidence from the experiment.

b. Where is the car moving at a constant speed? Justify your answer with evidence.

5C Gravity and Falling Objects

How does gravity affect the motion of falling objects?

Gravity causes objects to accelerate as they fall. An object is in *free fall* if it is moving under the influence of only gravity. For example, when you drop a ball, it is in free fall from the time it leaves your hand until it hits the ground. In this investigation, you will compare the motion of objects with different masses that are in free fall.

Materials

- Data Collector and 2 photogates
- Physics stand
- Plastic ball and steel marbles
- Piece of string
- Tape

1 Setting up the experiment

1. This investigation should be done with the physics stand on the floor and with the pole as perfectly vertical as you can make it. Tape the steel marble to a piece of string to make a plumb bob. Hold the string at the top of the physics stand and adjust the leveling feet until the string hangs down the center of the pole.
2. Attach photogate A near the top of the stand. The bottom of the "U" of the photogate should be against the pole.
3. Attach photogate B so it is 50 cm (10 holes) below photogate A.
4. Plug both photogates into the Data Collector.

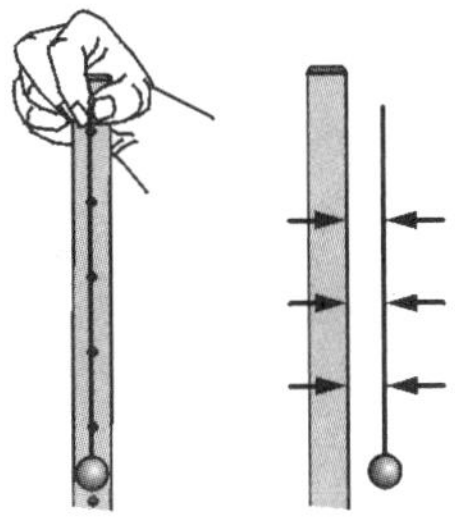

When the pole is vertical, the string is centered and parallel

2 Collecting data

1. Examine the photogates and find the two small rectangular openings where the infrared light beam is emitted and detected. Hold the steel marble (without the string) so it is above the top photogate, centered just above the openings.
2. Carefully drop the marble without giving it a push. Allow it to fall to the bottom of the stand where one group member should catch it. The marble should fall straight through the center of the two photogates. This may take some practice!
3. Record the time through photogate A (t_A), the time through photogate B (t_B), and the time from A to B (t_{AB}) in Table 1.
4. Repeat for a total of three trials with the steel marble.
5. Repeat with the plastic marble.

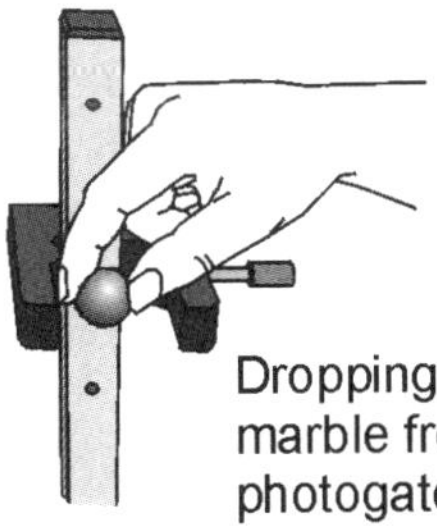
Dropping the marble from photogate A

Table 1: Photogate data

Marble	t_A (s)	t_B (s)	t_{AB} (s)
steel			
steel			
steel			
plastic			
plastic			
plastic			

3 Calculating acceleration

1. Calculate the speed of the marble through each photogate. The distance a marble moves when passing through the photogate is its diameter, 0.019 meter.
2. Use the two speeds and the time A to B to calculate the acceleration for each trial.

Table 2: Speed and acceleration data

Marble	speed through A (m/s)	speed through B (m/s)	acceleration (m/s^2)
steel			
steel			
steel			
plastic			
plastic			
plastic			

4 Analyzing the data

a. Calculate the average acceleration for your three trials for the steel marble. Use the average from each group in your class to find the class average for the steel marble's acceleration.

b. Calculate your average and the class average for the plastic marble's acceleration.

c. How do the accelerations compare? Are they about the same, or are they very different?

d. Which marble has the greater weight, and is therefore pulled on by a stronger force of gravity?

e. Suppose you wanted to move the steel and plastic marbles across the floor with the same acceleration. Which would require more force to accelerate?

f. Use your answers from questions d and e to explain why the accelerations of the two marbles in free fall compared as they did.

g. Why was it important to drop the marbles straight through the center of the photogate beam?

6C Collisions

Why do things bounce back when they collide?

Newton's third law tells us that when two objects collide, they exert equal and opposite forces on each other. However, the effect of the force is not always the same. What happens when you collide two Energy Cars that have unequal masses?

Materials

- Energy Car
- Data Collector and photogates
- Thick rubber band
- Electronic scale (or triple-beam balance)

1 Making a collision

Launching the Car

Rest your palm on the wood and pull the car against the screw with your finger on the tab nearest the far end of the car.

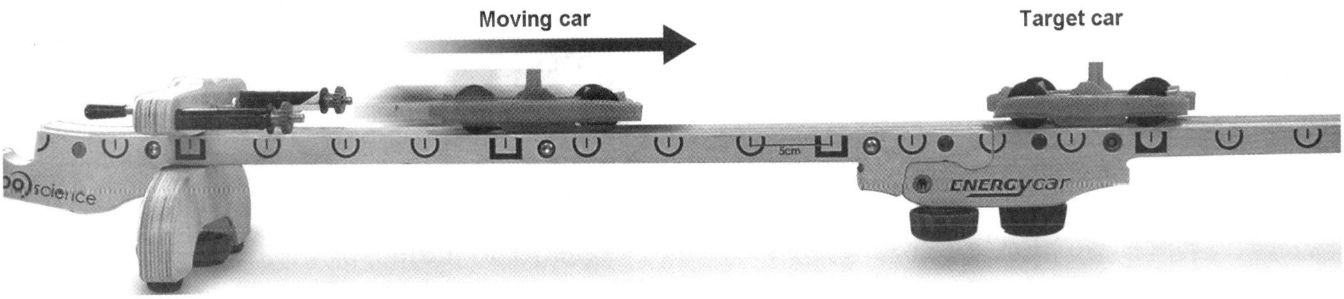

1. Set up the long straight track with a rubber band on one end and a clay ball on the other end. Use the bubble level to set the track level.
2. Place one steel marble in each car.
3. Wrap the thick rubber band around the moving car. Place both cars on the track so their noses are pointed toward the rubber band launcher.
4. Place the target car near the center of the track. Use the screw to launch the car using the same deflection of the rubber band each time. This means the same force is applied to each launch. You will use this car to create the collision.
5. To make a collision, release the moving car from the rubber band launcher. It will speed down the track, and hit the target car. This is an efficient way to produce collisions on the track.

a. Does the **moving** car bounce back after the collision?

b. Does the **moving** car keep going forward after the collision?

c. Does the **moving** car stop at the collision?

d. How does the target car behave?

2 Thinking about what you observed

a. Describe in words the motion of the two cars before and after the collision.

b. The target car must exert a force on the moving car to stop it. How strong is this force relative to the force the moving car exerts on the target car to get it moving? How could you use the photogates to provide evidence for your answer?

3 Gathering evidence

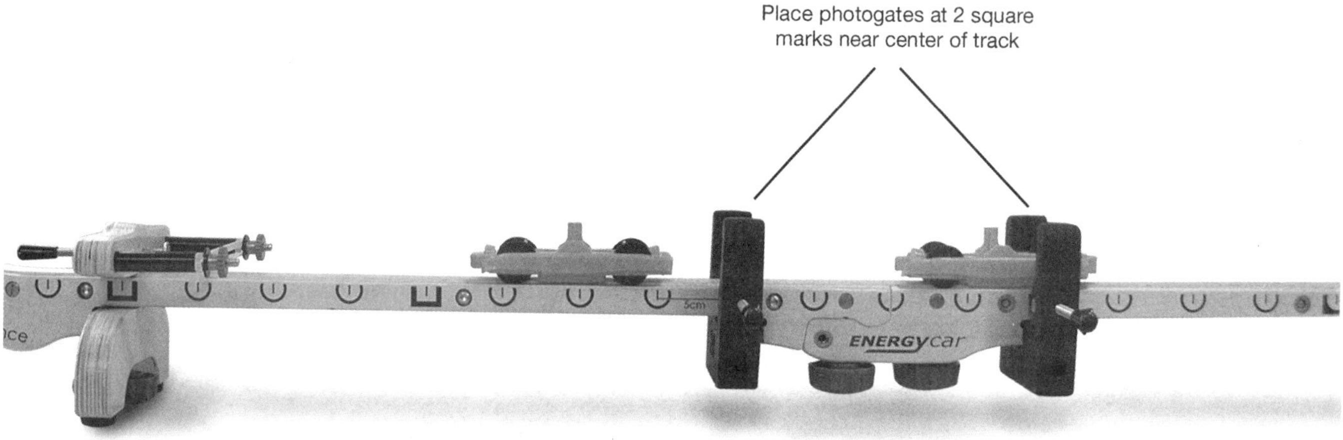

1. Try the experiment again, but now use two photogates to collect time data.
2. Place two photogates on the square marks near the middle of the track.
3. Put the target car on the track so it is near photogate beam B.
4. Release the moving car from the rubber band launcher as before and make a collision.
5. Repeat several times and record trial Times in Table 1.

Table 1: Collision Times

Collision Trial	Time for moving car to pass through A before collision (s)	Time for target car to pass through B after collision (s)
1		
2		
3		
4		
5		

a. Newton's third law tells us that when the moving car exerts a force on the target car, the target car exerts an equal and opposite force on the moving car. Does your data provide evidence for this? Explain.

b. You can compare times through A and B for each individual trial. How can using these times show there are equal and opposite forces at work?

4 Changing the masses

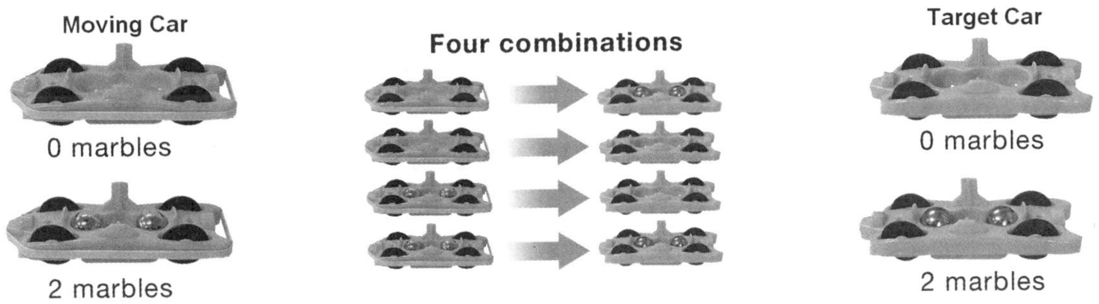

Try the experiment with the four combinations of mass shown above. You do not need to use photogates for this part of the investigation.

5 Applying what you have learned

a. Describe the motion of the two cars when the target car has more mass than the moving car.

b. Describe the motion of the two cars when the target car has less mass than the moving car.

c. Explain how your observations support the idea that there are action and reaction forces.

d. If the action and reaction forces are equal in strength, why does one car move at a different speed after the collision than the other car when the masses are unequal? Hint: the answer involves the second law.

7C Conservation of Energy

What limits how much a system may change?

A car launched up the hill at a given speed will never go higher than a certain point. A car rolling downhill will only reach a certain speed. Why? The answer is that nature keeps an exact balance of energy. Speed uses one form of energy and height uses another. This investigation explores the exchange of energy.

Materials

- Energy Car
- Physics stand
- Data Collector and one photogate
- Electronic scale (or triple-beam balance)
- Clay
- String
- Meter stick

1 Energy exchange from potential to kinetic

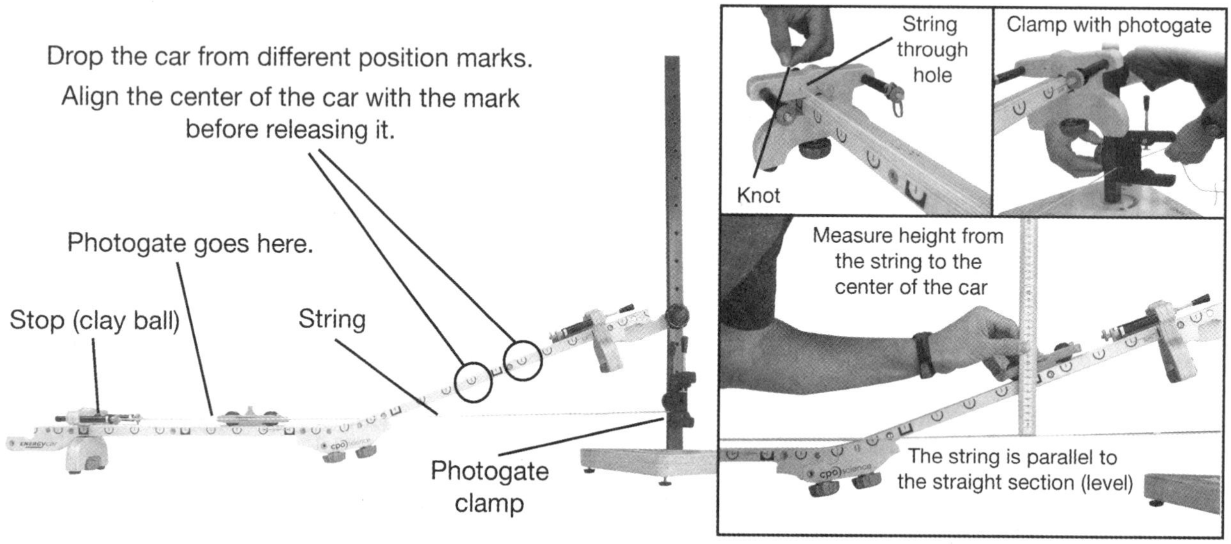

1. Set up the track with the steeper hill and a level section. Make the level section as level as you can. Attach a photogate at the bottom of the hill on the level section.

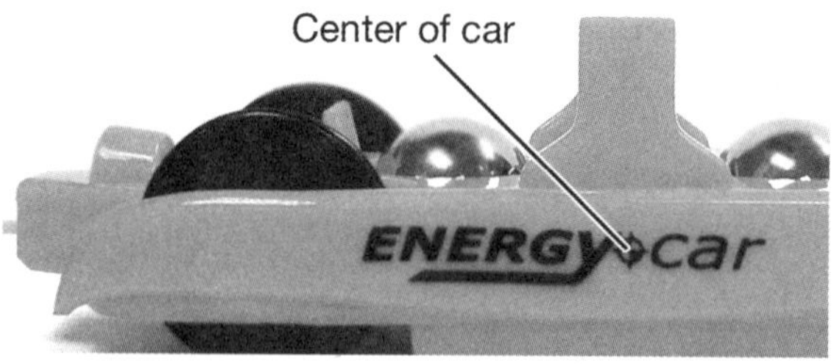

2. Thread a string though the hole in the lower stop and use a photogate to clamp the other end of the string to the stand. Adjust the string so it is parallel to the level section of the track.
3. Drop the car from each 5-cm mark on the hill and measure the speed with the photogate. Measure the height of the car from the string to the center of the car.

4. Measure the mass of the car and record it in the table.

Table 1: Height and speed data

Drop Height (m)	Mass of car (kg)	Photogate time (s)	Speed (m/s)

2 Thinking about what you observed

a. Graph the speed of the car vs. the height.

b. What does the graph tell you about the relationship between speed and drop height?

3 Analyzing the data

a. Use the formula for potential energy to fill in the second column of Table 2.

b. Use energy conservation to derive a formula for the speed of the car in terms of the energy it has at the start. (Hint: your formula should include only two variables, velocity and height.)

c. Use the formula you just derived to fill in the column for the predicted speed of the car.

d. Plot the curve for the predicted speed on the same graph as you made in part 2a above.

Table 2: Energy data and predicted speeds

Drop Height (m)	Potential energy (J)	Predicted speed (m/s)	Measured speed from Table 1 (m/s)

4 Thinking about what you observed

a. Explain the relationship between speed and height using the idea of energy conservation.

b. Explain any difference between the predicted and measured speeds. If there is a difference, what does it tell you about the energy of the car as it rolls along the track?

c. Let the car roll downhill, bounce off the rubber band and go back up hill again. Does it reach the same height as it was dropped from? Explain why or why not using the idea of energy conservation.

d. Challenge experiment. Use a rubber band to launch the car uphill so it goes through the photogate with the same speed as it had going down. You won't be able to get it precisely the same, but come as close as you can. If the speeds are the same, the car's kinetic energy is also the same. Does the car reach the same height on the hill that it was dropped from to get the same speed in part 1? Explain any difference using the idea of energy.

8C Pure Substance or Mixture?

How can observing the melting point identify a pure substance or a mixture?

Matter can be divided into two main categories: mixtures and pure substances. *Pure substances* are homogeneous throughout. They have the same chemical properties no matter where the sample is obtained or how large the sample is. *Mixtures* are combinations of two or more substances, with each substance retaining its chemical identity.

In this lab, you will obtain four test tubes containing unknown solids. The melting point, or the temperature at which the matter changes from a solid to a liquid, will be measured to determine if the matter is a mixture or a pure substance. Any given pure substance will always have the same melting point. Pure substances usually melt over a small temperature range while mixtures often melt over a very wide temperature range.

Materials

- 4 test tubes with unknowns
- Hot plate
- Stirring rod
- 1 250 mL beaker
- Water
- Safety goggles
- Lab apron
- Data Collector
- Temperature probe

WARNING — This lab contains chemicals that may be harmful if misused. Read cautions on individual containers carefully. Not to be used by children except under adult supervision.

1 Thinking about what you will do

Why do you think a pure substance will melt over a smaller range of temperatures than a mixture?

2 Doing the experiment

1. Attach the temperature probe to the Data Collector and set it to meter mode for the entire experiment. Record your sample label (A–D) in Table 1.
2. Add approximately 200 mL of cold water to a 250 mL beaker. Place the beaker on the hot plate, and turn the hot plate on to medium.
3. Place the first sample into the water bath and have one group member hold the temperature probe about halfway in the water.
4. Another group member should stir the contents of the test tube with a stirring rod as it heats.

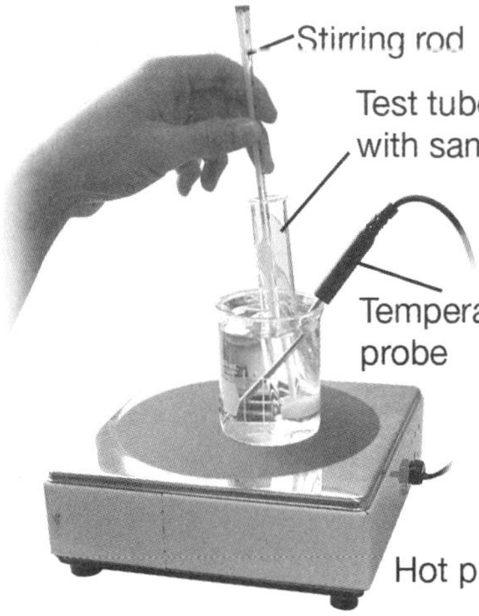

5. Watch the contents of the test tube carefully. At the moment the contents begin to melt, measure and record the temperature in Table 1.

6. Continue stirring the contents of the test tube and watching the contents until the entire sample is liquefied. Once the last solid particle melts, measure and record the temperature in Table 1.
7. Return your sample to your instructor.
8. Pour out the hot water in the beaker.
9. Repeat steps 2-8 for each of the remaining samples. Be sure to start with a fresh sample of cold water for the water bath.

Table 1: Melting data

Sample label	Temperature melting starts (°C)	Temperature melting ends (°C)
A		
B		
C		
D		

3 Analyzing the data

a. How would you know which test tubes contain pure substances and which contain mixtures?

b. If a sample contained 1 gram of pure substance A and another test tube contained 4 g of pure substance A, how would the melting points differ? How would the experiment differ?

c. Name two possible sources of error for this experiment. How would they affect your data?

4 Applying your knowledge

A white, waxy substance is heated in a test tube. Part of the substance melts almost immediately, and is poured off into a separate test tube after a minute of further heating, leaving a little more than half or the original sample behind.

a. Explain why you think the original sample was a pure substance or a mixture.

b. Is the melted portion that was poured off a pure substance or a mixture? What evidence do you base your answer on?

c. Can you be sure that the remaining portion left behind in the test tube is a pure substance or a mixture? How could you know for sure?

9C The Specific Heat of a Metal

How can you use specific heat to identify an unknown metal sample?

If you have ever walked barefoot on a concrete walkway or street during a hot and sunny day, you have felt its warmth on your feet. In fact, the thermal energy transferred to your feet may send you retreating to the grass or even a swimming pool. Why is the temperature of the concrete so different from the temperature of the grass or the swimming pool? Even though the sunlight shines on all three surfaces, it is easier to raise the temperature of concrete than the water in a swimming pool. A hot summer day may raise the temperature in a pool by only one or two degrees.

In this investigation, you will use a calorimeter, the specific heat of water, and the law of conservation of energy to determine the specific heat of a sample of copper and two unknown metal samples.

Materials

- Data Collector
- Temperature probe
- 3 Foam cups (with or without lids)
- Electronic scale (or triple-beam balance)
- Safety goggles
- Lab apron
- Metal samples (from Density cubes kit): Copper, 2 unknown samples
- Hot water
- Tongs

1 Doing the experiment

1. Place your sample of copper on the balance and record its mass in Table 1.
2. Make a calorimeter by nesting two foam cups.
3. Pour about 150 mL of room temperature water into the calorimeter. Calculate the mass of the water you add to the calorimeter (Density of water = 1g/mL). (Note: Make sure there is sufficient water in the calorimeter to submerge your metal samples COMPLETELY).
4. Fill a styrofoam cup 2/3 with hot water.
5. Place your sample of copper into the cup with hot water. Allow the copper sample to sit in the water for about a minute so it will get warmed by the hot water.
6. Use the temperature probe to measure the temperature of the hot water. Once it stabilizes, record this temperature as the initial temperature of the metal in Table 1.
7. Place the temperature probe in the calorimeter. When the temperature stabilizes, record it as the initial temperature of water. Leave the temperature probe in the calorimeter.
8. Use the tongs to remove the copper from the hot water and place the copper into the calorimeter. Try to transfer as little hot water as possible into the calorimeter when moving the copper. Go as quickly as possible but do not spill any hot water.
9. Once the temperature stabilizes, record the final temperature in Table 1.
10. Repeat steps 1–9 with each unknown metal provided by your teacher. You may be asked to conduct two trials with your unknown samples, if time allows.

Table 1: Mass and temperature data

	Copper	Unknown Metal # 1	Unknown Metal # 2
Mass of Metal (g)			
Mass of Water in Calorimeter (g)			
Initial Temperature of Water (°C)			
Initial Temperature of Metal (°C)			
Final Temperature of Mixture (°C)			

2 Processing the Data

a. Calculate the temperature change of the water.

b. Calculate the temperature change of the metal.

c. Calculate the heat gained by the water using the equation below:

HEAT EQUATION

$$E = mC_p (T_2 - T_1)$$

Where: E = thermal energy (Joules) lost or gained by the water in the calorimeter
m = original mass of the measured water in the calorimeter
C_p = specific heat of water (4.184 J/g °C)
$T_2 - T_1$ = change in temperature of the water (also referred to as ΔT)

d. The amount of energy released by the metal is equal to the energy absorbed by the water. Knowing the value of E for the metal, calculate the specific heat (C_p) of the metal. (Remember, since the energy is being released by the metal, change the sign of E from answer #3 to use for #4).

e. Identify your unknown metal(s) by comparing your calculated value of its specific heat to known specific heat values of common metals provided by your teacher.

f. Calculate the percent error for your unknown sample(s).

3 Thinking about what you observed

a. What did you determine was the identity of your unknown metal(s)?

b. Looking at your data/class data, were the experimental values too high or too low? Based on the experimental procedure, give an explanation for your observations.

c. How does the Law of Conservation of Energy allows us to make the calculations needed to determine the specific heat of the mystery metal?

d. The second unknown metal sample you tested is an alloy containing up to 80% of one of the other two metals you tested. The specific heat of this alloy should be almost the same as ***one*** of the other two metals. According to your results, which metal makes up 80% of the second unknown metal?

e. Water has a high specific heat. How does the fact that humans are largely made up of water help us regulate our body temperature?

10C Mountains and Earth's Crust

How can mountains float?

Solid objects float if they are less dense than the liquid in which they are placed. Objects sink if they are more dense than the liquid. You may have noticed that large ships are often made of steel. Ships made of steel float in water, even though steel is much more dense than water. Mountains, embedded in the continental crust, float in the upper mantle of the Earth. So how does a steel boat or a mountain float? The answer is in the concept of *apparent density*. You will soon discover how and why boats can be made of materials that are denser than water, and how a giant mountain can float.

Materials

- Displacement tank
- Balance
- Disposable cup
- Beakers
- Graduated cylinder
- GeoBox
- Buoyancy blocks
- Metric ruler

Unlike steel, concrete or clay, Earth's crust is not shaped like a boat. For the most part it is solid, without a hollowed out section like the boat you made, and yet it floats. Beneath Earth's crust is the mantle. The mantle is made of very hot rock.

1 Why Earth's crust floats

a. There are two types of Earth's crust; continental and oceanic. Oceanic crust is on average about 15% more dense than continental crust. How do you think the difference in density would affect how both types of crust float on the mantle?

b. What kinds of things would make part of Earth's crust thicker than another part?

c. Continental crust is usually much thicker than oceanic crust. Make a prediction; Would being thicker affect how continental crust floats on the mantle compared to oceanic crust? If so, how?

2 Testing your prediction

As mountains and mountain chains grow, the part of the crust they are on gets thicker and thicker. We are going to test your prediction of how getting thicker affects how the crust floats on the mantle by watching what happens when we build a mountain.

1. Fill the GeoBox with water up to the 5 cm line.
2. Measure the thickness of the largest mountain block and record your measurement in Table 3.
3. Place the largest mountain block in the water. This block represents Earth's continental crust.

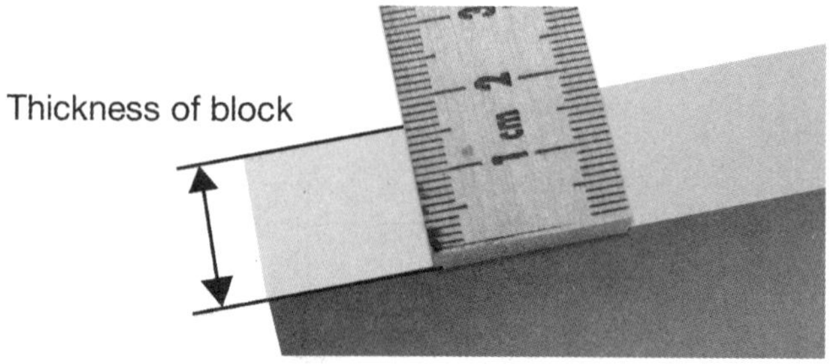

4. Measure how much of the block is above the surface of the water, and how much is below the surface of the water. Record your measurements in Table 3.

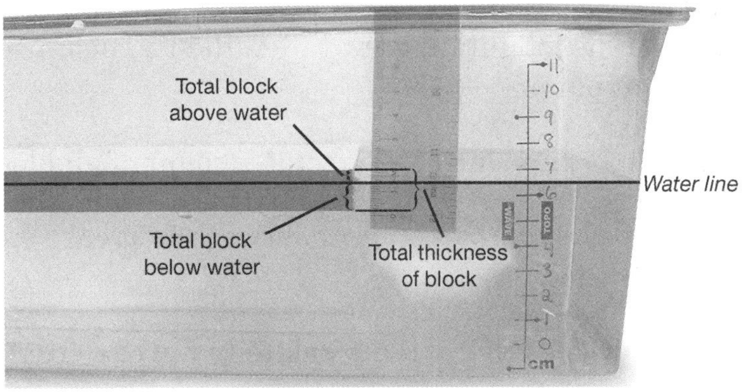

5. Measure the thickness of the second largest mountain block. Record your measurement in Table 3. Add this measurement to the thickness of the first block and record the total thickness of all blocks in Table 3.
6. Its time to start building your mountain. Place the second largest block on top of the block in the water. There are guide lines painted on each block. Align the corners of the smaller block with the guide lines on the larger block as you stack them up. This helps keep the entire stack balanced as it floats. When you add a block to the stack, align its corners with the guide lines painted on the block on top of the stack.
7. Measure the total amount of both blocks above the surface of the water, and the total amount of both blocks below the surface.

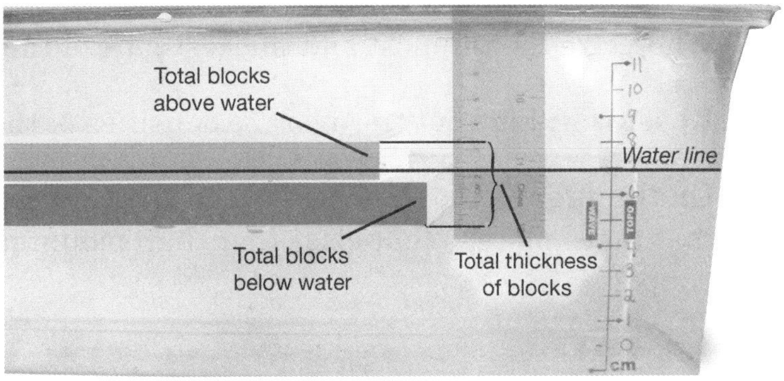

8. Repeat steps 5–7 until you've measured all the blocks and completed the mountain.

Table 1: Mountain blocks data

Block	Thickness (mm)	Total thickness of all blocks (mm)	Total amount above water (mm)	Total amount below water (mm)
first				
second				
third				
fourth				
fifth				

3 Thinking about what you observed

a. As the mountain grew taller, did the increase in height equal the thickness of each block you put on?

b. What happened to the total amount of all blocks underwater as the mountain grew taller?

c. Did the total amount underwater increase steadily as the mountain grew taller? Why or why not?

d. How does this activity relate to Earth's crust and mountain building?

e. Predict what you think happens to the crust as its mountains and mountain chains slowly erode over many years through weathering and erosion. Take your mountain apart one block at a time and see if your prediction is correct.

4 Exploring on your own

a. How thick on average are oceanic and continental crust?

b. Where are the thinnest and thickest part of the crust, and how thick are they?

c. **Research:** Find out the average densities of the mantle, oceanic and continental crust.

d. **Challenge:** As the thickness of the crust decreases, what do scientists call the process that takes place? What else other than mountains can cause this same process to happen?

11C Storms

How does Doppler radar work?

"The National Weather Service has just issued a severe thunderstorm warning for the following locations..."

Has a broadcaster ever interrupted your favorite television program with these words? In this investigation, you will learn about one of the important tools the National Weather Service uses to track storms: Doppler radar.

Materials

- Storm radar images (set of six)
- A metric ruler (with millimeter marks)

1 How can we use radar to track precipitation?

You may be familiar with the radar images used by weather forecasters on the news. Did you ever wonder how these images are produced?

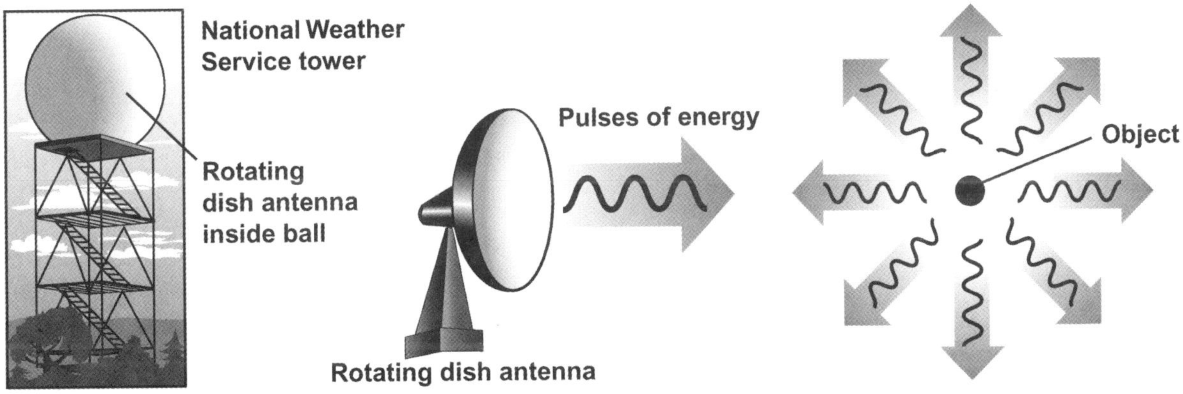

The National Weather Service has radar stations located across the United States. At each station there is a tower with what looks like a large soccer ball on top. Inside this protective ball is a 10-meter-diameter rotating dish antenna that sends out pulses of energy. If a pulse of energy strikes an object like a raindrop, the energy is scattered in all directions. A small part of the energy will bounce back to the antenna.

This reflected energy is then received by the radar station during its listening period. This process of sending a signal, listening for any returned signal, then sending the next signal, takes place very fast, up to around 1,300 times each second. Computers analyze the strength of the returned signal and the time it took to travel to the object and back. They use this information to create a color diagram showing the location, amount, and movement of precipitation falling over a region. In this reflectivity mode, four colors are typically used to show the rainfall rate per hour. For example, green may indicate traces of precipitation; yellow, a light rain; orange, heavy rainfall; and red, severe thunderstorms or hail. A color key is usually provided on the radar image.

a. Look at Radar Image Plate 1. Over which state is the most intense rain falling?

b. Name two states in which mixed precipitation is falling.

2 How can we use radar to detect a tornado?

Doppler radar, developed in 1988, has an advantage over earlier radar systems: It can detect the direction of air movement inside a storm. Doppler radar works by detecting the difference between signals bouncing off raindrops moving away from the antenna and signals bouncing off raindrops moving toward the antenna. In Doppler mode (also called velocity mode), the radar image is presented in two colors: green represents raindrops moving toward the antenna, and red represents raindrops moving away from the antenna. Doppler radar is very useful for locating rotating columns of air within a storm. One feature meteorologists pay close attention to is the appearance of a "J" hook-shape appearing near the back of the storm. This means that air is beginning to circulate in a pattern that often leads to the formation of a tornado.

Daphne Zaras NSSL/NOAA photo

Central Oklahoma tornado, May 3, 1999

a. Look at Radar Image Plate 2. Near which interstate highway intersection is a tornado possibly beginning to form?

b. The National Weather Service issues tornado watch and tornado warning bulletins when certain atmospheric conditions are observed. Use the Internet to find out what each notice means. How does Doppler radar help meteorologists issue these bulletins?

3 Recognizing insects, birds, or airborne debris on a radar image

Of course, the pulses of energy sent by the antenna can bounce off things other than raindrops or other forms of precipitation. Interference from objects on the ground can be easily recognized because those objects do not move. However, in the spring and fall, swarms of insects or flocks of migrating birds can show up on radar images. Meteorologists practice looking at images and recognizing the characteristic patterns these creatures create, so that they do not mistake the birds and bugs for rain.

In some cases, the interference is actually important to meteorologists. In Radar Image Plate 3, locate the white spot in the center of a very intense storm. The color white represents an area that is reflecting back a large number of energy pulses—too many pulses to be bouncing off raindrops. These energy pulses are bouncing back from debris in the air thrown there by a tornado. This white area helps meteorologists locate places where the tornado is causing damage.

a. Near which city is the airborne debris located?

b. What safety precautions does the National Weather Service recommend when conditions like these are observed?

4 How can we use radar to track a hurricane?

Satellite image of Hurricane Georges as it hit Puerto Rico on September 21, 1998.

Hurricane Georges, a category 3 hurricane, struck the island of Puerto Rico on Monday, Sept. 21, 1998. The next six radar images are from the National Weather Service's San Juan headquarters. The radar station is located in the Guavate Forest just south of the city of Cayey. All of the hurricane images are in reflectivity mode, depicting the intensity of the rainfall through a series of colors. The color key to the right of each image tells you that blue indicates light rainfall; green moderate rainfall; and yellow, orange, red, purple, and white indicate increasingly severe rainfall, often accompanied by thunderstorms and hail.

a. The hurricane's eye is the low-pressure center around which the storm rotates. Inside the eye, winds are calm and skies are blue. Look at Hurricane Image 1. Where is the eye located?

Airplane entering the eye of Hurricane Georges to take storm measurements. The sky above is clear and sunlight can be seen reflecting off clouds.

b. In the afternoon, Georges gathered strength because of the combination of the Caribbean Sea's warmer waters, the demise of some earlier wind shear, and the fact that as the storm moved over open sea between St. Croix and Puerto Rico, the wind circulating around the eye had no resistance from land objects. What evidence can you see in Hurricane Image 2 that the storm is gaining strength?

c. The National Weather Service reported that Georges made landfall on Puerto Rico at 7 p.m. local time (23:00 GMT). Landfall is defined as the time when the center of the hurricane's eye reaches land. Use the scale printed on Hurricane Image 3 to determine the diameter of the eye when it made landfall.

d. As the hurricane moved across Puerto Rico, intense rain and a possible tornado were spawned as the eye wall's heavy thunderstorms interacted with the mountainous terrain. Notice the strong bands of precipitation found in Hurricane Image 4. Why do you suppose it is harder to see the eye in this image?

USGS photo

Hurricane George's high winds tore the roof and exterior walls from this home in Puerto Rico.

e. Just after midnight on the morning of Sept. 22, 1998 (04:01 GMT), the eye of Hurricane Georges began to move off the west coast of Puerto Rico. Using the time stamps on Hurricane Image 3 and Hurricane Image 5, calculate the time it took for Georges to make its way across Puerto Rico.

Hurricane Georges' travel time (round to the nearest hour): __________

Then, using a ruler and the scale provided on the radar images, calculate the distance from the center of the eye in the third image to the center of the eye in the fifth image.

Hurricane Georges' travel distance (round to the nearest km): __________

Divide the distance traveled by the time taken in order to calculate the speed of the hurricane as it moved across the island.

Hurricane Georges' speed (round to the nearest km/hour): __________

f. By 2:51 a.m. local time (06:51 GMT) the eye of Hurricane Georges was again over water. Use information from Hurricane Image 5 and Hurricane Image 6 to calculate the speed of the storm as it moved off of Puerto Rico. Does the storm appear to be intensifying or dissipating? Explain your answer.

Distance __________ km ÷ **time** __________ hours = **speed** __________ km/hour.

12C Activity Series of Metals

How reactive are different metals?

An activity series is a list of elements arranged from most reactive to least reactive. When you place a solid piece of metal in contact with a solution, there may or may not be a reaction. In order for a reaction to proceed, the solid metal that is touching the solution must be more reactive than the metal that is part of the ionic compound (solute) dissolved in the water. In this investigation, you will examine several metals and rank them from most to least active based on how many reactions take place for each metal.

Materials

- Well plate (5x5) or overhead (lead acetate sheet)
- Copper solid
- Magnesium solid
- Zinc solid
- Lead solid
- Forceps
- Eyedroppers
- Goggles
- Apron

- $Mg(NO_3)_2$ solution
- $Zn(NO_3)_2$ solution
- $Pb(NO_3)_2$ solution
- $AgNO_3$ solution
- $Cu(NO_3)_2$ solution

1 Stop and think

Based on the metals given in the materials list, which do you think will be the most reactive?

2 Doing the experiment

1. You will need a 5x5 well plate. It will be arranged similar to the layout of Table 1. With an eyedropper, place 7 drops of each solution into each of four wells. For example, place 7 drops of $Cu(NO_3)_2$ solution in each of four wells across the top of the well plate
2. If you are using a clear plastic well plate, place a sheet of white paper underneath the it. It will allow you to see evidence of a potential chemical reaction more easily.
3. Place one sample of each metal into each solution. For example, place one sample of copper in the first well of the row containing the $Cu(NO_3)_2$ solution. Then place a sample of copper into the first well of the row containing $Zn(NO_3)_2$, and so on until the first well in each row has a sample of copper in it. Repeat this process with one sample of zinc placed in each of the second wells in each row, lead in the third, and magnesium in the fourth.

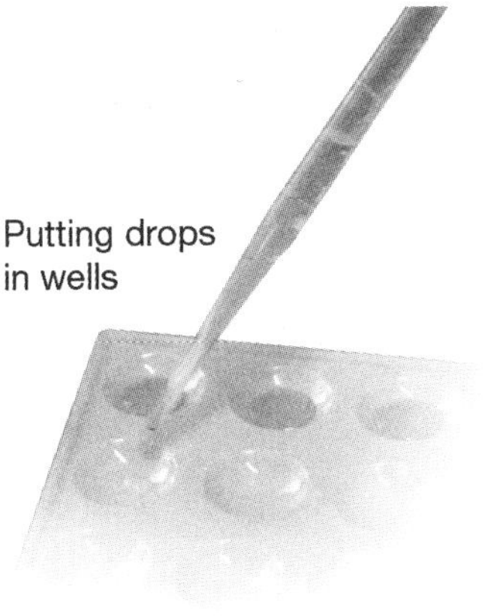

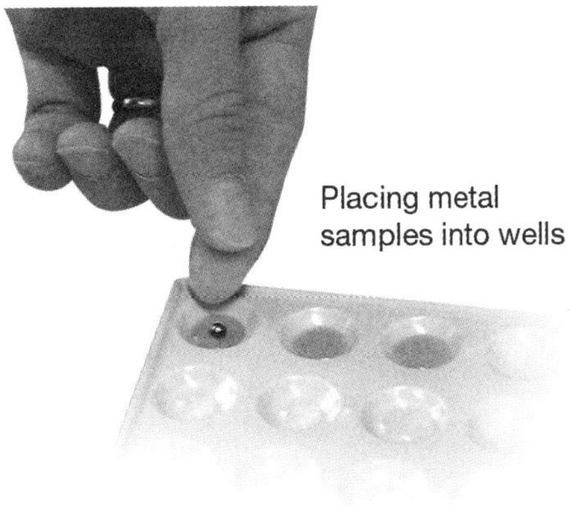

4. After 15 minutes, use forceps to examine each metal. Be sure to turn the metal over and look at the underside for evidence of a chemical reaction.
5. Place an "X" in any square of the data table where a reaction occurs.
6. Clean up according to your instructor's directions and be sure to wash your hands.
7. Add up the number of "X's" for each metal and write the total in the second data table.

Table 1: Reactivity data

	Cu	Zn	Pb	Mg
$Cu(NO_3)_2$				
$Zn(NO_3)_2$				
$Pb(NO_3)_2$				
$Mg(NO_3)_2$				
$AgNO_3$				

Metal	Cu	Zn	Pb	Mg
Number of Reactions				

3 Thinking about what you observed

a. What evidence did you observe to indicate a chemical reaction had taken place in this experiment?

b. Rank the solid metals used in the experiment from most reactive to least reactive.

c. Even though silver solid was not used in this experiment, make a judgement as to where it would rank compared to the other metals. Explain.

d. Compare the activity series you just constructed with one provided by your teacher. How do they compare?

13C Carbon and its Chemistry

What are some common molecules that contain carbon?

Living organisms are made up of a great variety of molecules, but the number of different elements involved is very small. Organic chemistry is the science of molecules that contain carbon and these are the ones most important to living organisms. This investigation will introduce you to some small organic molecules.

Materials

- Periodic Table Tiles
- Reference book of organic compounds

1 The chemistry of carbon

Carbon is the central element in the chemistry of living things. This is because carbon can make complex molecules. Each carbon atom can make four bonds because carbon has four electrons in the outer energy level, which can hold a total of eight. Carbon can also form double and triple bonds by sharing two or three electrons with a single atom. Because of carbon's flexible bonding ability, many molecular structures are possible.

Straight chains

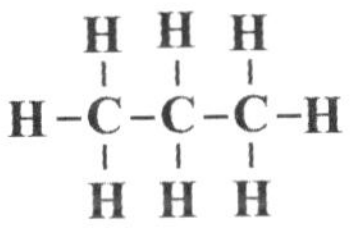

Propane
C_3H_8

Branched chains

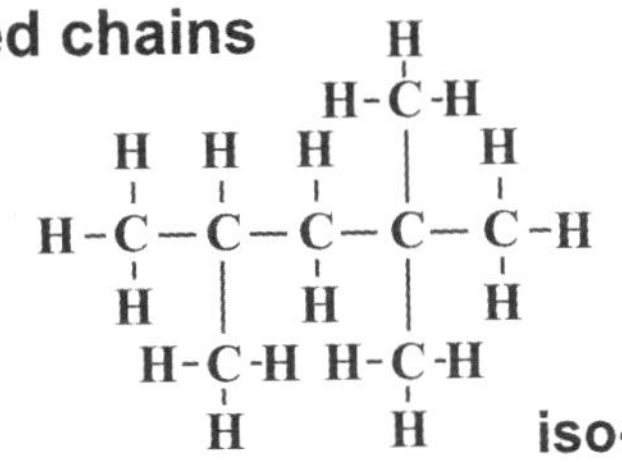

iso-octane
C_8H_{18}

Rings

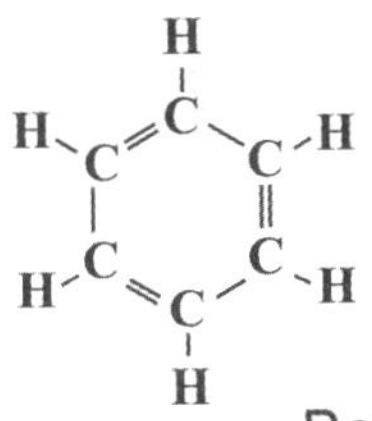

Benzene
C_6H_6

2 Single carbon molecules

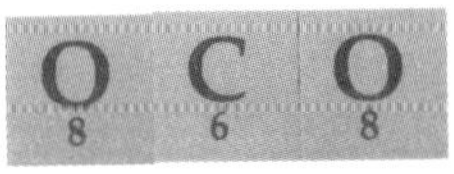

Carbon dioxide
CO_2

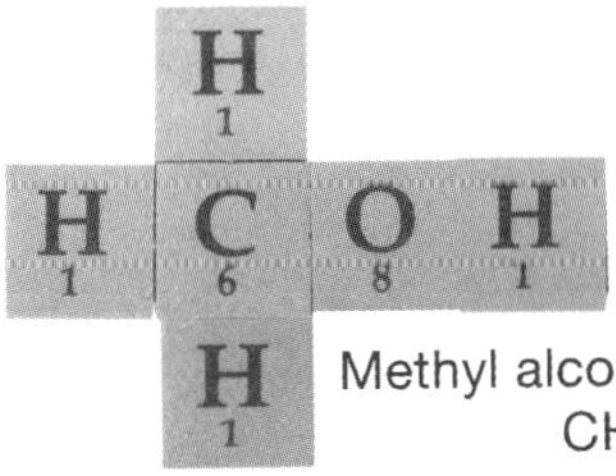

Methyl alcohol (methanol)
CH_3OH

Methane
CH_4

1. Build the three carbon molecules above using the periodic table tiles.
2. Build a model of carbonic acid: H_2CO_3.

3 Stop and think

a. Research and describe at least one use of methyl alcohol.

b. Research and describe the use and production of methane.

4 Molecules with two carbon atoms

Once there are two carbon atoms, the structures get more complicated. The carbon atoms may share a single, double, or triple bond between them.

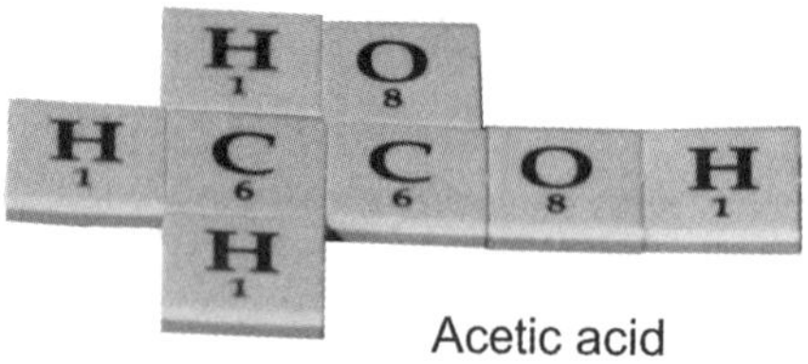

Acetic acid
CH_3COOH

Ethane
C_2H_6

1. Build the 2-carbon molecules above using the periodic table tiles.
2. Build a model of ethyl alcohol (ethanol): C_2H_5OH.
   (Hint: Each carbon atom makes 4 bonds and the oxygen atom makes two bonds)
3. Build a model of acetylene: C_2H_2.
   (Hint: there is a triple bond between the two carbon atoms)

5 Stop and think

a. Research and describe where acetic acid is found.

b. Research and describe at least one use of ethyl alcohol.

6 Biological molecules

Living organisms are constructed mainly of proteins, which are very large carbon-based molecules, such as hemoglobin (right). A single protein may contain 10,000 or more atoms. Proteins themselves are constructed of simpler units called amino acids. For example, the hemoglobin molecule contains 584 amino acids.

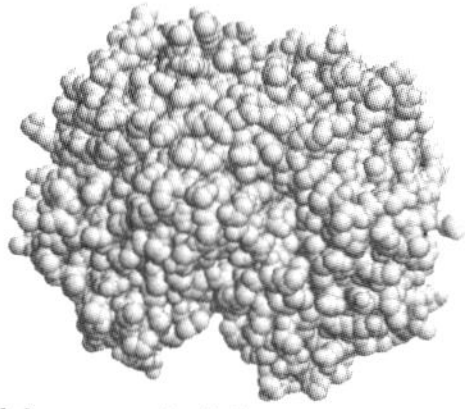

Hemoglobin molecule

Use the periodic table tiles to build models of the three smallest amino acids.

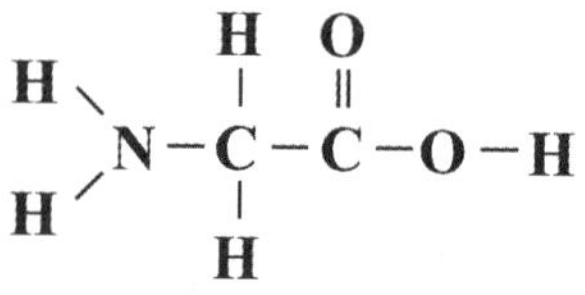

Glycine

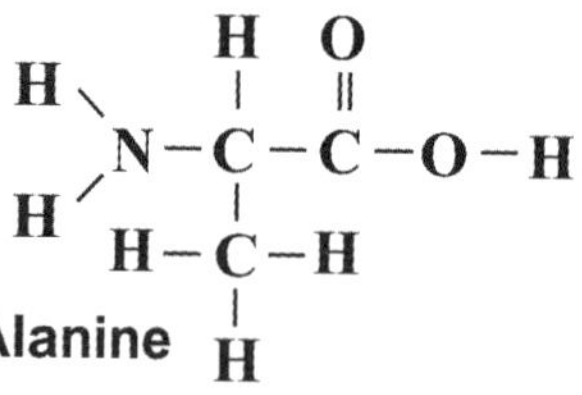

Alanine

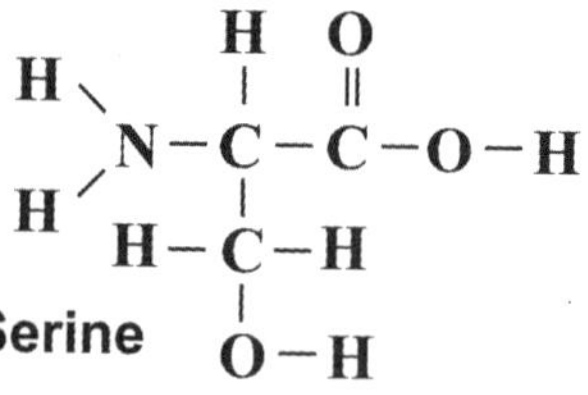

Serine

7 Stop and think

a. What is similar about all three amino acids?

b. Is the chemical in the diagram on the right an amino acid? Explain why you think your answer is correct.

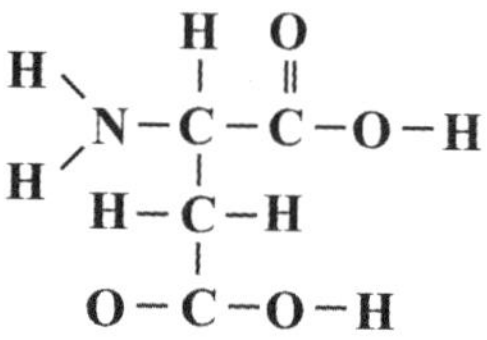

Is this an amino acid?

14C Types of Chemical Reactions

Can we measure the heat released/energy absorbed by instant hot and cold packs?

All chemical reactions are either exothermic (release energy) or endothermic (absorb energy). However, some physical processes such as dissolution (dissolving) can also release/absorb energy. This is the basis for commercially available instant hot packs and cold packs. Most hot and cold packs work by breaking a membrane that separates a solid and water. Once the membrane is broken, the solid dissolves in the water. Depending on the nature of the compound, heat is either released (hot pack) or absorbed (cold pack) during the process. In this investigation, you will examine the temperature changes of hot and cold packs and determine the energy absorbed or released during the chemical reactions that take place.

Materials

- 2 foam cups
- Temperature probe
- Data Collector
- Safety goggles
- Lab apron
- Water
- Calcium chloride (from an instant hot pack or ice melt)
- Ammonium nitrate (from an instant cold pack)
- Electronic scale (or triple-beam balance)
- 100 mL graduated cylinder

WARNING — This lab contains chemicals that may be harmful if misused. Read cautions on individual containers carefully. Not to be used by children except under adult supervision.

1 Stop and think

Calcium chloride is used to melt ice on winter roads. Would that process be exothermic or endothermic? Why?

2 Doing the experiment

Part A: Hot Pack - Record all measurements in Table 1.

1. Use a graduated cylinder to measure 50 mL of water. Record the mass of the water in Table 1 (HINT: 1 mL of water = 1g).
2. Pour the water into 2 nested foam cups and place the temperature probe into the water.
3. Use a balance to measure out about 10 g of calcium chloride ($CaCl_2$). Record the mass of the $CaCl_2$ in Table 1.
4. Select new experiment on the Data Collector and press Go. Wait for the temperature of the water to stabilize and record that temperature in Table 1 (initial temperature).
5. Pour the $CaCl_2$ into the water and stir with the temperature probe.
6. Watch the graph and the temperature reading as the reaction proceeds.
7. Watch for the highest temperature reading before the temperature starts to go down again. Record that temperature in Table 1 (final temperature).
8. Pour the solution down the drain and then rinse and dry your foam cups.

Table 1: Hot pack mass and temperature data

Mass of water (g)	Mass of $CaCl_2$ (g)	Initial temp. (°C)	Final temp. (°C)

Part B: Cold Pack

1. Use a graduated cylinder to measure 50 mL of water. Record the mass of the water in Table 2 (HINT: 1 mL of water = 1g).
2. Pour the water into 2 nested foam cups and place the temperature probe into the water.
3. Use a balance to measure out about 10 g of ammonium nitrate (NH_4NO_3). Record the mass of the NH_4NO_3 in Table 2.
4. Select new experiment on the Data Collector and press Go. Wait for the temperature of the water to stabilize and record that temperature in Table 2 (initial temperature).
5. Pour the NH_4NO_3 into the water and stir with the temperature probe.
6. Watch the graph and the temperature reading as the reaction proceeds.
7. Watch for the lowest temperature reading. Record that temperature in Table 2 (final temperature).
8. Pour the solution down the drain and then rinse and dry your foam cups.

Table 2: Cold pack mass and temperature data

Mass of water (g)	Mass of NH_4NO_3 (g)	Initial temp. (°C)	Final temp. (°C)

3 Analyzing the Data

a. Calculate the temperature change of the $CaCl_2$ and water reaction.

b. Calculate the heat gained by the solution using the equation below: Assume the specific heat of the solution is equal to the specific heat of water. The specific heat of water is 4.184 J/g °C.

HEAT EQUATION

$$E = mC_p(T_2 - T_1)$$

Where: E = heat lost or gained
m = mass of solution in the calorimeter
C_p = specific heat of substance
$T_2 - T_1$ = change in temperature (also referred to as ΔT)

c. Calculate the amount of energy (in Joules) released per gram of $CaCl_2$. Remember: Q is the opposite sign of the value you calculate.

d. Calculate the temperature change of the NH_4NO_3 reaction.

e. Calculate the heat lost by the solution using the heat equation: Assume the specific heat of the solution is equal to the specific heat of water.
Where: E = heat lost or gained
m = mass of solution in the calorimeter
C_p = specific heat of substance
$T_2 - T_1$ = change in temperature (also referred to as ΔT)

f. Calculate the amount of energy (in Joules) absorbed per gram of NH_4NO_3. Remember: Q is the opposite sign of the value you calculate.

4 Thinking about what you observed

a. Look at the reactions below. Place the energy value on the appropriate side of the two equations.

$$CaCl_2(s) \rightarrow Ca^{2+}(aq) + 2Cl^-(aq)$$

$$NH_4NO_3(s) \rightarrow NH_4^+(aq) + NO_3^-(aq)$$

b. When $CaCl_2$ dissolves in water the true value for $E = -747$ J/g. When NH_4NO_3 dissolves in water the true value for $E = 326$ J/g. Compare your experimental values by calculating the percent error for each reaction.

c. Describe why your skin feels cool when a cold pack is applied and warm when a hot pack is applied.

15C Natural Resources

How fast are we using nonrenewable resources?

As the world population and demand for energy increases, having a variety of energy sources becomes more and more important. We consume energy quickly, so we need sources of energy that are long-lasting, efficient, and as clean as possible.

In this investigation, you will model the depletion of resources over time at a constant rate of use. Then you will model the depletion of resources when the rate of use increases. Next, you will make a physical model of an oil seep. This will help you understand how we find oil in the ocean.

Materials

- 1 open container per two students
- 2 types of dried beans or nuts—100 beans per pair (90% one color/type, 10% another color/type)
- Blindfold
- Large clear cup or glass
- Small mixing bowl
- 2 mL cooking oil
- 10 mL sand
- 30 mL soil
- 1 stick of modeling clay
- Water

WARNING — This lab contains chemicals that may be harmful if misused. Read cautions on individual containers carefully. Not to be used by children except under adult supervision.

1 Stop and think

a. What is the difference between renewable and nonrenewable resources?

b. What are some examples of each of these resources?

c. What type of resource do you expect will be depleted first?

2 Activity A

In this activity, you will learn how nonrenewable resources get depleted over time, especially as rate of use increases.

1. Fill your container with 100 beans (90 of one color type, 10 of another color/type). The 90 beans represent the fact that the United States currently uses nonrenewable fossil fuels for 90% of our energy needs. The other 10% of our energy needs are met with renewable resources.
2. One student in each pair will put on the blindfold. This student will choose beans from the container.
3. For the first trial, the situation is as follows: The population is NOT growing and the demand for resources stays exactly the same from one year to the next. The blindfolded student will randomly pick 10 beans out of the jar. Any "renewable beans" can be put back in the jar. After replacing these beans, count how many beans remain

in the jar. Record this information in the "Year 1" column of Table 1. How many years do you think it will take to deplete the nonrenewable beans?

4. Repeat the process for year two. Continue until only renewable beans are left. Record all your data in Table 1. You may or may not use all of the columns available. You might even have to add more columns. Calculate the percentage left after each drawing. How many years did it take to run out of nonrenewable resources?

Table 1: Renewable and nonrenewable energy use

Consumption level	Year 1	2	3	4	5	6	7	8	9	10	11	12	13	14	15
Remove 10 beans each year (constant use)	10	10	10	10	10	10	10	10	10	10	10	10	10	10	10
# of beans remaining in container															
% renewable															
% nonrenewable															

5. Now, remove the blindfold. Put it on the other student. Place all the beans back in the container. This time, you will be modeling a situation where the population is increasing and so is the demand for energy.
6. Predict again how many years it will take to deplete the resources.
7. Proceed in the same way, except follow the table to see how many beans to pull out each year.

Table 2: Energy use increasing demand

Consumption level	Year 1	2	3	4	5	6	7	8	9	10	11	12	13	14	15
Remove 5 more beans each year (increasing demand)	10	15	20	25	30	35	40	45	50	55	60	65	70	75	80
# of beans remaining in container															
% renewable															
% nonrenewable															

8. Remove the blindfold. How many years did it actually take to run out of nonrenewable resources?
9. Make two graphs—one for each table. On the x-axis put the year. On the y-axis, put the percent renewable and nonrenewable remaining.
10. Answer the questions **a** through **d** in Part 4.

3 Activity B

In this activity, you will make a model of an oil seep, like those in the Santa Barbara Channel off the coast of California. There, cracks in rock layers of the ocean floor allow oil and gas to ooze or seep through the water to the surface. The gas dissipates in the air, but the oil floats on the ocean surface. Oil and gas are both fossil fuels.

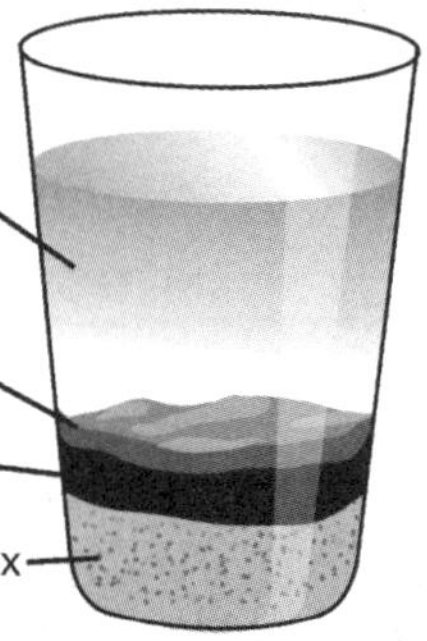

1. Pour the sand into the bottom of the glass or cup.
2. Pour the oil into the sand. Add 1 mL of water.
3. Now, get some soil and mix it with water so that it is very wet. Pack it tightly into the glass on top of the sand/oil mixture.
4. Take the clay and flatten it into a circle that can fit into the glass. Stick it into the glass, making a thin seal over the mud mixture.
5. Fill the rest of the glass with water.
6. Time how long it takes the oil to seep through the layers to the top of the water.
7. Now, answer questions **e** and **f** in Part 4.

4 Thinking about what you observed

a. Which graph in Activity A is more realistic? Why?

b. Which type of resource lasts for a longer time?

c. Compare your predictions for the number of years it would take to deplete the resources to the actual number of years that you measured. How close were you? How did you come to your prediction?

d. What did you learn about the rate of depletion of the two types of resources?

e. How long did it take for the oil to seep to the top of the water? Do you think it would take longer if there was more water on top of the clay? Why?

f. In the future, what should the United States try to do about its energy consumption? Why?

5 Exploring on your own

1. Try the oil seep experiment with a taller glass so that you can add more water. Also try it with salt water. Did it take more or less time for the oil to seep to the surface?
2. Research some ways that oil and natural gas are removed from the ocean floor.
3. How is coal created?

16C Electric Circuits

What are the different types of circuits?

A simple electric circuit contains one electrical device, a battery, and a switch. Flashlights use this type of circuit. However, most electrical systems, such as a stereo, contain many electrical devices connected together in multiple circuits. This investigation introduces two ways to connect multiple devices in a circuit.

Materials

- Electric Circuits kit
- Multimeter
- 2 "D" cell batteries

1 Series circuits

Build this circuit

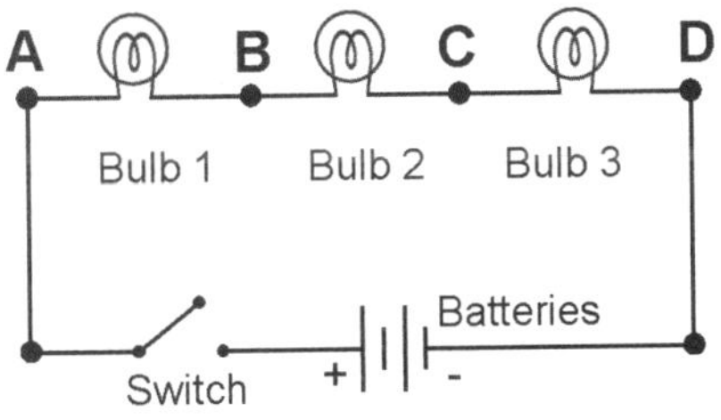

1. Using two batteries, build the simple circuit with three light bulbs and a switch as shown above.
2. Set the meter to DC volts. Close the switch and measure the voltage across the different places by touching the meter's leads to the bulbs' terminals. Record the voltages in Table 1.

Table 1: Voltage measurements (volts)

Between A and B (V)	Between B and C (V)	Between C and D (V)	Between A and D (V)

2 Thinking about what you measured

a. What relationships do you see among the voltage measurements in Table 1?

b. What do the voltage measurements tell you about the flow of energy in the circuit?

3 The current in series circuits

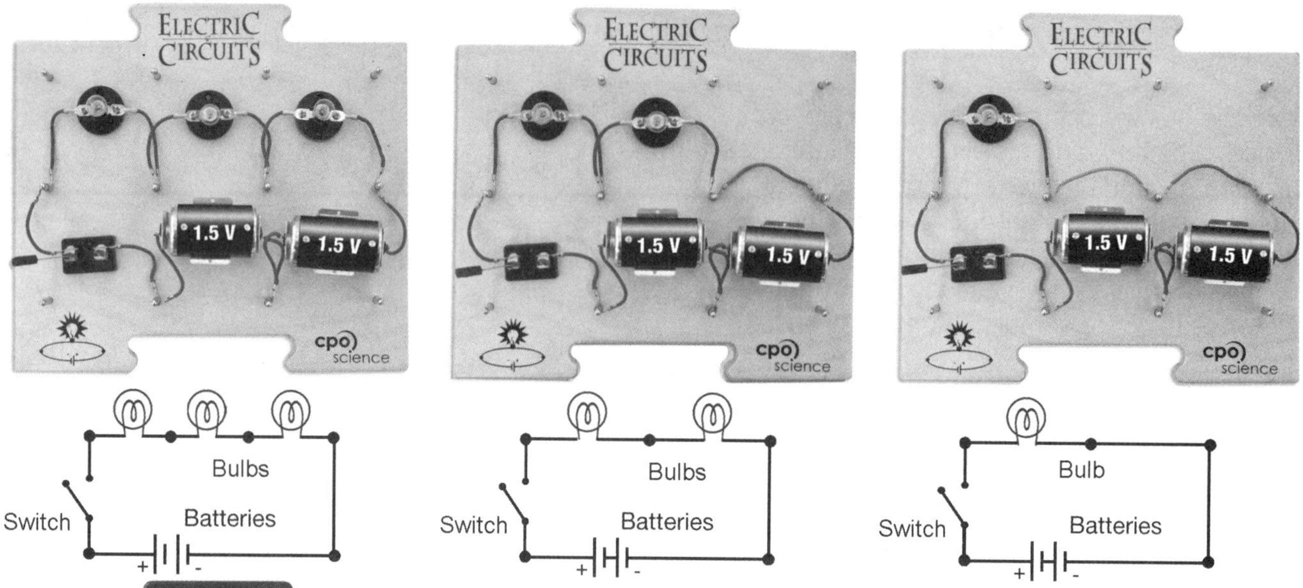

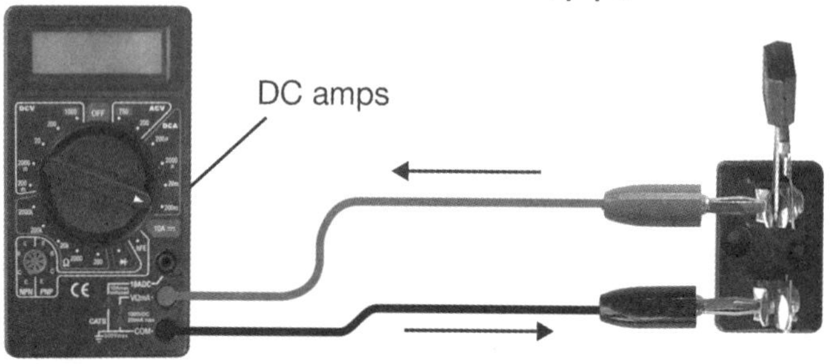

To measure current, touch the leads to the terminals of the switch with the switch open.

1. Set the meter to DC amps. Measure the current by opening the switch and touching the leads of the meter to the terminals of the switch in the three bulb circuit. Record your measurements in Table 2.
2. Remove one bulb and replace it with a wire. Measure and record the current for the two-bulb circuit.
3. Remove a second bulb and replace it with a wire. Measure and record the current again for the one-bulb circuit.

Table 2: Current Measurements (amps)

Three bulbs (A)	Two bulbs (A)	One bulb (A)

4 Thinking about what you observed

a. What happens to the current in the circuit as the number of bulbs is reduced? Explain why this occurs using Ohm's law and the concept of resistance.

b. What happens to the other two bulbs when one bulb is removed from the three-bulb circuit? Try it and explain why the circuit behaves as it does.

5 Short circuits

A short circuit is an easy (but dangerous) shortcut that current can travel through to avoid one or more of the electrical components in the circuit.

Circuit diagram

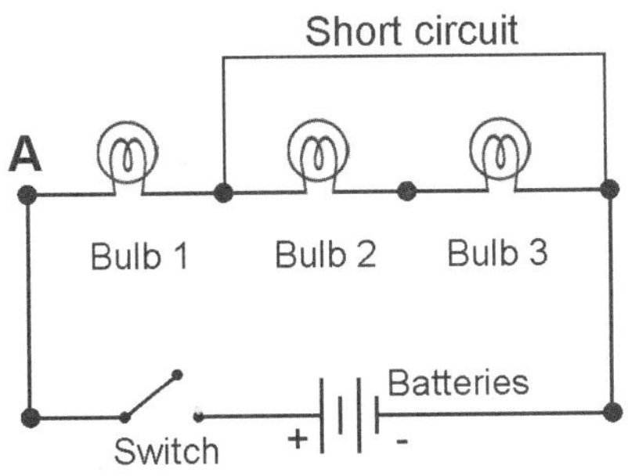

Add this wire to make the short circuit

1. Rebuild your three-bulb circuit with the switch open.
2. Check the current and observe which bulbs light and how bright they are.
3. Add a section of wire that bridges the last two bulbs in the circuit. This wire is the “short circuit.”
4. Complete the circuit (with the switch open) using the meter to measure the current. Observe which bulbs light and how bright they are.

6 Thinking about what you observed

Table 3: Short Circuit Current Measurements (amps)

Three bulbs in series (A)	Three bulbs with two short circuits (A)

a. Compare the current in the three-bulb circuit with the current when two bulbs are bypassed by a short circuit. Which is greater? Use Ohm’s law and the concept of resistance to explain why.

b. How does the current in the “short circuit” version compare with the current you measured in a one-bulb circuit? Explain why this should be true.

c. How does the resistance of a wire compare to the resistance of a bulb? Measure the resistances to test your answer. NOTE: Most meters cannot measure very low resistance and display “0.00” when the resistance is lower than 0.01 Ω.

d. Why would a short circuit be dangerous? Discuss (as a class) the consequences of very large currents in wires of different sizes.

7 Parallel circuits

Build this circuit

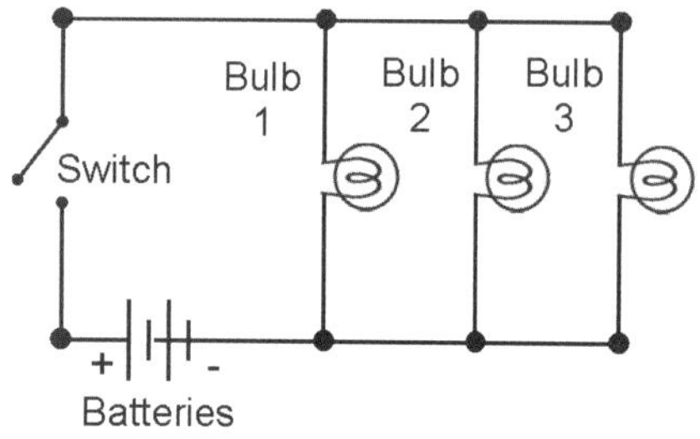

1. Build a circuit with two batteries, a switch, and three bulbs as shown in the diagram.
2. Close the switch and measure the voltage across the battery. All three bulbs are lit.
3. Measure the voltage across each bulb by touching the leads of the meter to the terminals of each bulb separately.
4. Set the meter to DC amps. Measure the total current in the circuit by opening the switch and touching the leads of the meter to the terminals of the switch.

Table 4: Voltage and current in a parallel circuit

	Total circuit	Bulb 1	Bulb 2	Bulb 3
Voltage (*V*)				
Current (*A*)				

8 Thinking about what you observed

a. Compare the brightness of the bulbs in the parallel circuit with the brightness in the series circuit.

b. Compare the total current in the single-bulb circuit, the three-bulb series circuit, and the three-bulb parallel circuit. Propose a relationship between the currents that agrees with the brightness of the bulbs.

c. Do the other two bulbs continue to light when the third bulb is removed from the parallel circuit? Try it. How does this differ from what happened with the series circuit?

d. Do you think the electrical outlets in your home are connected in a series or parallel circuit? Give two reasons why one type of circuit has an advantage over the other for connecting outlets.

17C Electromagnetic Forces

How does an electric motor work?

Electric motors are found in many household devices, such as a hair dryer, blender, drill, and fan. In this investigation, you will build a simple electric motor and see how it works. The concepts you learn with the simple motor also apply to other electric motors.

Materials

- 1 permanent magnet (from kit)
- 1 "D" cell battery
- Large rubber band
- Metric ruler
- 1/2 stick of modeling clay
- Sand paper
- Varnished magnet wire
- Paper clips
- Data Collector (optional)

1 Making the base

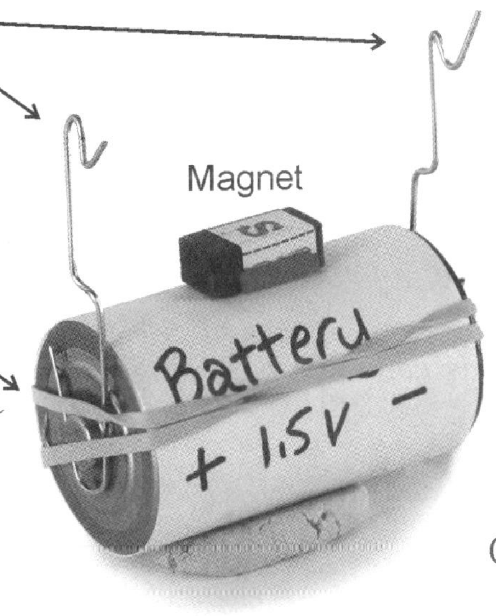

1. Bend the two paper clips so they look like the photo.
2. Fasten them with rubber bands so they contact the positive and negative terminals of the battery.
3. Break off a small lump of clay from the 1/2 stick. Set the battery on the small clay lump so it stays in one place without rolling around.
4. Use a tiny piece of clay to stick a magnet to the top of the battery.
5. Your motor base is complete!

2 Making the coil

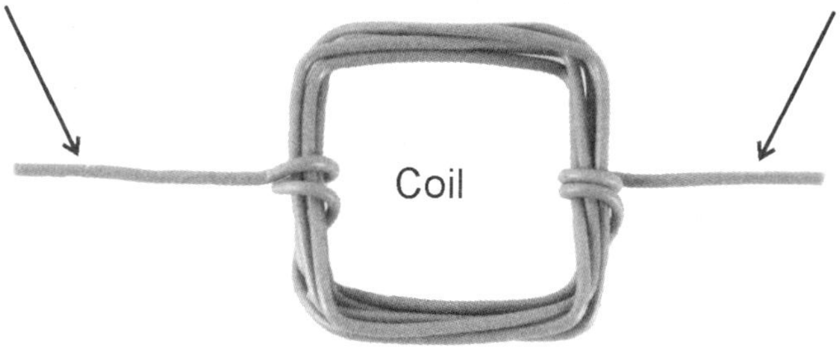

1. Cut 1 meter of magnet wire (also called varnished magnet wire). This wire has a painted insulation layer on the surface.
2. Wrap the wire around the square form of the modeling clay with one end sticking out 4 cm or so.
3. Keep wrapping until you have only 4–5 cm left.
4. Remove your coil from the form and wrap the ends of the wire a few turns around the sides of the coil to keep things together. There should be about 3 cm of wire on each side of the coil.
5. Take some sand paper and sand off all the varnish on one wire.
6. Sand off the varnish on ONE SIDE ONLY of the other end of the wire.
7. Adjust the wires until the coil balances as well as you can get it.
8. Your coil is done!

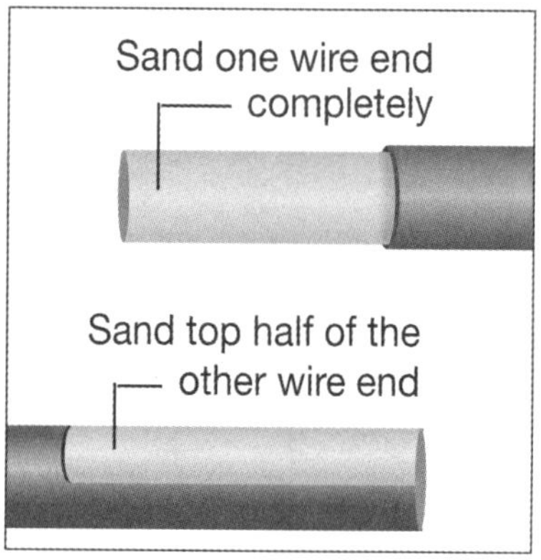

3 Making the motor work

1. Set the coil into the paper clips so it is free to spin.
2. Adjust the height of the paper clips until the coil rotates just above the magnet.
3. The motor should spin! Adjust the balance by bending the wires or paper clips.

Complete motor

Troubleshooting: the main problem areas to watch out for are

- unbalanced coil
- poor connections
- improperly sanded coil ends
- dead D-cell

4 Thinking about what you observed

a. When electricity runs through the coil of wire, what type of force is created around the coil?

b. What is the purpose of the permanent magnet?

c. What interactions cause the coil of wire to spin?

d. Try adding a second magnet. Does this make the motor go faster, slower, or about the same? What observations did you make that support your answer?

e. What else might make the motor spin faster? With your teacher's approval, try it!

5 Challenge

See if you can hold a photogate so the spinning coil breaks the light beam. With the Data Collector mode set on frequency, you will know how many breaks per second the spinning coil makes.

a. Describe how you set up the experiment to get frequency readings from the Data Collector/photogate equipment.

b. How did your spinning coil measurements compare to other groups' data?

c. Repeat the last 2 steps in the previous section (4d and 4e). Record your data.

d. Does the spinning coil data support your answers to 4d and 4e? Explain.

e. The basic parts of any simple DC motor like the one you made are: electromagnet, permanent magnet, commutator, and energy source. Draw a diagram of your simple motor and label these parts.

18C The Rock Cycle

How does the rock cycle create new rocks and change one type into another?

There are many regular repeating cycles in nature, like the seasons, or the water cycle. When it comes to rocks, the regularity of patterns does not apply. Igneous, metamorphic, and sedimentary rocks can be changed from one type to another by Earth's geologic processes. A rock's proximity to these processes is an important factor that impacts if and when any changes occur. In this investigation, you will learn about the formation of these 3 types of rock and the transformations they can make from one type of rock into a new type.

Materials

- 2 (or more) crayons or candles of different colors for original samples
- Scraper (a penny works well)
- Paper towel
- Aluminum foil
- Aluminum pie tin
- 3 cups of water (one with ice cold water, one with warm water, and one empty)
- Ice
- Candle for heating sample
- Textbook

1 Creating rock type #1

1. Use your scraper to scrape small pieces of all of your original samples (crayons or candles) into separate piles onto the paper towel.
2. Follow your teacher's instructions on how to move all your scrapings onto a piece of aluminum foil. Be sure your scrapings are in a neat pile on the foil.
3. Fold up your aluminum foil with the scrapings contained in a small, sealed pocket so they can't escape or leak out. Keep folding until your foil package is about 3 x 4 cm.
4. Place the foil package on the table. Place a textbook on top of the foil package and press down on the book with one hand. You should feel the package compress slightly.
5. Remove the textbook and carefully unwrap the foil package. Examine the sample inside the package and remove a small piece of the sample. Place it onto the blank piece of paper and write "Rock Sample Chart" at the top of the paper. This will be where you place and compare small samples of each rock formed during this investigation. Answer questions a-e below.

a. Examine and describe what your sample looks like.

b. What kind of rock does this sample represent? Use your answer to label the small piece of the sample you removed from the foil and placed onto the Rock Sample Chart.

c. What geologic process does the scraping of small pieces off of the original sample represent?

d. What geologic processes do the moving and placing of the scrapings onto the aluminum foil represent?

e. What geologic processes do the pressing on the foil package with the textbook represent?

2 Creating rock type #2

1. Reseal the rock you just made in the aluminum foil package. (Do not include the small sample on the Rock Sample Chart, it will remain there for the duration of the investigation, along with a small sample of each rock you form).
2. Place your textbook back on top of the sealed package.
3. Place both hands on top of the textbook and push down as hard as you can without damaging the textbook or the package.
4. Light your candle and follow your teachers instructions on how to hold the package over the lit candle for 5 seconds.
5. Set the package on the table and place the textbook back on top of it. Let the package cool with the textbook on top of it.
6. Once the package is cool, remove the textbook and carefully unwrap the package. Examine the sample, and carefully remove a small piece and place it on the Rock Sample Chart. Answer questions a-e below.

a. Examine and describe what your sample looks like.

b. What kind of rock does this sample represent?

c. Describe the transition in rock types that took place from start to finish in this part of the investigation.

d. Is your sample completely uniform in its consistency? If not, what was responsible for the inconsistency?

e. What does the heat from the candle represent?

3 Creating rock type #3

1. Prepare 3 cups; one with ice cold water (without ice cubes in cup), one with warm water, and one without water.
2. Place your sample into the aluminum pie plate
3. Heat the pie plate over your candle until the entire sample has melted. Try to prevent ovulating your sample. If it starts to emit smoke, raise the pan away from the candle's flame and heat more slowly.
4. Pour approximately equal amounts of the remaining melted sample from the pie plate into each of the 3 cups.
5. Let the samples cool for 1 to 2 minutes. Place and label the 3 samples onto the Rock Samples Chart.

a. Examine and describe what your 3 new samples look like.

b. What kind of rock do these samples represent?

c. Describe the transition in rock types that took place from start to finish in this part of the investigation.

d. What do the three different cups represent? How would actual samples of this type of rock differed if cooled in different ways?

4 Thinking about what you observed

a. Could these transitions continue, or do the transitions from one rock type to another have to stop at rock type #3?

b. Compare the sedimentary rock and metamorphic rock forming processes.

c. What is the difference between the process of forming igneous and metamorphic rocks? What do these processes have in common?

d. Does there seem to be any real pattern as to the order in which rocks transform from one type into another?

5 Geologic transformation scenarios

a. Can a rock change but remain the same type of rock? For example, could a sedimentary rock somehow become a different kind of sedimentary rock? Describe a scenario where this could occur. Include the names of different rocks and their types when possible.

b. Limestone is formed from the calcium-rich exoskeletons of trillions of deceased tiny sea creatures being deposited and compacted on the ocean floor over millions of years. Limestone can turn into marble when subjected to enormous heat over time. Explain what kind of rock this makes limestone and marble, and describe a geologic event or scenario that would turn limestone into marble.

c. Half Dome in Yosemite National Park is a structure made from granite. It was once a large plume of material from Earth's mantle that rose up towards the surface, but stopped rising when it was still thousands of feet below the surface. What kind of rock is granite, and what processes helped to finally expose this giant rock formation instead of keeping it buried deep underground forever?

Half Dome in Yosemite National Park

Photo courtesy of U.S. Geological Survey

d. What conditions may cause some rocks to be transformed from one kind of rock into another several times, while other kinds of rock remain unchanged for billions of years?

19C Evidence for Plate Tectonics

How are fossils useful evidence for continental drift?

Fossils are the remains or evidence of living organisms. Fossils come in different forms, including casts, molds, imprints, amber, and ice. Scientists can learn a great deal about life and the history of Earth using fossils. Fossils are useful evidence for understanding how organisms have evolved over time. They are also used to see how life has changed throughout Earth's history. Fossils have also been used to help scientists understand how the continents appeared millions of years ago. In this investigation, you will use several fossils to try to reconstruct how Earth's landmasses may have appeared approximately 250 million years ago.

Materials

- Bathymetric map
- Fossil data (included in investigation)
- Scissors
- Tape
- A piece of construction paper
- Pen or pencil

1 Setting up

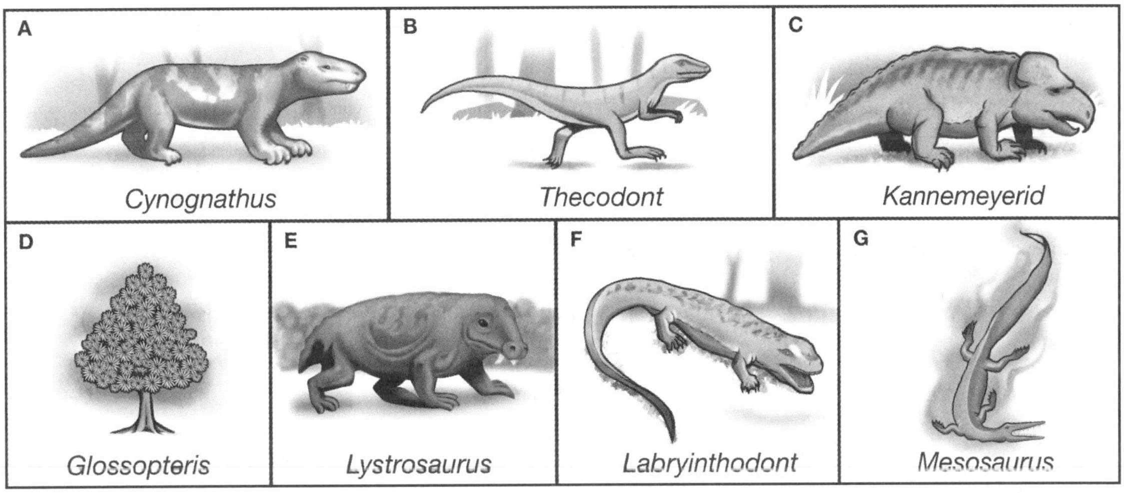

	Fossil name	Locations found
A	Cynognathus	South America, western Africa
B	Thecodont	Europe, eastern North America
C	Kannemeyerid	northern South America, Africa, India, Asia
D	Glossopteris	eastern South America, central Africa, India, Australia
E	Lystrosaurus	Antarctica, southern Africa, India
F	Labyrinthodont	Antarctica, central Africa, eastern Asia, Australia
G	Mesosaurus	southern South America, southern Africa

1. Using the information table above, write the letter of each fossil onto the correct locations on your bathymetric map. Note: the locations given in the table are only approximate locations.

2 Stop and think

a. Which of the fossils from the table were found in Antarctica? Why might this seem strange or unexpected?

b. Which of the fossils from the table were found both in North America and Asia?

c. What are some possible ways these organisms could have traveled from continent to continent when they were living?

d. If these animals and plants were not able to swim across large bodies of water such as oceans or sea, explain how else the pattern of fossil distribution can be explained.

3 Doing the experiment

1. Cut out each of the following continents and landforms from your map: North America, South America, Antarctica, Africa, Europe, Australia, and Asia.
2. Place these continents onto your piece of construction paper.
3. Using the shape of the continents' coastlines and the locations where each type of fossil has been found, reconstruct the world so that all the continents are connected to form a large "supercontinent."
4. When you have completed the previous step tape down the continents to your piece of construction paper in the same positions you arranged them in step 3.

4 Thinking about what you observed

a. *Glossopteris* is an extinct type of plant referred to as a seed fern. These plants most likely thrived in tropical climates. Do any of the locations where the fossils of the *Glossopteris* have been found seem strange? Explain your answer.

b. *Thecodont* was a small dinosaur. Where have fossils of this dinosaur been uncovered? Does it seem likely that this animal could have traveled between these two locations? Explain your answer.

c. How did the fossils of *Cynognathus* help you construct your map?

d. Where on your new map is Australia? To which continents is it connected? Which fossils did you use to help place Australia? How were they useful?

e. What other evidence might be useful for connecting the continents together into one giant landmass?

5 Exploring on your own

Pick two of the organisms from the fossil list you used in this investigation. Research this organism and write a one paragraph summary of what you found about each. Your research should include but is not limited to - when it lived, what it ate, how it behaved, its size, and the habitat in which it lived. Include a sketch of your organisms as well.

20C Igneous Rocks

How are igneous rocks classified?

Geologists divide rocks into three groups—igneous, metamorphic, and sedimentary—based on how they were formed. Igneous rocks are formed from the cooling of magma or lava either on or below Earth's surface. Many igneous rocks consist of mineral crystals. When magma cools, it usually includes the ingredients for more than one kind of mineral. As the magma cools, it begins to form mineral crystals. As the cooling continues, different mineral crystals form. Eventually the last of the cooled magma solidifies, locking all of the crystals in place, forming an igneous rock.

Some igneous rocks do not contain minerals with crystal structures. These rocks are formed when magma cools rapidly on the surface, not allowing the minerals to crystallize. Igneous rocks formed this way are known as volcanic glass. In this investigation you will learn how to classify igneous rocks. You will also determine how crystal size relates to the cooling rate of magma.

Materials

- Igneous rock samples: gabbro, granite, basalt, andesite, rhyolite, obsidian, tuff, pegmatite
- Hand lens
- 2 small plastic dishes or saucers
- Table salt
- Plastic beverage cup
- Plastic spoon

1 Examining igneous rocks

1. Arrange the igneous rock samples on your table so that they can easily be examined.
2. Rocks are classified based on their texture and color. You will be classifying the rocks on your table based on these properties.
3. Use the hand lens and classify each of the rocks according to its color. Some rocks may have more than one color. Be sure to classify all of the colors you see. Record your observations in Table 1 in part 2.
4. Rocks are also classified based on their texture. Geologists define a rock's texture as the shape, size and pattern of the mineral crystals that make up the rock.
5. Classify each of your rock samples based on its texture. Place a check mark in each of the appropriate columns in Table 1 to classify your rock's texture.

2 Rock classification

Table 1: Rock Classification

	Color(s)	TEXTURE						
		Fine Grained	Coarse Grained	Round	Jagged	Flat layers	Wavy	No Visible Grains
Granite samples								
Basalt								
Obsidian								
Andesite								
Rhyolite								
Gabbro								
Pegmatite								
Tuff								

3 Stop and think

a. Which of the rocks you observed were composed of larger, coarse-grained mineral crystals? Did these rocks have any other properties in common?

b. Which of the rocks you observed were composed of smaller, fine-grained crystals or no visible crystals at all? Did these rocks have any other properties in common?

c. Based on your observations and what you know about the formation of igneous rocks, which rocks do you think formed from magma that cooled quickly? Which of the rocks may have formed from magma that cooled very slowly?

d. Based on your answer to question **c**, predict which rocks may have formed below Earth's surface? Which rocks may have formed on or near the surface?

4 Doing the experiment

1. Add 4 heaping teaspoons of salt to a plastic beverage cup half-filled with water.
2. Stir this mixture for 1 minute. Not all the salt will dissolve because you have formed a saturated solution. This means the water has dissolved all the salt it can hold.
3. Pour a little of the salt solution into two dishes. Label one dish "warm" and the other "cool."
4. Place the dish labeled "warm" in a warm place such as over a radiator or near a sunny window. Alternatively, you can use a hair dryer to warm the liquid in this dish until it dries up. Operate the hair dryer on low speed at least 60 cm from the dish. You want to warm the salt water without blasting it out of your dish!
5. Place the dish labeled "cool" away from sources of warmth and light.
   Make a prediction:
   How do you think the crystals will look in each dish after the water has evaporated? Draw a picture in each of the circles below to illustrate your prediction.

How do you think the crystals will look in each dish after the water has evaporated?

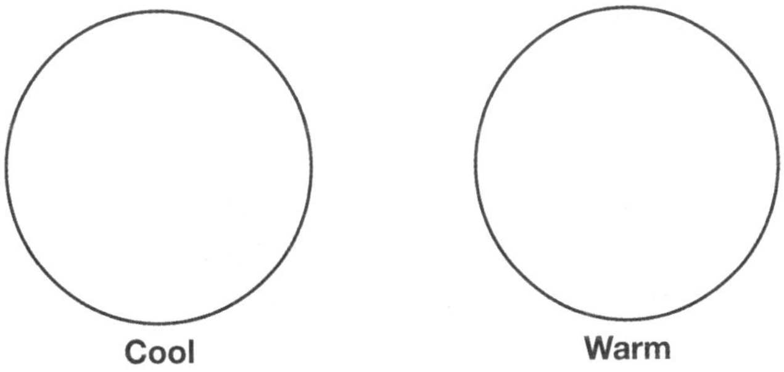

NOTE: Crystal growth is slow and will take more than one day.

6. Examine both dishes from time to time for the next few days until all the water has evaporated. You may want to record the temperature at each dish's location.
7. Make a table to record your observations.
8. Record the date, time, and the amount of water left in each dish at each observation. Describe the crystals in each dish. Use relative terms such as *slightly*, *more*, and *most* to record these observations. Measure and record the size of the largest crystal in each dish.
9. When the water has evaporated from one of the dishes, make a final inspection of both dishes. Both dishes will have large and small crystals, but one will have the largest crystals. Usually these largest crystals will form near the edges of the dish.
10. Record your final observations in the table you made in step 7 and 8.

5 Thinking about what you observed

Part 1

a. Which dish had more time to grow crystals? Which dish contained bigger crystals?

b. Write a conclusion sentence about the time available for crystals to form and the size of the crystals that form.

Part 2

The size of rock crystals depends on the amount of interaction among the materials in the magma that forms the rock. More time allows more interaction. As a result, slow-growing crystals tend to be bigger than fast-growing crystals. Anything that might slow the cooling of magma will provide that time.

a. List two natural events or conditions on Earth (other than the methods you used in this investigation) that might slow the cooling of magma.

b. List two natural events or conditions on Earth that might speed up the cooling of magma.

6 Exploring on your own

a. Go back to the rock samples you classified earlier in this investigation. Compare the appearance of each sample and look at the observations you made about each rock's texture in Table 1. Based on your work in this part of the investigation, which rocks formed slowly and which rocks formed quickly? Explain your answer to this investigation based on what you learned about crystal growth.

b. Igneous rocks that form on Earth's surface are *extrusive* and those that form underground are called *intrusive*. Based on your observations of the igneous rocks, which of the rocks are likely to be extrusive? Which are likely to be intrusive? Explain your answer.

21C Acids, Bases, and pH

What is pH?

Life exists inside a certain range of pH values. A pH value describes whether a solution is acidic, basic (alkaline), or neutral by describing the concentration of hydronium ions in a solution. An acid is a substance that produces hydronium (H_3O^+) ions when dissolved in water, and a base (or alkali) is a substance that produces hydroxide (OH^-) ions when dissolved in water. Neutral solutions have equal numbers of H_3O^+ and OH^- ions.

In this investigation, you will learn the pH of several common solutions by making a pH scale using a pH indicator and chemicals of known pH. You will also identify two 'mystery' chemicals with your pH scale.

Materials (per group)

- 2 or 3 well plates depending on the size (need 12 wells)
- Permanent marker
- Eyedroppers or pipettes
- 5 mL of red cabbage juice (a pH indicator)
- 12 each red and blue litmus papers (pH indicators)
- 2 mL each of the 12 solutions listed in the data table
- Two 250 mL beakers – one contains warm water for rinsing pipettes, one empty beaker for discarded rinse water

WARNING — This lab contains chemicals that may be harmful if misused. Read cautions on individual containers carefully. Not to be used by children except under adult supervision.

1 Stop and think

Look at the table in the investigation. Using the pH values listed, predict whether the antibacterial soap will be acidic or basic and whether the apple juice will be acidic or basic. Of the solutions in the list, to which are they most closely related?

2 Doing the experiment

A. Make a pH scale using indicators

1. To create your pH scale, you will be using solutions 1 to 7 in the table below. Place the following labels for these solutions in order on a well plate. If you don't have seven wells in a row on one well plate, place two plates side by side. The labels should describe the solution and its pH: Lemon – pH 2, Vinegar – pH 3, Seltzer water – pH 4, Red cabbage juice (the control) – pH 6.5, Baking soda – pH 8.5, Bar soap – pH 10, and Ammonia – pH 11.
2. Using a pipette, place three drops of red cabbage juice in each of the seven labeled wells.

3. Using a pipette, add two drops of each of the solutions to the appropriately labeled well. Use a different eyedropper or pipette for each solution. However, if you must use the same dropper or pipette, thoroughly rinse it in fresh water after each solution before using it for a new solution. Record the color changes in your data table. The color series you see on the plate(s) represents a pH scale. We will refer to it as the pH test plate. You will use it to identify the pH of other solutions.
4. Dip the red litmus paper and the blue litmus paper into each well of the pH test plate. Record the results according to the directions in the data table.

Name of solution	Color when mixed with red cabbage juice	Red litmus paper: if paper turns blue, write "base" or make an "x"	Blue litmus paper: if paper turns red, write "acid" or make an "x"	pH
1. Lemon				2
2. Vinegar				3
3. Seltzer				4
4. Red cabbage juice				6.5
5. Baking soda solution				8.5
6. Bar soap solution				10
7. Ammonia				11
8. Green tea				
9. Antibacterial cleaner				
10. Apple juice				
11. Mystery solution A				
12. Mystery solution B				

B: Using pH indicators to measure unknown pH

1. Repeat steps A2 through A4 for solutions 8 to 12. Use another well plate for these five solutions. The labels should describe the solution. At this point, you do not know the pH of these solutions.
2. Identify the pH of solutions 8 to 12. Compare the color reactions and the litmus paper results for solutions 8 to 12 with the pH test plate.

3 Thinking about what you observed

a. What is the role of a pH indicator? What is the range of pH measured by each of the indicators you used (red cabbage juice, red litmus paper, blue litmus paper)?

b. Which of your solutions has the highest concentration of H_3O^+ ions? Which has the highest concentration of OH^- ions? Explain your reasoning.

c. The red cabbage juice used in the investigation has two roles. It is the pH indicator and, in the series on the pH test plate, it is a control. Why is a control needed on the pH test plate?

d. Mystery solutions A and B are identical to two other solutions you used in this lab. Use your results to identify these solutions. What is the identity of mystery solution A? What is the identity of mystery B? List evidence to support your claims.

e. List the pros and cons of using red cabbage juice and litmus paper as pH indicators.

f. Various professions use pH indicators. For example, photographers traditionally used and still sometimes use stop bath in developing, and swimming pools are maintained using information from pH indicators. Find out how these pH indicators work in these (or other) situations, and what the color changes mean.

22C Global Winds and Ocean Currents

How do temperature and salinity cause ocean layering

Did you know that there are rivers in the ocean? In some places, there are two rivers running side-by-side—in opposite directions. Now imagine underwater waterfalls and underwater springs. All of these exist in the world's oceans as a result of differences in water temperature and *salinity* (saltiness). How do temperature and salinity differences cause these underwater rivers, waterfalls, and springs? *Density* changes are the key. In this investigation, you will discover how temperature and salinity create currents, underwater waterfalls, and springs in the ocean.

Materials

- A clear plastic cup
- Two foam coffee cups
- Eyedropper
- Pipette with a barrel longer than an eyedropper's
- Salt
- Measuring spoon—teaspoon size
- 10 cm square of single ply cardboard
- Pencil
- Scissors
- 20 staples
- Newspapers
- Cafeteria tray or paper plate
- Source of hot and cold water
- Food coloring: red, green, and blue

The ocean contains:

1 Density and ocean currents

As global winds push ocean currents around the planet, the ocean water undergoes several changes. As the current moves nearer to the equator, the water warms up. As it moves toward the poles, the water cools down. These changes affect the density of ocean water.

When the current moves through a warm area, there is an increase in evaporation. Since evaporation removes fresh water and leaves salt behind, the salinity of the current increases. This increases the current's density. Fresh water is added back to the ocean through melting ice, rivers, and rain. Adding fresh water to the salty ocean water decreases its density.

These changes in density cause ocean currents to float and sink at different points in their journeys. In this investigation, you will model underwater rivers, waterfalls, and springs. Then you will use your observations to help you understand the movements of the Atlantic gyre, an ocean current system. You will also discover how these density differences play an enormous role in the life of the world's most important fishing grounds.

a. Which do you think is more dense: warm or cold ocean water? Why?

b. Explain why dissolved salt increases the density of ocean water.

2 Observing salinity-dependent layering

This investigation requires care and preparation to keep your cleanup quick and easy. Spread newspaper over your work area to catch drips. Keep any water-filled cups over a tray or paper plate. Wipe up any spills before they get tracked around. Discard any water in the bucket provided or in the sink.

1. Make a pouring stick by cutting a cardboard circle about 4 centimeters in diameter with your scissors. Press the point of a pencil into the center of the disk so that it is stuck firmly onto the pencil.
2. Fill a clear plastic cup half-full with cool water. Add 1 teaspoon of salt to the water. Add 2 drops of green food coloring. Stir until the salt dissolves.
3. Fill a foam cup half-full with cool water.
4. Have a team member hold the pouring stick at its top, near the eraser. Lower the pouring stick into the middle of the clear cup so that the cardboard disk is just under the surface of the green water. Have a second team member hold the lip of the foam cup up to the pouring stick.

5. Tip the foam cup so that the cool water flows slowly down the pouring stick. The first team member must move the pouring stick upward as the second team member pours so that the cardboard disk remains at the surface of the water. Continue to add water until the clear cup is almost full, and then gently remove the pouring stick. You have created two ocean layers, separated by their salinity.

a. Try slightly tipping the clear plastic cup. Are the layers stable? Do they resist mixing?

b. Tear off a small piece of foam cup. Press some staples into the foam, and place it on the surface of the clear water. Remove the foam and add more staples to it, one at a time, until the foam bit sinks. Where did the foam bit end up? Why?

c. Explain why the clear water floats over the saline water.

3 Exploring temperature-dependent layering

1. Fill a clear plastic cup half-full with cool water. Add 2 drops of blue food coloring. Stir to mix.
2. Fill a foam cup almost full from the hot-water source. Add 2 drops of red food coloring. Stir to mix.
3. Add hot red water to the clear cup using the pouring stick as you did in salinity-dependent layering. You have again created two ocean layers, this time separated by their temperatures.

a. Try tipping the cup slightly. Are the layers stable? Do they resist mixing?

b. Explain why the hot red water floats over the cool blue water.

4 Creating an underwater waterfall

1. Fill a clear plastic cup nearly full with cool water.
2. Fill a foam cup half-full with hot water. Add a pinch of salt. Add 6 drops of red food coloring. Stir until the salt dissolves.
3. Place the eyedropper into the hot red water to warm it up. After a minute, fill the dropper barrel with the water.
4. Hold the dropper so that it lies at a flat angle at the surface of the clear water with the tip just under the surface. Gently squeeze out a layer of hot red water onto the surface of the clear water.
5. After a short cooling time, the red layer will form little waterfalls that sink through the clear water. They may even form little smoke-ring-like structures as they fall. If this does not happen within a few minutes, add a little more salt to the hot red water, stir, and try again.

a. Explain why the red water floats at first.

b. Explain why the red water eventually sinks.

5 Observing underwater springs

1. Fill a clear cup three-quarters full with cool water. Add a heaping teaspoon of salt to the water. Stir until the salt dissolves.
2. Fill a foam cup half-full with cool water. Add 6 drops of blue food coloring. Stir to mix.
3. Fill the eyedropper with cool blue water.
4. Gently lower the dropper into the salt water so that the tip is near the bottom. Gently squeeze the dropper so that a small stream of blue water is released.

a. Where did the blue water go? Why?

b. In this model, the blue water was less salty than the surrounding water. Think of another difference you could use to create an underwater spring. Write your own procedure, test it, and explain what happened.

6 Applying your knowledge

You can use these experiments to understand the North Atlantic gyre, a system of currents that occupies the North Atlantic Ocean basin. After each description, fill in the experiment that applies (salinity-dependent layering, temperature-depending layering, underwater waterfall, or underwater springs), and then determine whether the current will float or sink.

We will start with the part of the North Atlantic gyre off the coast of Africa. We will assume that the temperature of the water is cool and its salinity is low. The low salinity is dominant and the current is floating on the surface.

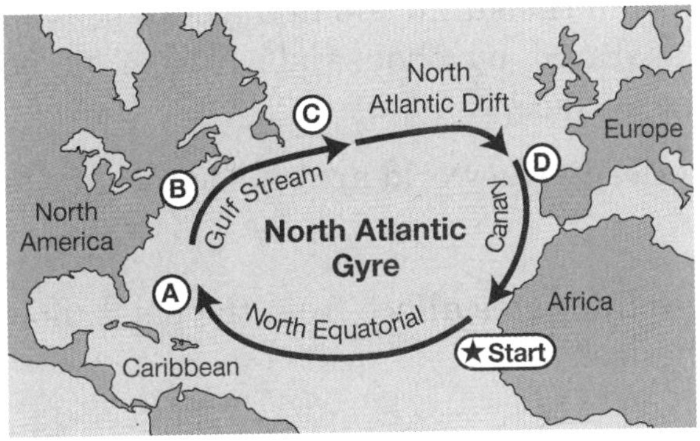

a. From there, the North Equatorial Current flows westward toward the Caribbean Sea. Then it turns north and becomes the beginning of the Gulf Steam. During this trip along the equator, the intense sun warms the current and evaporates a lot of water. This makes the water both warm and highly saline. The high temperature is dominant.

Experiment: ____________________ Float or sink: ____________________

b. After turning northward, the Gulf Stream flows along the United States' Atlantic Coast. The water is highly saline, but it remains warm. The high temperature remains dominant.

Experiment: ____________________ Float or sink: ____________________

c. As the Gulf Stream leaves the U.S. coast, evaporation is not as great in the cold northern regions, but the cooling effect is very significant. The low temperature is dominant.

Experiment: ____________________ Float or sink: ____________________

d. Now part of the Gulf Stream becomes the North Atlantic Drift. Fresh water from ice melt may mix with these cold waters off the coast of Europe. Lower salinity is dominant.

Experiment: ____________________ Float or sink: ____________________

7 Extension: Thermohaline currents and the ocean food chain

Back in the 1960s, it was popularly believed that the key to feeding the world was the bountiful harvest that could be taken from the seas. Today, we are faced with the collapse of fisheries on both sides of the Atlantic. Why has the ocean proven to be such a modest food source?

The food paradox of the oceans is based on the nutrient cycle. For new creatures to grow, the nutrients from the old creatures must be recycled. Unfortunately, when ocean creatures die, they take their nutrients to the deep bottom. The photosynthetic plankton (phytoplankton) that do the recycling must live in the sunlit top-600 feet of ocean, so recyclers and the needed nutrients are hopelessly separated by thousands of feet of ocean unless something can transport the nutrients to the surface.

a. Two of the biggest fisheries in the world are off the Canary Islands and Peru. Can you explain why?

b. If global climate change eliminates all ice from the poles, how might this affect ocean currents and world fisheries?

23C Sedimentary Rocks and Relative Dating

What can sedimentary rock tell us about its age?

Sedimentary rocks are formed from the compaction and cementation of separate particles called sediment. Sediment can consist of materials such as sand, clay, silt, pebbles, and gravel.

As time passes, sediment that is weathered from rock and/or eroded from the land gets deposited somewhere else. In certain environments, where weathering and erosion are slow, layers of sediment can build up faster than they are removed.

In this investigation, you will model the formation of sedimentary rock, observing how sediment size plays a part in its development, and gain an understanding of how scientists are able to place events in the sequence that they occurred, using these rock layers.

Materials

- Two 1-liter plastic soda bottles (labels removed)
- Scissors
- Permanent marker
- Soil mixture (one part topsoil to one part sand)
- Newspaper
- Beaker
- Metric ruler
- Spoon
- Sedimentary rock samples (sandstone, shale, conglomerate)

1 Setting up

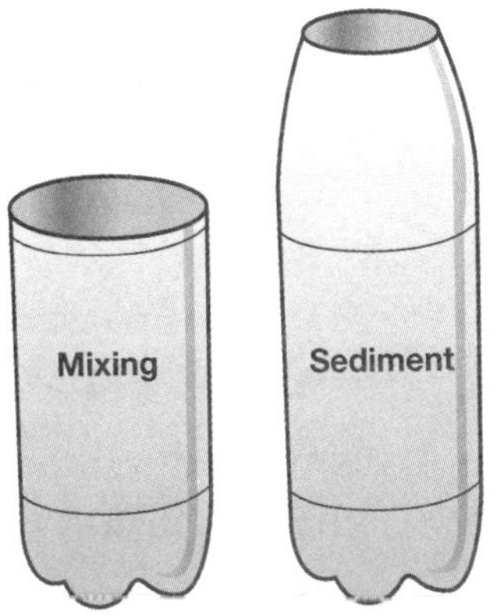

1. Gather the items needed for this investigation. Cover your work area with newspaper. At the end of the investigation, you can wrap up the newspaper and throw it away to ease your clean-up.
2. Ask your teacher to make a slit in your first plastic bottle just below the bottle's shoulder. Then use your scissors to cut all the way around the bottle. You should have an open cylinder with straight sides. This bottle will be used for mixing.
3. Your teacher should make a slit in the second bottle just above the shoulder. Then use your scissors to cut all the way around this bottle. It will be used for holding your sediments. See the diagram to the right before you cut your bottles.
4. Label the first bottle “mixing” and the second bottle “sediment.”
5. Add water to the sediment bottle to a depth of five centimeters.
6. Add the soil mixture to the mixing bottle to a depth of three centimeters.
7. Add water carefully into the mixing bottle and mix it to make a “mud slurry.” Add just enough water to make it the thickness of a thick milk shake.
8. Pour your mud slurry into the sediment bottle and mix it into the water.
9. Allow the sediment bottle to stand undisturbed for two minutes or until you can see layers forming.

2 Stop and think

a. Once you see layers in the sediment bottle, make a sketch of the layers in the first bottle below, labeled "first slurry." Be sure to draw and label the following in your diagram: floating debris, fine particles, and coarse particles.

b. Where in your bottle did you find the coarsest sediment? Where is the finest sediment?

c. Is there a change in color from bottom to top? Explain why or why not?

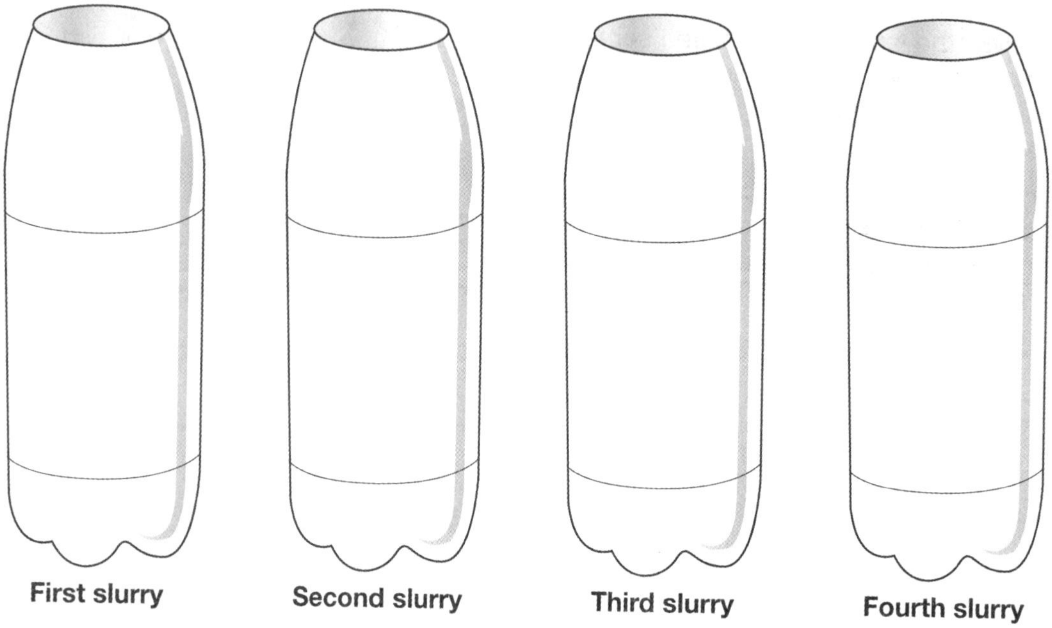

3 Doing the experiment

1. Now you will be adding more sedimentary layers to your bottle. The procedure is slightly different than the first trial.
2. Examine the water level in your sediment bottle. You need only five centimeters of water over the settled slurry. Carefully pour off excess water into a wastewater container.
3. Add 1 centimeter of sediment to the mixing bottle. Then, add water until the new mud slurry is much thinner than the first slurry you made.
4. Have one of your team members continuously stir the slurry for twenty seconds. Do the next two steps quickly.
5. Pull your spoon through the water in the sediment bottle just over the surface of the settled slurry, a little off-center.
6. Immediately add the new slurry to the sediment bottle while the water is still swirling.
7. After two minutes, sketch the layers you see. Use the bottle diagram labeled "second slurry." Pay special attention to thickness, sediment size, and the angle in which layers formed (i.e. slanted, straight across, etc.).

8. Repeat steps 2 through 7 until you have added three slurries to the original for a total of four slurries added to the sediment bottle. Stop the process if your sediment bottle gets too full.

4 Thinking about what you observed

a. Wipe off the sides of your bottle. Carefully examine the sediment in your bottle. Is each slurry you made represented by a distinct layer? If so how many layers do you see?

b. Are the sedimentary layers in your bottle the same thickness?

c. Which size sediments settled into the bottle faster: larger or small particles? Explain your answer.

d. What will happen to your sedimentary layers when the water evaporates?

e. What kinds of environments would cause sediment layers to be deposited over time on top of each other?

5 Exploring further

a. Label the sedimentary layers in your sketch from oldest to youngest. How do you think geologists can figure out the age of sedimentary rock layers in areas such as the Grand Canyon?

b. Do you think the age of the sedimentary rock layers from the Grand Canyon that geologists figure out would be the relative age (the age in reference only to the other rock layers in the area) or absolute age (the exact age)?

c. What might be a way of figuring out the absolute, or exact age of the rock layers? Hint: what might be embedded in those sedimentary layers that the geologists could use to get an exact date?

6 Exploring on your own

a. Examine the sedimentary rock samples. Classify these sedimentary rocks according to color, texture, and particle size. What types of particles do you think make up each of the rock samples?

b. Based on the results of this investigation, where would you expect to find each of these sedimentary rock samples in a river system? Upstream near the source of the river or farther downstream near the mouth of the river?

c. Go for a "rock walk" around your school or around your home. See if you can find samples of sandstone, shale, or conglomerate in these areas. Collect any other rocks that may seem interesting to you as well. Take these rocks back to your classroom and classify them according to their color, texture, and mineral size.

24C Sound as a Wave

How can we observe sound as a wave? How can we use the speed of sound and certain frequencies to build a basic instrument based on wavelength?

Waves are a traveling form of energy because they can cause changes in the objects they encounter. Sound waves carry energy that causes human eardrums to vibrate, which the brain interprets as sound. Because it is a wave, sound has both frequency and wavelength. In this investigation, you will use a sound's frequency and its speed to calculate desired wavelengths and use your calculations to build a musical instrument.

Materials

- 10 ft. length 1/2 inch PVC pipe
- Metric ruler
- Pipe cutter (or a saw)
- Sandpaper
- Permanent marker
- Data Collector
- Temperature probe

1 Thinking about sound as a wave

You have learned several important properties of waves. Answer the questions below to review what you have learned.

a. What property of waves describes the distance from any point on a wave to the same point on the next cycle of the wave?

b. What property of waves describes how often something repeats, expressed in hertz?

c. How are the frequency and wavelength of a wave related to its speed?

d. How is the wavelength of a sound wave related to the pitch it produces?

2 Making a prediction

You have been given a long length of pipe. In part three, you will cut two lengths of pipe, one twice as long as the other. You will then use the pipe to produce a sound by blowing over the top of the pipe with your hand covering the bottom, or by tapping the bottom of the pipe on the palm of your hand. Both methods produce a sound with the same pitch. The air inside the pipe vibrates to make the sound you hear when the pipe is played. You may have to practice to get a repeatable sound.

a. Make a prediction about how the sound produced by the two different lengths of pipe will compare. Describe the sound in terms of pitch.

b. Explain the reasoning behind your prediction.

3 Cutting two sections of pipe

1. A good rule to follow when cutting anything is "Measure twice, cut once". From the end of the pipe measure a distance of 10.5 cm and use a pen or pencil to mark the distance with a small straight line. Depending on the cutting tool you have, follow your teacher's directions on how to cut the pipe as close to 10.5 cm as possible.

2. Once it is cut, use some sandpaper to remove any rough edges from the cut end of the pipe. It does not need to be perfectly smooth, but there should be no sharp edges.
3. When your first pipe is cut and sanded, measure 21.0 cm from the cut end of the pipe and make a straight line mark. Double check your measurement and once again follow your teacher's directions on how to cut the pipe.
4. Sand both ends of the 21.0 cm pipe.

4 Using sound as an observation

You are ready to use your pipes to make some sounds. Make some sounds with both pipes and listen to the pitch each one makes. After you have compared the sounds created by each, answer the questions below.

a. Which pipe had a higher pitch, the short pipe or the long pipe?

b. Was your prediction about the pitch produced by two different lengths of pipe correct?

c. How does the pitch relate to the frequency of the sound created by both pipes?

d. How does the pitch relate to the wavelength of the sound created by both pipes?

e. How would the pitch of a pipe longer than 10.5 cm and shorter than 21.0 cm compare to these two pipes?

f. If you were to cut increasingly longer pipes, how would the sound produced by each longer pipe be different from the previous shorter length? Describe the change in terms of frequency and wavelength.

5 Wavelengths based on the speed of sound and frequency

The two pieces of pipe you cut have different pitches. Each different pitch is called a note. Certain notes sound pleasant to the human ear, and notes that sound pleasant can be arranged together to create a musical scale. By using what you know about the relationship between the speed of sound, frequency, and wavelength, calculate the wavelength of the sound wave that will make each note in a series of musical scales. The first task is to calculate the speed of sound in your classroom. The speed of sound depends mainly on the temperature of the air in your classroom. Use the formula to calculate the speed of sound in air for your classroom based on the temperature in degrees Celsius.

$$\textit{Speed of sound} = 331\,\text{m/s} + (0.6\,\text{m/s/}^\circ\text{C} \times {}^\circ\text{C})$$

Speed of sound m/s	=	331 m/s	+(0.6 m/s/°C	x	°C in classroom)
	=	331 m/s	+(0.6m/s/°C	x	°C)

Each note in all musical scales corresponds to a particular frequency. These notes and their frequencies have been musically established for thousands of years. Once you have calculated the speed of sound in your classroom, you can begin to calculate the wavelength of the sound waves needed to create the frequencies that make the musical scales we will use. Fill in your speed of sound on Table 1, and use the frequencies provided to calculate the wavelengths.

Table 1: Calculating wavelengths needed based on speed of sound and frequency

Speed of sound (m/s)	÷	Frequency (Hz)	=	Wavelength (m)
	÷	349	=	
	÷	392	=	
	÷	440	=	
	÷	466	=	
	÷	523	=	
	÷	587	=	
	÷	659	=	
	÷	698	=	
	÷	784	=	
	÷	880	=	
	÷	892	=	
	÷	1049	=	
	÷	1174	=	
	÷	1318	=	
	÷	1397	=	

6 Creating sound from a pipe with one closed end and one open end

The two pipes you've made have one closed end and one open end when they make sound. The air inside the pipe vibrates to make the note you hear when the pipe is played. Air in a tube like this vibrates in a specific way. The vibrations of the sound wave resonate in the pipe, amplifying their perceived volume. The length of the sound wave created by each pipe is actually 4 times the length of the pipe, as shown in the diagram below.

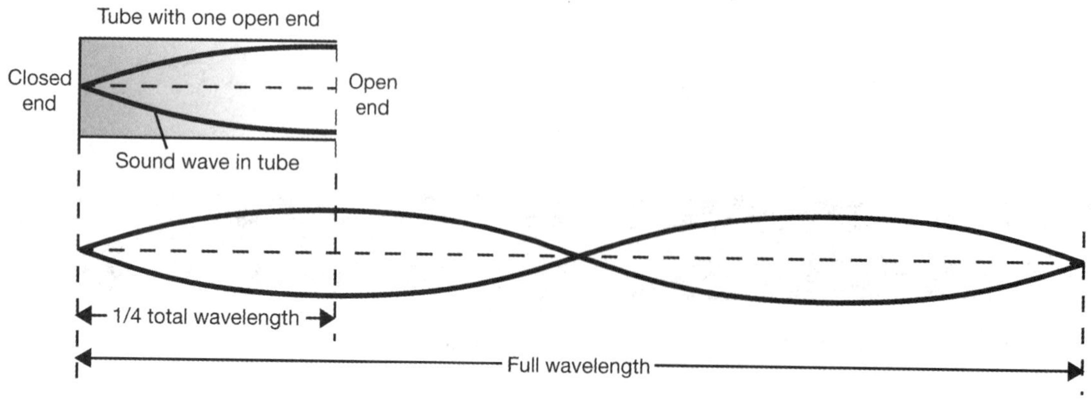

a. How would you calculate the length of pipe needed that would create a note of a specific frequency based on your calculated wavelengths from the previous section?

b. If the temperature in your room went up by 10 °C, how would that change the length of pipe you need to cut to make a note of 440 Hz compared to the current temperature of your classroom?

7 Cutting the remaining pipe

Fill in the wavelengths of sound you calculated from Table 2 to make the frequencies needed at your classroom's temperature. The frequencies are now also labeled with the notes they make in the musical scale. Use your answer from question a. above to calculate the length of pipe needed to make each note in the scale. Once you have calculated the length of pipe needed in meters, convert your calculation to cm and fill in the last column of Table 3. Your teacher will assign some notes to your group and you will cut those specific lengths of pipe. Use the length of pipe calculations in cm to measure and cut the lengths of pipe assigned to your group. You may be able to use the two pipes you made earlier. They may need to be trimmed to fit the particular lengths you need. Follow the instructions of your teacher to cut the pipe. Use a permanent marker to label each pipe with the corresponding note it creates.

Table 2: Calculating the length of pipe for the required frequencies

Note in scale	Frequency (Hz)	Wavelength of note (m)	Length of pipe (m)	Length of pipe (cm)
F	349			
G	392			
A	440			
BFlat	466			
C	523			
D	587			
E	659			
F	698			
G	784			
A	880			
BFlat	892			
C	1049			
D	1174			
E	1318			
F	1397			

8 Playing some music

Your teacher will have some scripted music that you can play. Look through the song to see what notes will be played. That way you can collect all the pipes you will need during the song and have them ready to go. With members each playing one or two pipes, your group should be able to cover all the notes in a song. Melody and harmony are different parts of the song that work together to make the song more interesting to listen to. Get your pipes ready and follow the beat of the conductor. Good Luck.

9 Challenge

a. There are some notes on the scale that have the same name but different frequencies. What is the relationship between these notes in terms of their frequencies and their wavelengths?

b. What do we call the group of 8 notes that are arranged in increasing or decreasing pitch, starting with one note and ending with the note of the same name but different frequency? (For example, from F 349 to F 698).

c. How does the concept of different lengths of pipe creating different notes relate to musical instruments, like the trombone?

d. What is the difference between how air in a tube with one open end vibrates and how a plucked string vibrates?

e. **EXTRA CHALLENGE:** The diagram in part 6 shows a sound wave with a node at one end and an antinode at the other end resonating in the tube. This represents the 1st harmonic of the air column in the tube. The column of air vibrates at several harmonics at once, but the 1st one is the one that creates the most sound you hear. Each harmonic has a node at one end and an antinode at the other due to one end being open and one end being closed. What would the next harmonic wave look like, what harmonic would it be, and what would its wavelength be in reference to the length of the tube?

25C Magnification and Mixing Pigments

How is the magnification of a lens determined? What happens when you mix different colors of pigments?

In optics, lenses are optical devices that use refraction to bend light. Some lenses can be used to produce images that are larger than the object they are collecting light from. This process is called magnification and it is used in equipment like microscopes and telescopes to investigate objects that are difficult to see with the naked eye. In this Investigation you will look at how distance can affect magnification, and also how mixing materials colored by pigments, paints, or dyes are different than mixing different colors of light.

Materials

- Optics with Light & Color
- Metric ruler
- Cyan, Magenta, and Yellow clay

1 Finding the magnification of a lens

1. Set your light blue lens directly on the graph paper and count the number of *unmagnified* squares that cross the diameter of the lens. In the example, the lens is 10 squares wide.
2. Next, examine a section of graph paper with your lens held above the paper. Move the lens closer and farther away until you have the biggest squares you can still see clearly in the lens.
3. Count the number of *magnified* squares that cross the diameter of the lens. For example, the picture shows 4 1/2 squares across the lens.
4. The magnification can be calculated by dividing the number of *unmagnified* squares by the number of *magnified* squares. In the example, you see 10 *unmagnified* squares and 4.5 *magnified* squares. The magnification is 10 /4.5, or 2.22.
5. Try the experiment again using a ruler to measure the distance between the lens and the paper. Notice that the magnification changes with different distances.

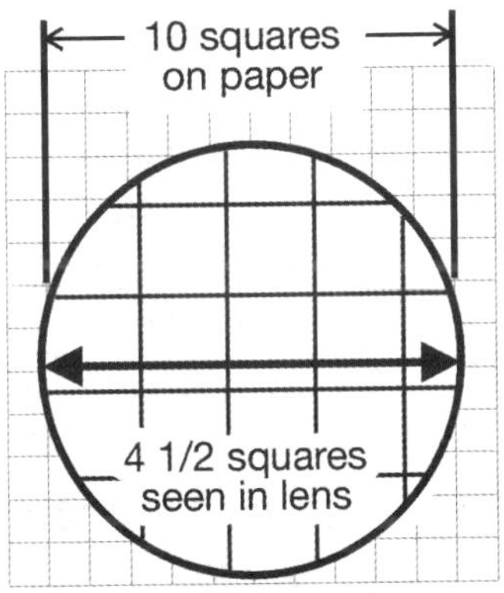

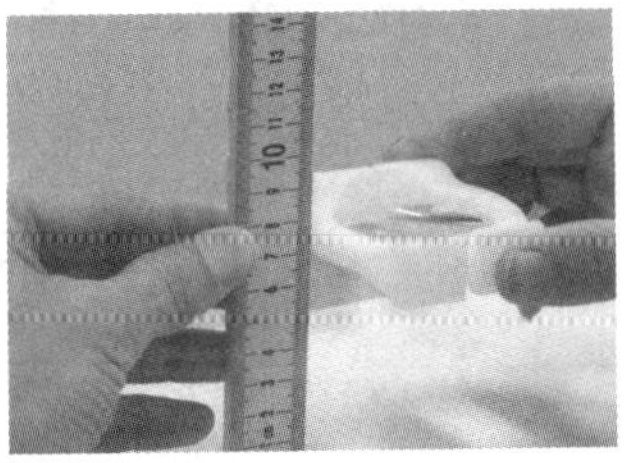

Measuring the distance from the lens to the paper

6. Fill in the table by measuring the magnification of your lens for at least four different distances. The number of squares on the graph paper will be the same for all distances.

Table 1: Magnification of a lens

Distance to paper	# of squares on graph paper (unmagnified squares)	# of squares in lens (magnified squares)	Magnification

2 Thinking about what you observed

a. Is the image in a magnifying glass inverted or upright?

b. At what distances will the lens act like a magnifying glass? What happens when the object is more than a focal length away?

c. Describe something that looks completely different under a magnifying glass than when seen with the un-aided eye.

d. Try the same activity with the dark blue lens. What happens to the image in the lens when you lift it up from the paper?

3 The subtractive color model (CMYK)

1. You have three colors of clay: cyan, magenta, and yellow. Take a portion the size of your fingertip of the both cyan and the magenta. Mix them together. What color do you get?
2. Mix equal amounts of cyan and yellow. What color do you get?
3. Mix equal amounts of yellow and magenta. What color do you get?

The subtractive color model (CMYK)

	Cyan	Magenta	Yellow	Black
Absorbs	Red	Green	Blue	Red, Green, Blue
Reflects	Blue, Green	Blue, Red	Red, Green	None

Cyan Magenta Yellow

Mix equal amounts of the three subtractive primary colors
(two colors at a time)

4 Thinking about what you observed

a. Explain how the mixture of magenta and cyan makes its color when seen in white light.

b. Explain how the mixture of cyan and yellow makes its color when seen in white light.

c. Explain how the mixture of yellow and magenta makes its color when seen in white light.

d. Why don't the mixed colors produce full red, green, or blue?

e. What color would you see if you looked at a mixture of magenta and cyan under a lamp that only made blue light?

f. Research how printers make colors. Do they use red, green, and blue (RGB) or cyan, magenta, yellow, and black (CMYK)? Explain why printed pictures need to use one or the other.

g. Research how computer monitors and televisions make colors. Do they use red, green, and blue (RGB) or cyan, magenta, yellow, and black (CMYK)? Explain why TV's and computer screens need to use one or the other.

h. Explain why mixing the primary colors of light is referred to as the "Additive Color Mixing Process", while mixing materials colored with pigments, paints or dyes is referred to as the "Subtractive Color Mixing Process".

26C Solar System

How big is the solar system?

It is difficult to comprehend great distances. For example, how great a distance is 140,000 kilometers (the diameter of Jupiter) or 150,000,000 kilometers (the distance from the Sun to the Earth)? An easy way to compare these distances is to create a scale model. For instance, a globe is a scale model of Earth and road maps are scale models of geographic regions. These scale models help us visualize the true sizes of objects and the distances between them.

In this investigation, you will compare an astronomical distance—like the distance from the Sun to Neptune—to a measurable distance—like the length of a football or soccer field. Using proportions, you will make a scale model of the distances of the planets from the Sun.

Materials

- Access to a large space with a length of 100 meters (a soccer field is ideal)
- A metric tape measure or a trundle wheel
- A variety of objects to represent the relative sizes of the planets (examples: softball, soccer ball, bowling ball, small plastic balls, inflatable beach balls). To represent the Sun and some of the larger planets may require imagination.
- Blank paper for making signs

1 Using proportions to determine scale distances

Neptune is an average distance of 4.5 billion kilometers from the Sun. We can use a proportion to determine a scale distance for our model. Assume the largest distance you can measure is 100 meters. The length of a soccer field is usually between 90 and 120 meters long. For this investigation, we will use 100 meters as the scale distance between the Sun and Neptune.

$$100 \text{ m} = 4{,}500{,}000{,}000 \text{ km}$$

If the distance from the Sun to Neptune equals 100 meters, where would you find the other planets? You can answer this question by setting up the following proportion where **x** is the distance from the Sun to any planet, in meters:

$$\frac{x}{\text{Distance of the Sun to planet}} = \frac{100 \text{ m}}{4{,}500{,}000{,}000 \text{ km}}$$

Mercury is 58,000,000 kilometers from the Sun. Using our proportion, we can find the scale distance:

$$\frac{x}{58{,}000{,}000 \text{ km}} = \frac{100 \text{ m}}{4{,}500{,}000{,}000 \text{ km}}$$

Cross-multiply and rearrange the variables to solve for x:

$$x = \frac{100 \text{ m}}{4{,}500{,}000{,}000 \text{ km}} \times 58{,}000{,}000 \text{ km} = 1.29 \text{ m}$$

Mercury is 1.29 meters from the Sun using this scale.

2 Determining scale distances for the other planets

Based on the example in Part 1, you would place Mercury 1.29 meters or 129 centimeters from the Sun in your 100-meter scale model. Use this example to help you calculate the placement of the other planets. Write the distance in meters for each planet in Table 1.

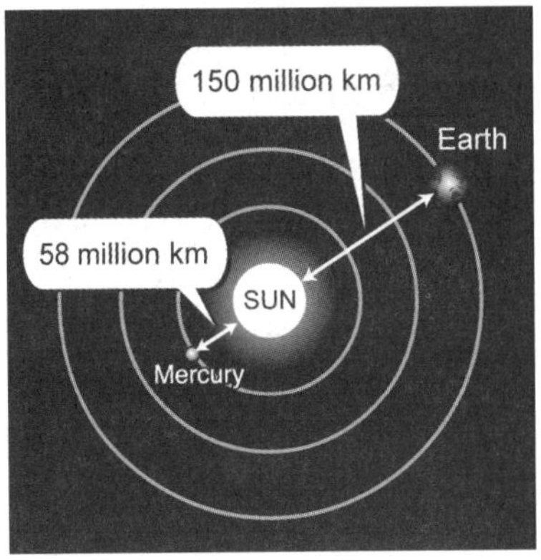

Table 1: Distance from the Sun

Planet	Actual distance to Sun (km)	Proportional distance from the Sun (m)
Mercury	**58,000,000**	**1.29**
Venus	108,000,000	
Earth	150,000,000	
Mars	228,000,000	
Jupiter	778,000,000	
Saturn	1,430,000,000	
Uranus	2,870,000,000	
Neptune	4,500,000,000	

3 Setting up the scale model

1. To begin, make signs for each of the planets and one for the Sun. In your scale model, a student in your class will hold the sign at each position of the planet.
2. In an area that is at least 100-meters long, identify the location of the Sun. A student will stand in this position with a sign that says "Sun."

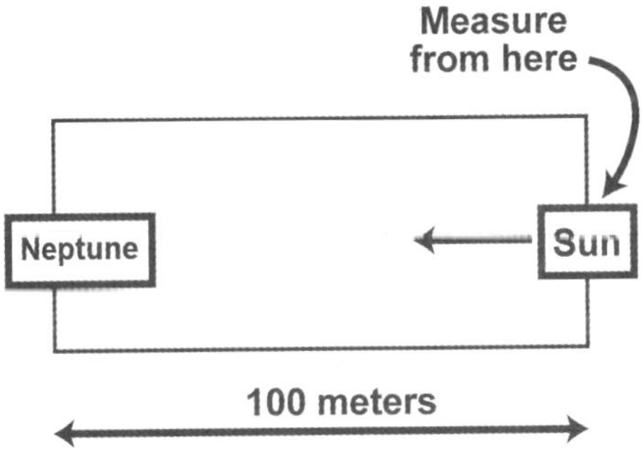

3. Measure 100 meters from the position of the Sun. At the 100-meter mark, a student will stand with a sign that says "Neptune." In this model, 100 meters is the scale distance from the Sun to Neptune.
4. Now, use the scale distances from Table 1 to find the locations of each planet. At the location of each planet, a student will stand with the appropriate sign. Then, answer the questions.

a. After constructing a model of it, what is your impression of our solar system?

b. Describe some disadvantages and advantages to this model of the solar system.

c. Alpha Centauri is the closest star to Earth at 274,332 AU. One astronomical unit is equal to 150 million kilometers. Where would you place this star in the 100-meter scale model?

d. The diameter of the Milky Way galaxy is known to be about 100,000 light years. One light year is 63,000 AU. How does the Milky Way compare using the 100-meter scale model?

4 Determining scale sizes of the planets

Mercury has a diameter of 4,880 kilometers. How big would Mercury be in your 100-meter scale model? You can use the same method to determine the scale diameter of Mercury that you used in Part 2:

$$\frac{x}{4{,}880\text{ km}} = \frac{100\text{ m}}{4{,}500{,}000{,}000\text{ km}}$$

Cross-multiply and rearrange the variables to solve for x:

$$x = \frac{100\text{ m}}{4{,}500{,}000{,}000\text{ km}} \times 4{,}880\text{ km} = 0.000108\text{ m}$$

Based on the example above, the diameter of Mercury in a 100-meter scale solar system would be 0.000078 meters or 0.078 millimeters. For comparison purposes, a single human hair is about 0.1 millimeters in diameter or one-tenth of a millimeter.

Use the above proportion to calculate the diameters of the other planets as well as the Sun and Earth's moon. Write these values in units of meters in the third column of Table 2.

Table 2: Diameters of the planets, our moon, and Sun

Planet	Actual diameter (km)	Scale diameter (m)	Scale diameter (mm)
Sun	1,391,980		
Mercury	4,880	**0.000108**	
Venus	12,100		
Earth	12,800		
Moon	3,475		
Mars	6,800		
Jupiter	142,000		
Saturn	120,000		
Uranus	51,800		
Neptune	49,500		

Now, answer these questions.

a. How big is the Sun in this model in units of centimeters?

b. How much larger is the Sun's diameter compared with Earth's? How much larger is Earth's diameter compared with the moon's?

c. The smallest object that the human eye can see without magnification is 0.100 millimeters. Given this information, which planets would be visible to the human eye? Would you be able to see the Sun or the moon on this 100-meter scale model of the solar system?

d. What is your impression of how the size of the planets and the Sun compare with the size of the solar system?

5 Extension: Making a larger scale model of the solar system

In this part of the investigation, you will use common objects to compare the diameters of planets, the Sun, and Earth's moon in our solar system. For example, an Earth globe can represent the scale size of Earth. The diameter of the globe we will use is 30 centimeters.

1. If an Earth globe is used to represent the size of Earth, what would the sizes of the Sun and the other planets be? How big would the moon be? Use what you have learned in this investigation to calculate the scale diameters of the other planets, the moon, and the Sun. Fill in the third column of Table 3 with these values.
2. What objects could be used to represent each of the planets, the moon, and the Sun? Fill in the fourth column of Table 3 with your answers to this question.

Table 3: A scale model of the solar system

Planet	Actual diameter of planet (km)	Scale diameter of Sun or planet (cm)	Representative object and its diameter or length (cm)
Sun	1,391,980		
Mercury	4,880		
Venus	12,100		
Earth	**12,800**	**30 cm**	**Earth globe, 30 cm**
Moon	3,475		
Mars	6,800		
Jupiter	142,000		
Saturn	120,000		
Uranus	51,800		
Neptune	49,500		

Now, answer these questions:

a. How many times greater is 30 centimeters than 0.284 millimeters? These are the diameters of Earth for the two scale models you created.

b. Using your answer to question 5a, what would be the distance between the Sun and Neptune on this larger scale? Come up with a way to explain or model this distance.

c. Why is it challenging to make a scale model of the solar system that includes the distances between planets and the Sun and the sizes of the planets?

27C Efficiency of a Solar Cell

How can we use energy from the sun to generate electricity?

The sun produces 3.9×10^{26} watts of energy every second. Of that amount, 1,386 watts fall on a square meter of Earth's atmosphere and even less reaches Earth's surface. This energy can be used to generate electricity without producing pollution or dangerous wastes. Photovoltaic (PV) cells convert sunlight *directly* into electricity and are used to run small appliances such as calculators and outdoor light fixtures. Many PV cells can be wired together to form *panels* that can be used to run larger devices such as irrigation pumps, radar stations, and even refrigerators. How much power does a PV cell produce? How efficient is a PV cell at converting the sun's energy into power?

Materials

- Solar cell with electric motor
- Small box
- Light fixture or lamp
- 100 W Light bulb
- Multimeter
- Electric Circuits kit

1 Setting up

1. Gather the following materials: PV cell, electric motor with fan, wires, circuit grid, digital multimeter, protractor.
2. Measure the length and width of your PV cell (do not include the base) and calculate its area in square centimeters. Record this value.
3. Build a circuit with the PV cell and the electric motor with fan on the circuit grid. Use the wires to make connections.
4. Bring this setup outside into direct sunlight.
5. Set the circuit grid on level ground and hold it so that the sun casts no shadow over it.
6. Vary the angle of the electric circuit board by tipping one end of it up or down. Record your observations of what happens to the speed of the fan at different angles.
7. Record the angle at which you get the fastest fan speed.

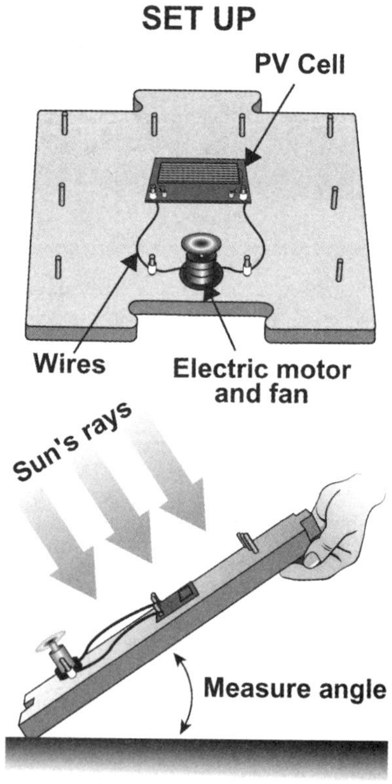

a. How does changing the angle of the PV cell affect the speed of the fan?

b. At which angle is the fan's speed the fastest?

c. Why do you think the angle of the PV cell affects the speed of the fan?

2 Measuring electrical quantities in the circuit

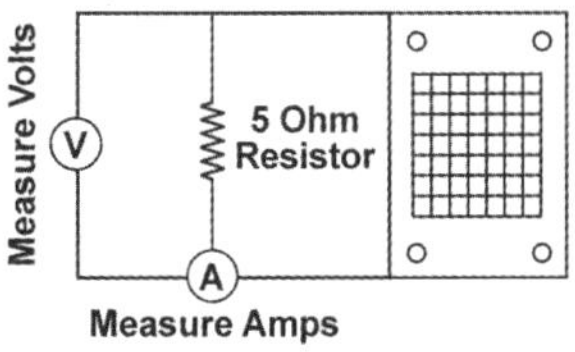

1. Readjust the angle of the electric circuit board until the fan's speed is its fastest. Record the angle.
2. Replace the fan with a 5-ohm resistor.
3. Use the digital multimeter to measure **voltage** across the resistor.
   Record the voltage.
4. Disconnect the circuit at point A and measure **current** in the circuit.
   Record the current.

3 How efficient is your photovoltaic cell?

In this part of the Investigation, you will determine how much of the energy that is reaching your PV cell is being converted into **power**. To do this, you will use your data from Parts 1 and 2.

a. Use the formula below to calculate the **power output** of your PV cell in watts/cm^2.

$$\frac{\text{voltage} \times \text{current}}{\text{area}} = \text{watts/cm}^2$$

Record your result.

b. Multiply your result by 10,000 to convert the value to watts/m^2. Record your result.

The amount of the sun's energy that reaches the edge of Earth's atmosphere is known as the **solar constant**. While the solar constant varies slightly, the average value is 1,368 watts per square meter (W/m^2). To visualize this amount of energy, imagine the energy of thirteen 100-watt light bulbs spread over a single square-meter surface.

How much of this energy actually reaches Earth's surface on a sunny day? This amount varies according to the time of year. The following values are estimates for how much energy reaches Earth's surface on a sunny day, according to the time of year:

- 1000 watts/m^2 on a sunny summer day.
- 900 watts/m^2 on a sunny autumn or spring day.
- 700 watts/m^2 on a sunny winter day.

Depending on the time of year, one of the values above is the *power input from the sun* that is converted into electrical energy by your PV cell.

c. Calculate the efficiency of your photovoltaic cell using the formula below.

$$\% \text{ efficiency} = \left(\frac{\text{power output of your PV cell}}{\text{power input from the sun}} \right) \times 100$$

Record your result.

d. Most PV cells have efficiencies between 5 and 20 percent. How does yours compare?

4 Applying your knowledge

a. Besides angle, what other factors do you think will affect the energy output of your PV cell?

b. PV cells are found on satellites, space probes, and the space shuttle. Do you think the power output of your PV cell would be greater or less just beyond Earth's atmosphere? Explain your answer.

c. With your group, design and conduct an experiment that addresses one of the questions below. Prepare a lab report that explains your hypothesis, procedure, data analysis, and conclusions.

- How does the *distance from a light source* affect the power output of a PV cell?
- How does the *color of light* affect the power output of a PV cell?
- How does *temperature* affect the power output and efficiency of a PV cell?

28C The Time Machine

How long does it take for starlight to reach Earth?

Light travels very fast at 300,000 kilometers per second. However, the universe is very large. Light that reaches your eyes from distant stars or galaxies has been traveling for a long time. Suppose you look through a telescope and see a galaxy that is one million light-years away. Because it takes one million years for the light to reach you, you are actually seeing the galaxy as it was one million years ago! Looking out into the universe is like looking back in time. The farther away you look in space, the farther back you see in time.

Materials

- Astronomy reference materials

1 How fast is fast?

When you travel in a car or a jet, your speed is usually measured in miles per hour or kilometers per hour. It takes about six hours to fly across the country on a modern airliner, which travels about 1000 km/hr. This is about as fast as a person can travel today, but it is nothing compared to how fast light travels.

Complete Table 1 below and compare the speeds of these fast moving objects to the speed of light (rounded to the nearest 300,000 km/sec). To calculate the percent of speed of light;

$$\frac{\text{Speed of object}}{\text{Speed of light}} \times 100 = \text{Percent of speed of light}$$

Table 1: Comparing speeds

Object	Speed in km/hr	Speed in km/sec	Time to reach the moon 384,400 km away (hours and days)	% of speed of light
Diving falcon	250			
Mag-Lev Train	430			
Top fuel dragster	535			
Commercial jet	1000			
Space shuttle	28,000			
Comet	200,000			

a. Earth is about 150,000,000 km from the sun. How long does it take the light from the sun to reach Earth?

b. Jupiter is 778,000,000 km from the sun. How long does it take the light from the sun to reach Jupiter?

c. Pluto is 5,900,000,000 km away from the sun, the farthest planet away. How long does it take light from the sun to reach Pluto?

2 How far is far?

When we looked the size of the solar system we saw that the distances between the planets were enormous in comparison with the distances we experience in our everyday lives here on Earth. In the study of the universe scientists often use the term light year as a measure of distance. This represents the distance light travels in one year, and since light is moving at 300,000 km/sec the distance adds up quickly. How quickly? Fill in Table 2 to find out.

Table 2: Light year data

Unit	Speed of light (in km/sec)		Time in seconds	Distance (in km)
1 light second	300,000	x	1	300,000
1 light minute	300,000	x	60	
1 light hour	300,000	x		
1 light day	300,000	x		
1 light week	300,000	x		
1 light year	300,000			

3 Looking back in time

The light from the sun that we see took several minutes to get here, even at the amazingly fast speed that light travels. But the light we see from stars takes much, much longer. Scientists measure the distances to objects like stars and planets outside our solar system in units of light years because the distances between objects in space is so huge. In fact, the closest star to Earth (other than our sun) is Alpha Centauri, which is 4.3 light years away. The light from that star took 4.3 years to get to us, so it is like looking back a little over four years into the past. The further objects are from Earth the longer the light takes to get to us, and the further back in time we are looking. When an astronomer looks at Alpha Centauri she is not seeing Alpha Centauri as it is right at this second, she is seeing it as it was 4.3 years ago.

Most of the stars we see in the night sky reside right here with us in the Milky Way galaxy. The ones that are not are actually galaxies that are very, very far away.

Research these night sky objects and see if you can locate them to look back in time.

Table 3: Night sky objects

Object	Distance from Earth (light years)	Years looking back in time
Sirius		
Vega		
Arcturus		
Canopus		
Polaris		
Betelgeuse		
Rigel		
Andromeda Galaxy (M31)		

SETUP

Equipment Setups

Physics Stand

You will need:

- Physics stand pole
- Physics stand washer
- Physics stand base
- Physics stand bolt

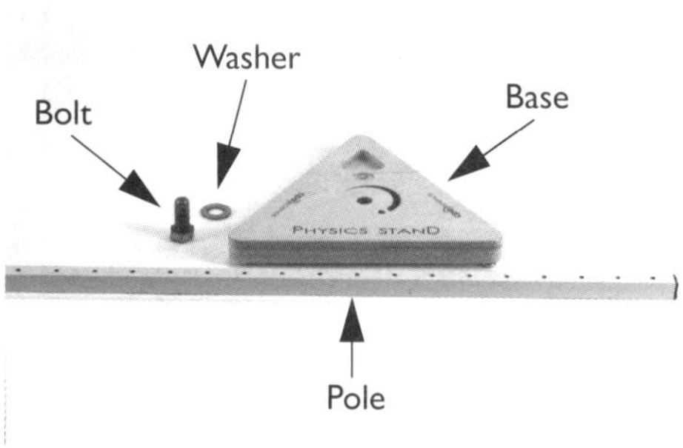

1. **Identifying the parts of the physics stand**

   There are four parts to the physics stand; the base, the pole, the washer and the bolt.

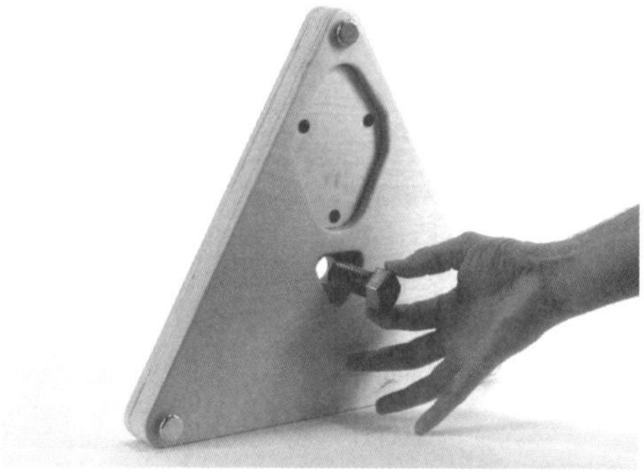

2. **Placing the bolt into the base**

   From the bottom of the base slide the large bolt through the hole.

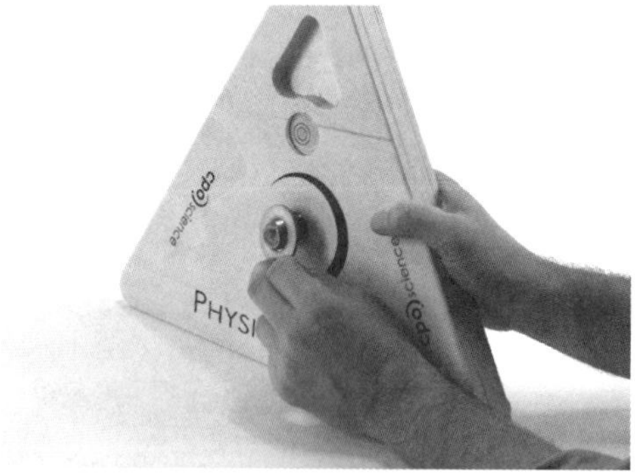

3. **Sliding the washer onto the bolt**

   Be sure to push the bolt into the cut-out triangle on the bottom of the stand. The washer will fit over the threaded part of the bolt.

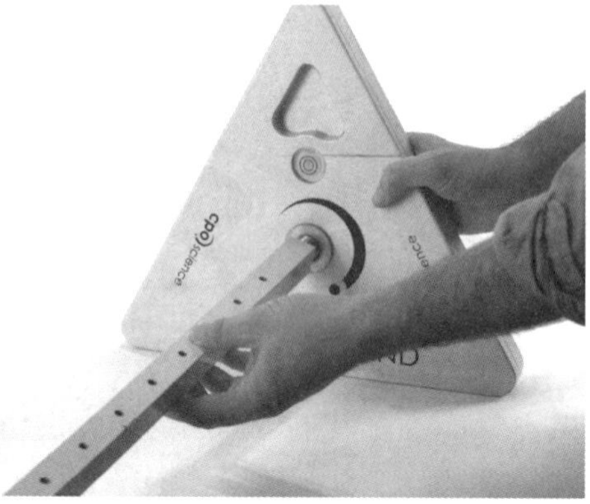

4. **Attaching the pole to the base**

   The pole screws onto the threads of the bolt. The bolt will not spin when you hold the bolt into the cut-out triangle on the bottom of the base. Spin the pole until it screws down snug onto the washer.

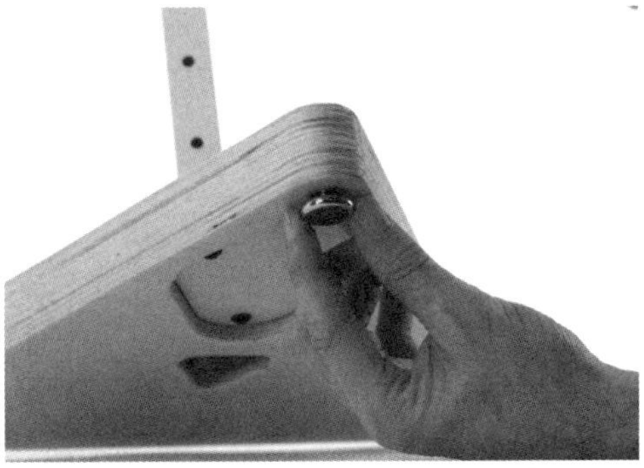

5. **Leveling the stand**

There are three adjustable feet on the bottom of the base. These feet screw into the base. They can be extended by unscrewing them a few turns.

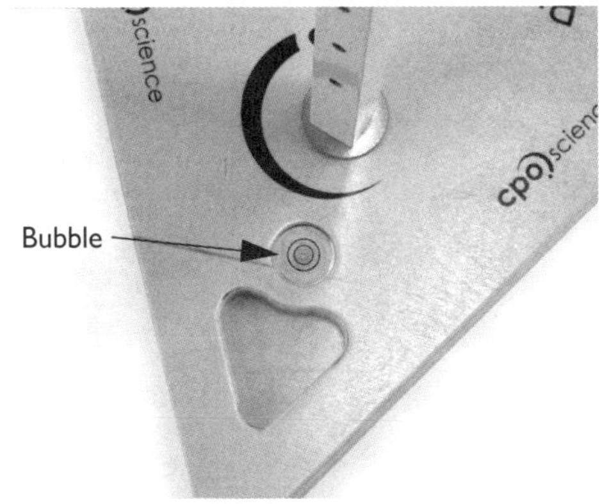

6. **Using the leveling bubble**

When the bubble is directly in the center of the small circle, the stand is level. By adjusting the feet on the bottom of the base by small amounts, the stand can be brought into level.

Energy Car

You will need:

- 2 sections of track
- Rubber bands
- Some string
- A small bubble level
- Three blue knobs
- 2 Energy Cars
- Some modeling clay
- 3 steel balls (3/4" dia.)
- 2 Stops
- 1 Sled
- Physics Stand

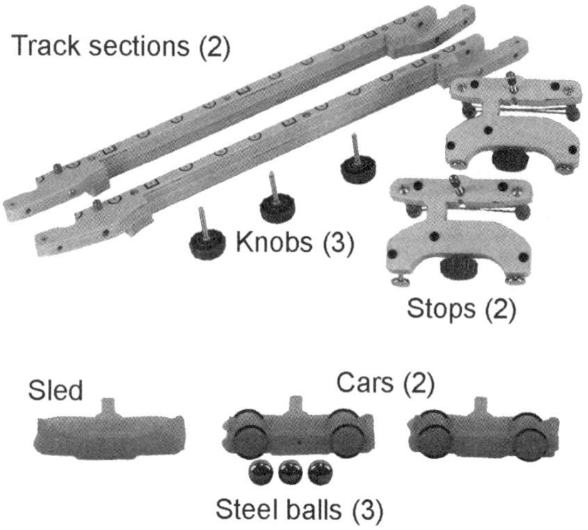

1. **Parts of the Energy Car set**

You will also need some rubber bands. #33 rubber bands work well but other sizes will also work. A small ball of modeling clay about 2 cm in diameter is also necessary.

The sled is the same as the energy car, but without wheels.

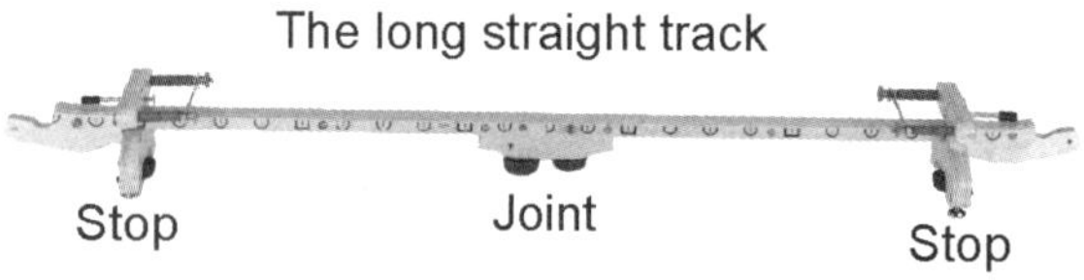

2. **Setting up the long straight track**

The long straight track is used for many experiments.

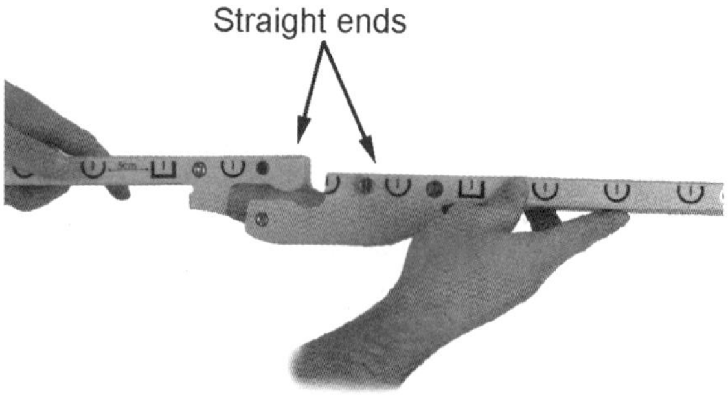

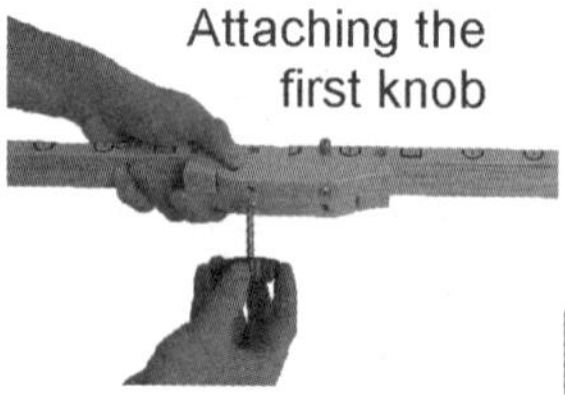

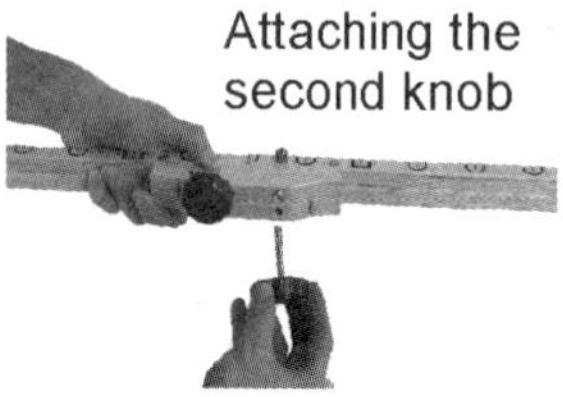

3. **Joining the track sections**

The two sections of the track join together with two blue knobs.

Notice that the ends of the track sections are different. One end is straight and the other has a slight curve. Join the two straight ends together to make the long straight track. The curved ends are for making a hill.

Screw one blue knob into each of the two holes on the bottom of the joint where the two track sections meet.

4. Attaching the stop and making a car launcher

The stop is used to both start and stop the car on the track. There are two stops, one for each end. They are identical.

To make the stop into a launcher, loosen the two thumbscrews by a few turns. Stretch a rubber band between the two screws. The rubber band should fit behind the washer. Give the rubber band one twist so it makes an "X" between the posts. The X helps provide even force when launching the car.

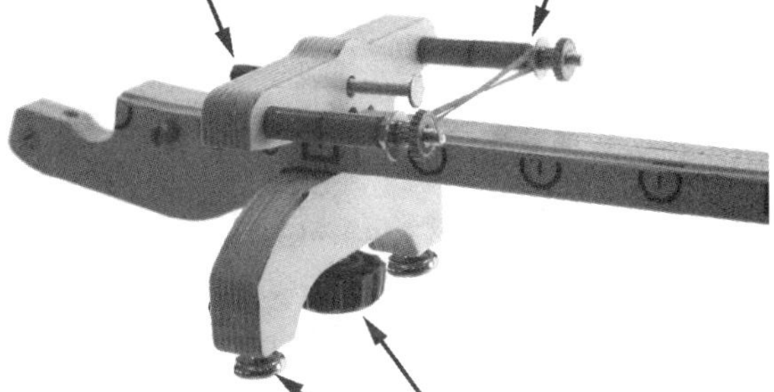

5. Launching the car

To launch the energy car, rest your hand on the top of the launcher and catch the thumb stop on the car with your finger (diagram).

Pull the car back until it hits the screw. Change the adjustment of the screw to get different speeds.

Flick your finger back and off the thumb stop with a quick motion to launch the car. With practice you can get speeds that are repeatable to within 1%.

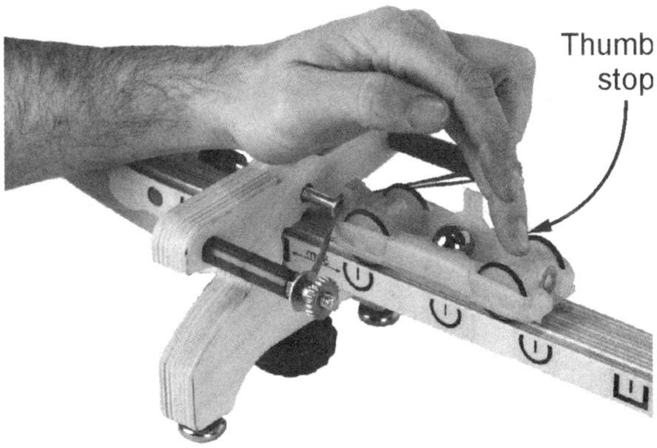

6. Measuring the force on the car

(1) Tie a knot on the end of a length of string. Thread the string through the small hole in the end of the car so the knot is on top the car.

(2) Thread the string through the notch just below the screw on the wooden stop. Tie a small loop on the free end of the string about 20 cm away from the wooden stop.

(3) Use a ruler to set the distance from the front of the screw to the front of the rubber band. Distances from 1 cm to 5 cm can be obtained.

(4) Attach a force scale to the loop on the string and pull the car back until it just touches the screw. The scale reads the force on the car at the measured deflection of the rubber band.

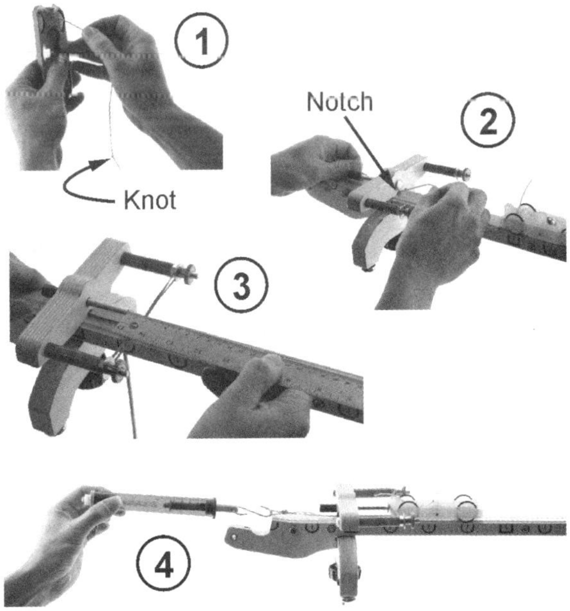

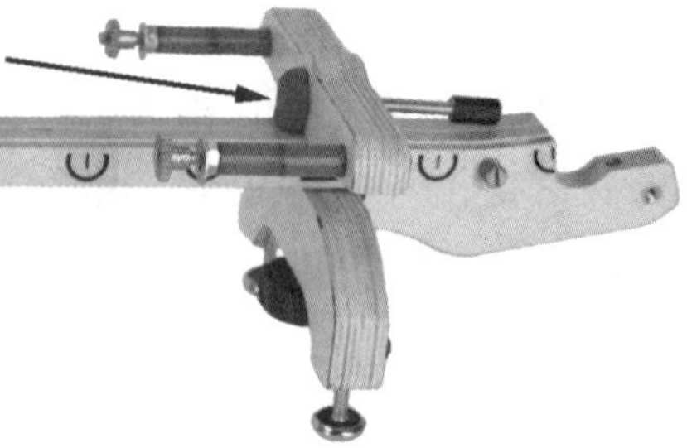

7. Stopping the car

Squish a small ball of modeling clay on the end of the screw to make a stop for the car. The clay prevents the car from bouncing.

8. Leveling the track

Some experiments require that the track be level. A small bubble-level works very well for this purpose. Adjust the leveling feet until the bubble is exactly in the center between the marking lines on the level.

Attaching the photogate

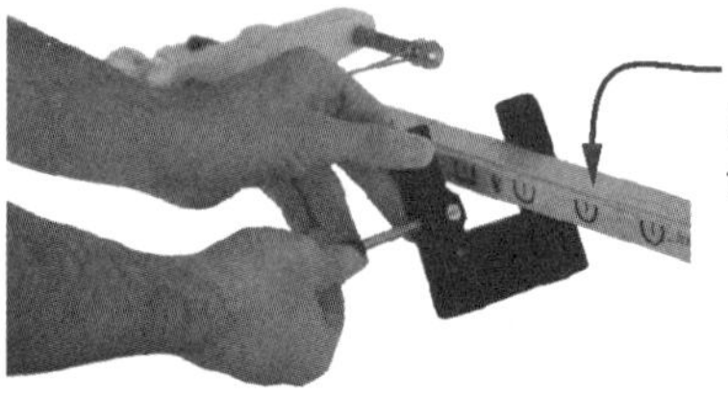

9. Using the photogates

The car has a small flag on top that breaks the light beam in the photogates. The flag is one centimeter wide.

Attach photogates as shown in the diagram. The flag breaks the light beam when the photogate is snug against the bottom of the track.

Add mass to the car so the steel balls are evenly spaced around the center.

0

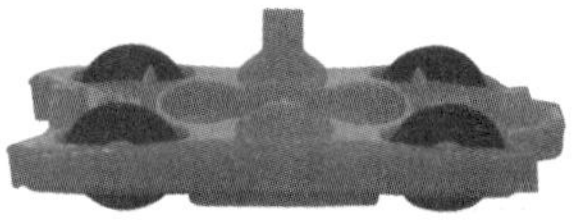

1

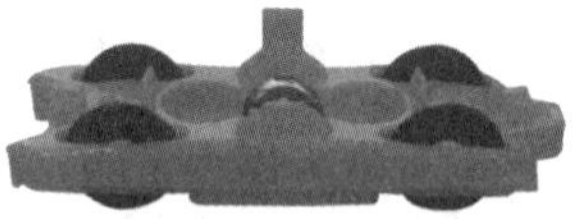

2

3

10. Adding mass to the car

The steel balls add mass to the car. Each ball is 50% of the mass of the empty car. That means adding one ball increases the mass by 50%. Adding two balls doubles the mass.

The car is designed so the center of mass stays in the same place if the balls are added symmetrically around the center (diagram).

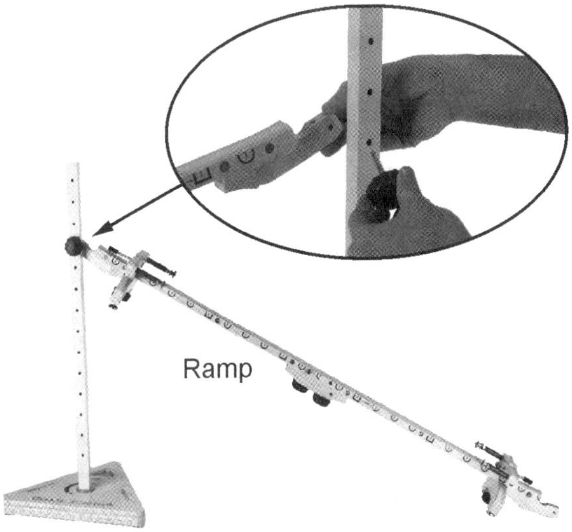

11. Making a straight ramp

The long straight track can be attached to the Physics Stand to make a ramp. Insert a blue knob through a hole in the stand and screw the threaded end into the end of the ramp.

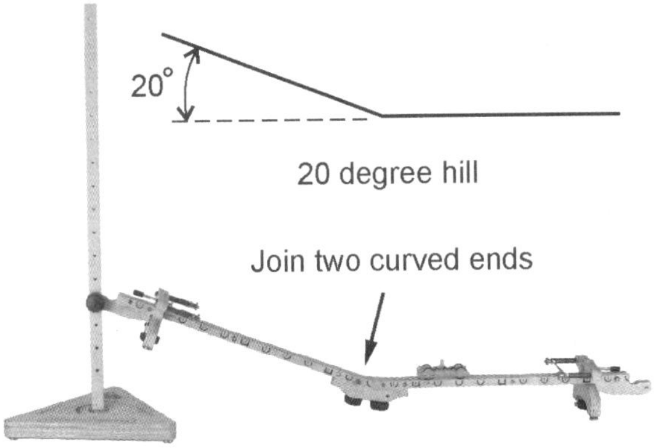

12. Making a flat section with a hill

You can join the two sections of the track to make a 10 degree hill or a 20 degree hill. To make the 10 degree hill join one curved end to one straight end. To make a 20 degree hill joining both curved ends together.

Use the Physics Stand to support the track at the top of the hill.

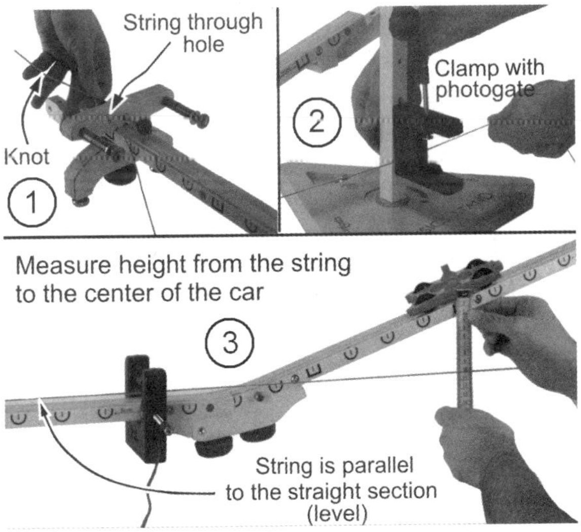

13. Measuring height on the hill

A level string can be used as a reference to measure the height of the car on the hill.

(1) Make the horizontal section of the track level. Tie a knot in one end of a string and thread the string through the small hole in the stop.

(2) Use a photogate and a knob to clamp the other end of the string against the Physics Stand. Adjust the position until the string is parallel to the horizontal section of the track.

(3) The string is now level and at the same height as the center of mass of the car! You can easily make height measurements on the hill by measuring the vertical distance between the center of the car and the string.

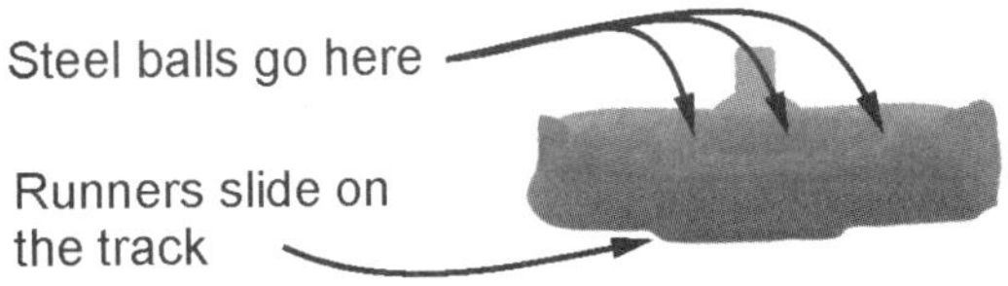

14. Using the sled

The sled is just like the car, but without wheels. The sled is used for friction experiments.

Atom Building Game

You will need:

- Atom game board
- 1 tube yellow marbles
- Laminated periodic table
- 1 tube blue marbles
- Nuclear Reactions cards
- Game booklet
- 1 tube red marbles
- Photons and Lasers cards

1. Identifying the parts of the game

The Atom Building Game comes with an atom game board, blue, red, and yellow marbles, game cards, a laminated periodic table, and an instruction booklet.

The game cards include Nuclear Reactions cards and Photons and Lasers cards. You will learn how to use these cards in the Investigations.

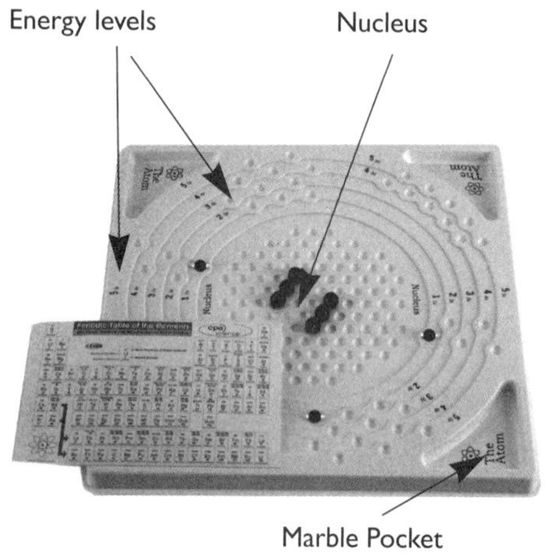

2. Using the atom game board

The Atom game board is designed to sit on a table top with four players (or teams) around it. Each player (or team) is assigned to one of the four marble pockets.

The center of the board represents the nucleus. This is where the protons (red marbles) and neutrons (blue marbles) are place during the activities.

The steps around the nucleus represent the energy levels that are occupied by the electrons (yellow marbles).

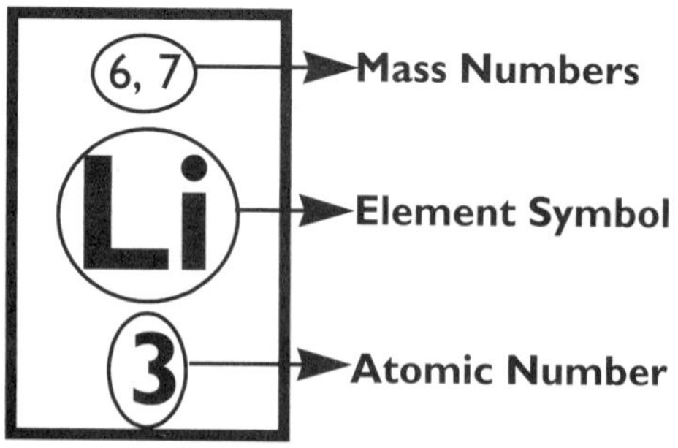

3. Using the periodic table

The periodic table is used for many of the activities. The atomic number is the number of protons (red marbles) in the nucleus.

The atomic number determines what element the atom is. The mass number is the total number of particles (protons plus neutrons) in the nucleus.

Isotopes are atoms with the same number of protons but different numbers of neutrons. You can figure out the number of neutrons by subtracting the atomic number from the mass number. For example, lithium-6 has 3 protons and 3 neutrons ($6 - 3 = 3$).

Neutral atoms have the same number of electrons as protons. The atom in step 2 is lithium-6.

Electric Circuits

You will need:

- Electric circuits table
- 6 light bulbs and 3 holders
- 2 brown (long-length) wires
- 1 red (20-ohm) resistor
- 1 potentiometer
- 6 green (short-length) wires
- 2 green (5-ohm) resistors
- 2 knife switches
- 2 battery holders
- 2 blue (medium-length) wires
- 1 blue (10-ohm) resistor

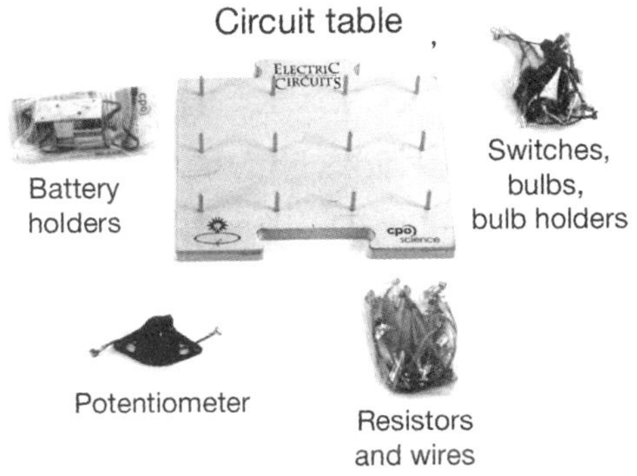

1. Identifying the parts of the electric circuits kit

The electric circuits table is a wooden platform with brass posts for securely assembling electric circuits. There are 12 brass posts used in making connections between different components of the circuit

The posts are not connected underneath and there are no hidden wires. All connections are made using wires that come in a separate pack.

Circuits are made with wires, batteries, bulbs and holders, resistors, and switches.

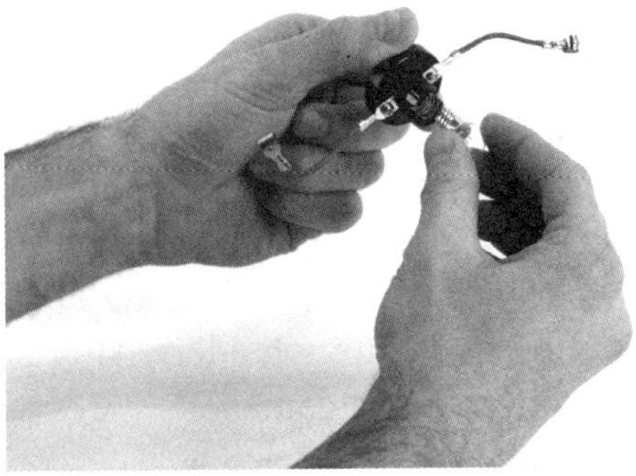

2. Assembling the components

Place D batteries (1.5 volts) into the battery holders.

Place a light bulb into each bulb holder.

3. **Adding wires and circuit elements**

Each wire has a circular connector at both ends called a hoop connector. To add a wire to the board, just place the hoop around the post and push down on the hoop. It will slide down the post, like a jewelry ring over a finger. If you need to add one or more wires to the post, simply push the first wire down the post to make room for another hoop. You can add up to 4 hoops to a post.

NOTE: Solid contact is made at any position on the post. It is not necessary to slide every wire to the bottom of the post.

A circuit element is any item that uses or affects electricity in a circuit. This includes batteries, light bulbs, resistors and switches. Each circuit element that comes with the kit has the same type of hoop connectors as the wires. To connect a circuit element to a post just place the hoop on the post and push down, sliding it down the post.

4. **Closing the circuit**

A circuit is made when wires and elements are connected together making a path for electricity. Shown at left is an example of a simple circuit with a battery, a light bulb, a switch, and some wires.

5. **Avoiding short circuits**

Circuits should always include a “resistor.” The term “resistor” refers to a device like a light bulb or one of the resistors that comes with the kit and provides a substantial resistance to the flow of electricity. A wire alone in a circuit provides very little resistance and is not considered a resistor. A circuit without a resistor, or one in which a branch bypasses a resistor, is called a short circuit. A *short circuit* causes unsafe heating of connecting wires, batteries, and battery holders. This could result in burns and irreparable damage to the equipment. Avoid short circuits at all times!

Pendulum

SETUP

You will need:

- Physics stand assembly
- Pendulum bob and string assembly
- Pendulum face
- 10 washers
- 2 blue knobs with threaded stud
- Timer and photogates

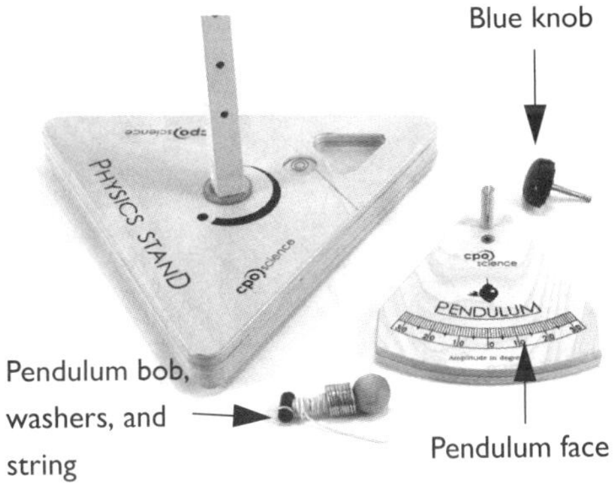

1. Identifying the parts of the pendulum

The pendulum experiment allows you to change three variables: the length of the string, the mass of the pendulum, and the angle of the swing (amplitude). The length of the string can be varied from 15 cm to nearly 1 meter. The pendulum has a hardwood face with an angle scale for easy determination of the amplitude. Washers can be added to or subtracted from the pendulum bob to change the mass.

2. Attaching the pendulum face to the physics stand

Slide the threaded stud on the blue knob through the desired hole in the physics stand.

Turn the blue knob to thread the stud into the back of the pendulum face, securing it to the physics stand.

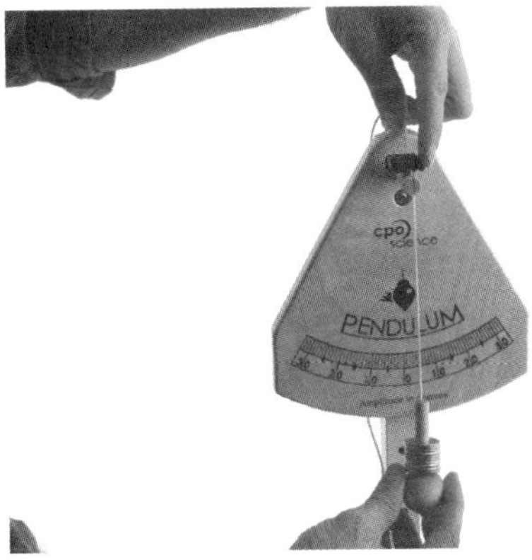

3. Attaching the pendulum bob and string

Select the length of string for the pendulum bob by sliding the string into the slot in the peg on the pendulum face. Check the length of the string by measuring from the bottom of the slotted peg to the bottom of the stack of washers on the pendulum bob. You can add or subtract washers from the pendulum bob to change the mass.

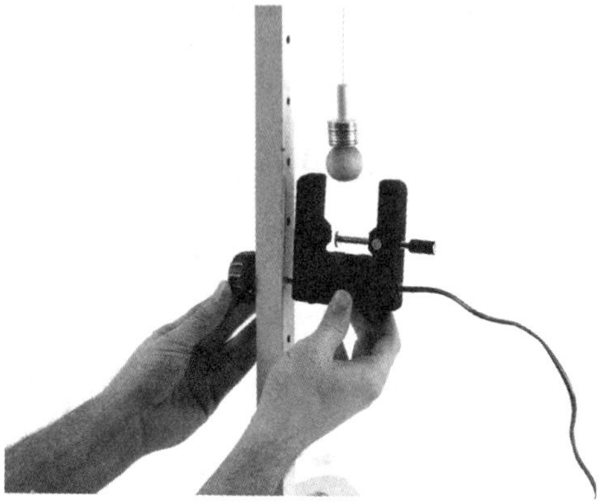

4. **Mounting the photogate to the physics stand**

To mount the photogate to the physics stand, you will need another blue knob with threaded stud. Place the outer edge of the photogate to against the hole that allows you to align the opening with the pendulum bob. Do not overtighten. Make sure the pendulum bob breaks the beam in the photogate each time it swings.

5. **Aligning the photogate**

Be sure to align the two small holes in the photogate with the center of the round portion of the pendulum bob.

Attach the photogate to slot A in the timer using the red or blue wire. Be sure the "A" light is on and that the timer is set to period mode.